Iwasawa Theory and Its Perspective

Volume 1

Mathematical
Surveys
and
Monographs

Volume 272

Iwasawa Theory and Its Perspective

Volume 1

Tadashi Ochiai

AMS AMERICAN MATHEMATICAL SOCIETY
Providence, Rhode Island

IWASAWA RIRON TO SONO TENBOU by Tadashi Ochiai © 2014, 2016 by Tadashi Ochiai. Originally published in 2014 and 2016 by Iwanami Shoten, Publishers, Tokyo. This English edition published 2023 by the American Mathematical Society, Providence, Rhode Island, by arrangement with Iwanami Shoten, Publishers, Tokyo.

2020 *Mathematics Subject Classification.* Primary 11R23, 11R42, 11R34, 13C05.

For additional information and updates on this book, visit
www.ams.org/bookpages/surv-272

Library of Congress Cataloging-in-Publication Data

Names: Ochiai, Tadashi, 1972- author.
Title: Iwasawa theory and its perspective / Tadashi Ochiai.
Other titles: Iwasawa riron to sono tenbō. English
Description: Providence, Rhode Island : American Mathematical Society, 2023- | Series: Mathematical surveys and monographs, 0076-5376 ; volume 272 | "Originally published in 2014 and 2016 by Iwanami Shoten, Publishers, Tokyo." | Includes bibliographical references and index. | Contents: Volume 1
Identifiers: LCCN 2022056149 | ISBN 9781470456726 (paperback: volume 1) | ISBN 9781470473259 (ebook: volume 1)
Subjects: LCSH: Iwasawa theory. | AMS: Number theory – Algebraic number theory: global fields – Iwasawa theory. | Number theory – Algebraic number theory: global fields – Zeta functions and L-functions of number fields. | Number theory – Algebraic number theory: global fields – Galois cohomology. | Commutative algebra – Theory of modules and ideals – Structure, classification theorems.
Classification: LCC QA247 .C32 2023 | DDC 512.7/4–dc23/eng20230317
LC record available at https://lccn.loc.gov/2022056149

10 9 8 7 6 5 4 3 2 1 28 27 26 25 24 23

Contents

Preface

Iwasawa theory began in the late 1950s with a series of investigations by Kenkichi Iwasawa on ideal class groups in the cyclotomic tower of number fields and their relation to p-adic L-functions. Iwasawa and others then developed the theory for ideal class groups during the 1960s and 1970s. Among all, the most important guiding principle in Iwasawa theory should be the "Iwasawa main conjecture" proposed by Iwasawa. The Iwasawa main conjecture in the most classical situation of class groups was first solved by Mazur–Wiles in the early 1980s using modular Galois representations. Then Rubin gave another proof using the Euler system of cyclotomic units. These monumental results form the foundation of the initial stage of cyclotomic Iwasawa theory for ideal class groups and are already well laid out in self-contained texts such as [**13**], [**51**], [**69**], and [**117**], which provide various good pathways to the heart of the theory.

There is a second stage of Iwasawa theory brought forward by Mazur putting it in the context of elliptic curves and modular forms. Coates, Greenberg, Kato, Perrin-Riou, and others continued further generalizations, giving rise to cyclotomic Iwasawa theory of p-adic Galois representations.

Unfortunately, there are no guiding textbooks that provide a compass for this world of developments. For cyclotomic Iwasawa theory of p-adic Galois representations, we can only find [**79**], [**90**], and [**101**], but they are limited to some specialized topics. Hence, this book was motivated by the need for a total perspective of Iwasawa theory that includes the new trends of generalized Iwasawa theory. Another motivation of this book is an update of the classical theory for class groups, as a long time has passed since [**13**], [**51**], [**69**], and [**117**] were published and the point of view on the classical theory has changed.

The book consists of two volumes, the first one for the classical Iwasawa theory and the second one for the generalized Iwasawa theory. Our goal in these two volumes is to focus on the "Iwasawa main conjecture" and its generalizations. Some important themes of Iwasawa theory that are not regarded as part of the Iwasawa main conjecture are left out of the main part of the text so as to keep a coherent flow and focus. However, for the interested reader, we include comments on these other themes at the end of each chapter.

The book is divided into three parts (Parts I, II, and III). Part I is in the first volume; Parts II and III are in the second volume. Let us explain briefly the contents of the first volume (Part I) and the second volume (Parts II and III) of this book.

In Part I (Chapter 1, Chapter 2, and Chapter 3) of this book, we explain "cyclotomic Iwasawa theory of class groups".

In Chapter 1, we discuss the motivation and the importance of Iwasawa theory in relation to the mysterious meaning of the special values of the L-function, the analogy between the number field and the function field, applications to Fermat's last theorem, etc.

In Chapter 2, we prepare some background knowledge. In Iwasawa theory, a Galois extension whose Galois group is isomorphic to $\mathbb{Z}_p$ plays an important role. Such an extension is called a $\mathbb{Z}_p$-extension. We cover some basic topics on $\mathbb{Z}_p$-extensions of a number field. In Iwasawa theory, the ring of one-variable formal power series $\mathcal{O}[[T]]$ over the ring of integers $\mathcal{O}$ of a p-adic field also plays an important role. Such an algebra is called an Iwasawa algebra. We spend four subsections discussing the Iwasawa algebra from four different angles. Then, we study a module over the Iwasawa algebra which is called an Iwasawa module. Some of the standard topics such as the structure theorem of Iwasawa modules are formulated over a one-variable Iwasawa algebra in classical textbooks. In this book, they are discussed in a more general setting over a general normal Noetherian domain. It is important that the setting is general enough to cover rings that appear in the theory of Galois deformation and the Hida theory since we will discuss a generalization of Iwasawa theory over a deformation ring later.

In Chapter 3, we discuss the Iwasawa theory of ideal class groups. As is explained Chapter 1, Iwasawa theory is divided into three main parts: the algebraic aspect; the analytic aspect; and the Iwasawa main conjecture, which is a bridge between the algebraic aspect and the analytic aspect. We first discuss the algebraic aspect. The first important topic on the algebraic aspect is Iwasawa's class number formula which describes the behavior of the orders of the ideal class groups of intermediate fields K_n of a given $\mathbb{Z}_p$-extension. We give a proof of Iwasawa's class number formula that is based on the result of the behavior of the orders of specialization $M/\omega_n M$ of a given Iwasawa module M with certain specific elements ω_n of the cyclotomic Iwasawa algebra given in the previous chapter. We also introduce some known problems and open conjectures on the Iwasawa module given by the ideal class group over the cyclotomic tower. For the analytic aspect, the main topic is the construction of the Kubota–Leopoldt p-adic L-function, which is a p-adic counterpart of the Dirichlet L-function. In light of its importance, we give two different constructions of the Kubota–Leopoldt p-adic L-function: Iwasawa's construction based on the Stickelberger element and the Coleman construction based on Coleman power series. Finally the cyclotomic Iwasawa main conjecture is formulated in two equivalent versions ($(+)$-type and $(-)$-type). We give two completely different proofs of the Iwasawa main conjecture: one is the proof due to Mazur–Wiles and Wiles using the method of the Eisenstein ideal and the other is the proof due to Rubin using the method of an Euler system of cyclotomic units. The proof using an Euler system was already discussed in classical textbooks such as [**13**], [**69**], and [**117**], but the proof using an Eisenstein ideal was never explained earlier in classical textbooks. Another new point in the treatment of Chapter 3 is that we discuss the cyclotomic Iwasawa main conjecture for ideal class groups over totally real number fields and complex multiplication (CM) number fields, which were not treated in classical texts.

In Part II (Chapter 4, Chapter 5, and Chapter 6) of the book, we will explain "cyclotomic Iwasawa theory of p-adic Galois representations". In Part III (Chapter 7 and Chapter 8) of the book, we will explain "Iwasawa theory of p-adic Galois deformations".

Acknowledgment. This book is a translation from a book originally written in Japanese. I would like to thank Yoshitaka Hachimori, Takashi Hara, Shinichi Kobayashi, Kazuo Matsuno, Tatsuya Ohshita, Takeshi Saito, Kazuma Shimomoto, Takae Tsuji, Atsushi Yamagami, and Seidai Yasuda who provided me a lot of valuable comments and pointed out a lot of corrections. Finally, I thank my parents, my wife, and my daughters for constant support for my work.

CHAPTER 1

Motivation and utility of Iwasawa theory

The aim of this book is to explain both the current state of Iwasawa theory, and how it should evolve from classical Iwasawa theory for ideal class groups as discovered by Kenkichi Iwasawa. We feel that it is very important to set the stage for later developments by first discussing this classical Iwasawa theory. This is done in Part I of the book. Chapter 1 is devoted to a brief historical tour without proofs, and Chapters 2 and 3 then develop the classical theory in detail.

1.1. Fermat's last theorem and ideal class groups

Fermat's last theorem asserts that, for any integer $n \geq 3$, triples of integers (x, y, z) can satisfy $x^n + y^n = z^n$ only when $xyz = 0$. Clearly it suffices to prove the theorem for the cases when $n = p$ is an odd prime, or $n = 4$. Fermat made the first great progress, proving the assertion for $n = 4$, by his revolutionary idea of infinite descent. Subsequently, generalizing Fermat's method, Euler and Lagrange proved the case $n = 3$, and Dirichlet the case $n = 5$. After Fermat's death in 1665 until Wiles proved the conjecture in 1994, there were many efforts to extend Fermat's proof and, although these were not successful, it is fair to say that the modern theory of algebraic numbers grew out of these attempts.

The most important of these efforts is due to Kummer, whose key idea was to enlarge the field $\mathbb{Q}$ of rational numbers to the field $\mathbb{Q}(\mu_p)$, which is obtained by adjoining to $\mathbb{Q}$ the p-th roots of unity, for p an odd prime. He then proposed to study the decomposition:

$$\prod_{i=0}^{p-1}(x + \zeta_p^i y) = z^p$$

in $\mathbb{Q}(\mu_p)$. However, as Kummer quickly realized, a major obstruction to this programme was the fact that the "uniqueness of prime decomposition" which holds for the rational field $\mathbb{Q}$ sometimes fails for the larger field $\mathbb{Q}(\mu_p)$. In terms of the modern language of algebraic number theory, for each finite extension field K of $\mathbb{Q}$, the obstruction to the uniqueness of prime decomposition is measured by a finite abelian group $\mathrm{Cl}(K)$, called the ideal class group of K. Using the ideal class group, Kummer succeeded in proving the following result in the middle of the 19th century.

THEOREM (Kummer). *When the order of the ideal class group* $\mathrm{Cl}(\mathbb{Q}(\mu_p))$ *is prime to* p*, Fermat's last theorem is valid for* $n = p$.

We say a prime p is a **regular prime number** when p does not divide the order of $\#\mathrm{Cl}(\mathbb{Q}(\mu_p))$, and an **irregular prime number** when it is not regular. Kummer himself showed that $p = 37, 59, 67$ are the only irregular primes less than 100, and thus established Fermat's last theorem for all prime exponents p less than 100 which are different from these three irregular primes. Moreover, building on more remarkable work of Kummer, which will be discussed in the next section, it was shown in the second half of the 19th century that there are infinitely many irregular primes. However, it has never been proven there are infinitely many regular primes, although numerical data strongly suggests that about 60 percents of all primes should be regular (see [**117**, §5.3]).

For a given algebraic equation f in one variable, we have a formula of the roots of f by means of root symbols if and only if Galois group of the equation f is solvable. Looking back the history of algebra, we remember that Galois group first appeared as an "obstruction" of solving algebraic equations. Then Galois groups became gradually important as one of main characters in elementary algebra and solving algebraic equation went out of our main interest of research. In fact, we obtain a lot of mathematical applications knowing and studying Galois groups.

As a similar paradigm shift, ideal class group, which first appeared as an "obstruction" to the uniqueness of prime factorization of algebraic numbers gradually became the main interest of algebraic number theory. For a number field K, we have an explicit algorithm to determine the structure as abelian group of the ideal class group $\mathrm{Cl}(K)$ by using the theory of "geometry of numbers" by Minkowski. Also, according to Dirichlet's class number formula, the order h_K of $\mathrm{Cl}(K)$ is expressed as

$$h_K = -\lim_{s\to 0} \frac{\zeta(K,s)}{s^{r_1+r_2-1}} \times \frac{w_K}{R_K}$$

where $\zeta(K,s)$ is Dedekind zeta function, r_1 (resp. $2r_2$) is the number of real embeddings (resp. complex embeddings) of K into $\mathbb{C}$, w_K is the number of roots of unity in K, and R_K is the regulator of K.

For example, let us take an imaginary quadratic field $K = \mathbb{Q}(\sqrt{a})$ where a is a negative integer without any square factors. For this K, we have $r_1 + r_2 - 1 = 0$, $R_K = 1$. Hence, Dirichlet's class number formula for K amounts to

$$h_K = -w_K \zeta(K,0). \tag{1.1}$$

It is known that we have an expression

$$\zeta(K,0) = -\frac{1}{a^*} \sum_{i=1}^{a^*} i \left(\frac{i}{a^*}\right) \tag{1.2}$$

where we set

$$a^* = \begin{cases} |a| & a \equiv 1 \bmod 4, \\ |4a| & a \equiv 2,3 \bmod 4, \end{cases}$$

which is a special case of the formula

$$\zeta(\chi,0) = -\frac{1}{C(\chi)} \sum_{i=1}^{C(\chi)} i\chi(i)$$

for a nontrivial Dirichlet character χ of conductor $C(\chi)$ (see §3.2.1 for the definition of Dirichlet character χ and its conductor $C(\chi)$). In fact, we can prove the equation (1.2) using the method of contour integrals which will be explained in §3.2.2. By combining (1.1) and (1.2), we obtain

$$h_K = \frac{w_K}{2a^*} \sum_{i=1}^{a^*} i \left(\frac{i}{a^*}\right). \tag{1.3}$$

In this way, we can calculate the class number of an imaginary quadratic field by repeating simple additions of integers.

In general, when we are given a single algebraic number field K, we have an explicit algorithm to calculate the class group $\mathrm{Cl}(K)$. However our understanding is still unsatisfactory in the following sense.

(i) We have very few general results which hold for the ideal class groups $\mathrm{Cl}(K)$ of certain inifinite families of algebraic number fields K (not only for each single K).

(ii) Note that $\mathrm{Cl}(K)$ is equipped with the natural action of the Galois group $\mathrm{Gal}(K/\mathbb{Q})$. We have very few general results which say something for the Galois module structure of $\mathrm{Cl}(K)$ (not only for the structure as an abelian group).

As an example of the problem (i), there exists a conjecture of Gauss which states that there are infinitely many real quadratic fields whose ideal class groups are trivial[1].

Though the problem of type (i) is difficult as the above example shows, Iwasawa proved the following remarkable result (which will be restated as Theorem 3.1 and will be proved in §3.1.1 of the book):

THEOREM A (Iwasawa's class number formula). *For each prime p, there exist invariants $\lambda, \mu \in \mathbb{Z}_{\geq 0}$, $\nu \in \mathbb{Z}$ such that the largest power of p dividing $\mathrm{Cl}(\mathbb{Q}(\mu_{p^n}))$ is equal to $p^{\lambda n + \mu p^n + \nu}$ for all sufficiently large natural numbers n.*

As for the problem (ii), Iwasawa theory provids a fairly good understanding of $\mathrm{Cl}(K)$ as $\mathrm{Gal}(K/\mathbb{Q})$-module when K is an abelian extension of $\mathbb{Q}$. For example, the Iwasawa main conjecture which is explained later is one of the important answers. We have a lot of important open problems[2].

1.2. Important meaning of special values of Riemann's zeta function

As is well-known, special values $\frac{\zeta(r)}{\pi^r}$ and $\zeta(1-r)$ are rational integers for every positive even integer r. More precisely, the following statements hold:

(1) $\zeta(1-r) = (-1)^{\frac{r}{2}} \frac{(r-1)!}{2^{r-1}} \frac{\zeta(r)}{\pi^r}$.

(2) $\zeta(1-r) = -\frac{B_r}{r}$ where $B_r \in \mathbb{Q}$ is the r-th Bernoulli number[3].

Let us look at the special values of $\zeta(s)$ at several of the first odd negative integers:

$$\begin{array}{lll} \zeta(-1) = -\frac{1}{2^2\cdot 3}, & \zeta(-3) = \frac{1}{2^3\cdot 3\cdot 5}, & \zeta(-5) = -\frac{1}{2^2\cdot 3^2\cdot 7}, \\ \zeta(-7) = \frac{1}{2^4\cdot 3\cdot 5}, & \zeta(-9) = -\frac{1}{2^2\cdot 3\cdot 11}, & \zeta(-11) = \frac{691}{2^3\cdot 3^2\cdot 5\cdot 7\cdot 13}, \\ \zeta(-13) = -\frac{1}{2^2\cdot 3}, & \zeta(-15) = \frac{3617}{2^5\cdot 3\cdot 5\cdot 17}, & \cdots\cdots\cdots \quad . \end{array}$$

[1]Since the language of algebra such as "algebraic number field" or "ideal class group" did not exist at the time of Gauss, Gauss stated the conjecture by means of quadratic forms in integer coefficients and the number of equivalence class.

[2]For example, there are well known conjectures such as Vandiver's conjecture and Greenberg's conjecture as well as a semi-simplicity conjecture of the action of $\mathrm{Gal}(\mathbb{Q}(\mu_{p^\infty})/\mathbb{Q})$ on the p-part of the ideal class group on $\mathbb{Q}(\mu_{p^\infty})$, which will be explained in §3.1.4.

[3]Later, in §3.2.2, we give the definition of Bernoulli number, state the formula of special values in a more general situation for Dirichlet L-function and state a brief idea on the proof.

As is observed in the above numerical examples, the prime numbers which appear in the prime factorization of the denominator of $\zeta(1-r)$ are smaller than or equal to $r+1$. This phenomenon holds always true since there is a well-known theorem by von Staudt and Clausen which says that the rational number $B_r + \sum_{(p-1)|r} \frac{1}{p}$ (In this summation, p runs through only prime numbers) is an integer[4]. On the other hand, in the prime factorization of the numerator of $\zeta(1-r)$, sometimes there appear large prime numbers such as $p = 691, 3617$ of the above example, in an unexpected manner. Naturally, we might ask the meaning of these prime numbers.

By applying Dirichlet's class number formula, which already appeared in the previous section, to $K = \mathbb{Q}(\mu_p)$, we obtain the following theorem[5]:

THEOREM. *For any prime number $p \geq 5$, we have the following equivalence:*

$$p | \#\mathrm{Cl}(\mathbb{Q}(\mu_p)) \Longleftrightarrow p \textit{ divides either of } \zeta(-1), \cdots, \zeta(4-p).$$

Further, there is a refinement of the above theorem which specifies which one of $\zeta(-1), \cdots, \zeta(4-p)$ is divisible by p. Let us define the Teichmüller character $\omega : \mathrm{Gal}(\mathbb{Q}(\mu_p)/\mathbb{Q}) \longrightarrow \mathbb{Z}_p^\times$ to be the canonical character obtained by taking the composite of the canonical isomorphism $\mathrm{Gal}(\mathbb{Q}(\mu_p)/\mathbb{Q}) \xrightarrow{\sim} (\mathbb{Z}/p\mathbb{Z})^\times$ and the canonical injective homomorphism $(\mathbb{Z}/p\mathbb{Z})^\times \hookrightarrow \mathbb{Z}_p^\times$ of Teichmüller representatives.

The refinement of the above theorem, mentioned above, is as follows:

THEOREM (Herbrand–Ribet). *Let $p \geq 5$ be a prime number, and let r be an even integer satisfying $1 \leq r \leq p-1$. Then we we have the following equivalence:*

$$p | \zeta(1-r) \Longleftrightarrow \mathrm{Cl}(\mathbb{Q}(\mu_p))[p]^{\omega^{1-r}} \neq \{0\}$$

where the symbol $[p]$ denotes the subgroup which consists of elements annihilated by multiplication of p.

As is discussed later in §3.2, special values of Riemann's zeta function $\zeta(s)$ at negative integers have a p-adic continuity, in the sense that as natural numbers r and r' get closer to each other in the p-adic topology, the special values $\zeta(1-r)$ and $\zeta(1-r')$ get closer in the p-adic topology. It is known that we have the following analogy:

(1.4) holomorphic functions on $\mathbb{C}$

$\longleftrightarrow$ bounded analytic functions on a p-adic unit disc in $\overline{\mathbb{Q}}_p$

Though the above analogy is not precisely formulated, a p-adic analog of a holomorphic function in the above analogy (1.4) is called sometimes an **Iwasawa function**. It is known that Iwasawa functions satisfy an analog of the Identity theorem for holomorphic functions (see Proposition 2.21).

[4]For the proof of the theorem of von Staudt and Clausen, the reader can refer to [**117**, Theorem 5.10] for example.

[5]We refer to Theorem 4.17, Theorem 5.16, and Theorem 5.34 of [**117**] for the detailed proof on how to deduce the following theorem from a well-known version of Dirichlet's class number formula.

Regarding this analogy, a p-adic analog of Riemann's zeta function is known to exist as Iwasawa function (see Theorem 3.30 for a more precise statement of the following theorem):

THEOREM B (Kubota–Leopoldt, Iwasawa, Coleman). *For any odd integer satisfying $0 < i < p-1$, there exists an Iwasawa function $\zeta_p^{(i)}$ characterized by the following interpolation property*[6]*:*

$$\zeta_p^{(i)}((1+p)^{1-r}) = (1-p^{r-1})\zeta(1-r)$$

(r runs through natural numbers such that $r \equiv 1-i \bmod p-1$).

1.3. Mysterious analogy between function fields and number fields

Finite algebraic number fields have a lot of mysterious analogies with algebraic function fields of one variable over a finite field $\mathbb{F}_q$. For example, for any prime v of a finite number field K, the completion K_v is a locally compact topological field, and this is also the case when K is an algebraic function field of one variable over a finite field $\mathbb{F}_q$. Also, class field theory to control abelian extensions of K holds true when K is a finite algebraic number field and when K is an algebraic function field of one variable over a finite field, and it has a unified way to state the main theorem and to prove it which is valid for both cases[7].

While we have a strong analogy between finite algebraic number fields and algebraic function fields of one variable over a finite field $\mathbb{F}_q$, it is also true that the theory is not always completely parallel between two cases. For example, Fermat's last theorem for the function field case is much easier than the number field case. Similarly, the abc conjecture is also much more difficult in the number field case. By following the philosophy of the analogy between number fields and function fields and by trying to fill such "gaps", we might often encounter some interesting ideas or useful conclusions.

When K is a finite number field or an algebraic function field of one variable over a finite field $\mathbb{F}_q$, the ring of integers $\mathfrak{r}_K$ is a Dedekind domain in either case[8]. We can define a zeta function $\zeta(K,s)$ for K with a complex variable s to be the following infinite product:

$$\zeta(K,s) = \prod_{\mathfrak{m} \in \mathrm{Spec}(\mathfrak{r}_K)-\{0\}} \frac{1}{(1-\#(\mathfrak{r}_K/\mathfrak{m})^{-s})}$$

where $\mathfrak{m}$ runs through maximal ideals of $\mathfrak{r}_K$. The infinite product converges at $\mathrm{Re}(s) > 1$ and $\zeta(K,s)$ is meromorphically continued to the whole $\mathbb{C}$-plane. Both for the number field case and for the function field case, $\zeta(K,s)$ multiplied by "infinite factor" satisfies a similar functional equation with respect to the transformation $s \leftrightarrow 1-s$.

The Iwasawa main conjecture is one of the most important themes of Iwasawa theory. Since the Iwasawa main conjecture appears as an analog in the number field case of the determinant expression of zeta functions which is known in the function field case, we will briefly review this determinant expression.

---

[6]Exceptionally, when $i=1$, $\zeta_p^{(1)}$ has a simple pole at $1 \in \mathbb{Q}_p^\times$.

[7]See [**121**] for example.

[8]The ring of integers $\mathfrak{r}_K$ depends on the choice of "point at infinity" in the case of an algebraic function field of one variable over a finite field $\mathbb{F}_q$.

Concerning the arithmetic of the function field, we have the following one-to-one correspondence:

$$\{\text{(isom. classes) of smooth projective alg. curves } C \text{ over } \mathbb{F}_q\}$$
$$\updownarrow$$
$$\{\text{(isom. classes) of one-var. alg. function fields } K \text{ whose constant field is } \mathbb{F}_q\}$$

Here, to pass from up to down, we assign its function field K_C to a given curve C. To pass from down to up, we assign its model C_K to a given algebraic function field K .

Now, let J_K be the Jacobian variety of C_K and let $J_K(\overline{\mathbb{F}}_q)$ be the group of $\overline{\mathbb{F}}_q$-valued points on J_K where $\overline{\mathbb{F}}_q$ is an algebraic closure of $\mathbb{F}_q$. For each prime number ℓ, we can define a $\mathbb{Q}_\ell$-vector space $V_{K,\ell}$ to be

$$V_{K,\ell} := (\varprojlim_n J_K(\overline{\mathbb{F}}_q)[\ell^n]) \otimes_{\mathbb{Z}_\ell} \mathbb{Q}_\ell.$$

The vector space $V_{K,\ell}$ is equipped with a continuous action of $\mathrm{Gal}(\overline{\mathbb{F}}_q/\mathbb{F}_q)$ and is of dimension $2g_K$ over $\mathbb{Q}_\ell$ where g_K is the genus of the algebraic curve C_K. The following theorem gives a presentation of the zeta function $\zeta(K,s)$ as a certain determinant associated to a Frobenius operator $\mathrm{Frob}_q : x \mapsto x^q$ acting on $V_{K,\ell}$.

THEOREM. *For any prime number ℓ different from p, we have the following equality:*

$$\zeta^*(K,s) = \left.\frac{\det(1 - \mathrm{Frob}_q t; V_{K,\ell})}{(1-t)(1-qt)}\right|_{t=q^{-s}}.$$

Here since $\zeta^*(K,s)$ as is defined above depends on the choice of "infinity" to remove, $\zeta^*(K,s)$ denotes the zeta function obtained by completing the contribution of "infinity" to $\zeta^*(K,s)$.

We recall that Weil proved that the complex absolute values of all the eigenvalues of the action of Frob_q on $V_{K,\ell}$ equal to $\sqrt{q}$. This result of Weil tells us that all the zeros of the function $\det(1 - \mathrm{Frob}_q t; V_{K,\ell})|_{t=q^{-s}}$ on complex variable s is on the line $\mathrm{Re}(s) = \frac{1}{2}$, which gives an analog of Riemman hypothesis for $\zeta^*(K,s)$ by combining with the above theorem.

Let us go back to the situation where K is an algebraic number field. A presentation of $\zeta(K,s)$ as a determinant associated to a certain operator acting on a certain space is not known yet. An important idea of Iwasawa theory is to work rather with a p-adic analog of $\zeta(K,s)$ than $\zeta(K,s)$ itself[9]. When K is an algebraic function field of one variable over a finite field $\mathbb{F}_q$, we considered an infinite extension $K\overline{\mathbb{F}}_q/K$ obtained by extending the constant field $\mathbb{F}_q$ to its algebraic closure $\overline{\mathbb{F}}_q/\mathbb{F}_q$. The Galois group $\mathrm{Gal}(K\overline{\mathbb{F}}_q/K) \cong \mathrm{Gal}(\overline{\mathbb{F}}_q/\mathbb{F}_q)$ is topologically cyclic and its generator Frob_q played an important role as we saw in the above theorem on the determinant expression of $\zeta^*(K,s)$.

When K is an algebraic number field, K has no "constant field" and it does not make sense to "extend the constant field of K". However, coming back to the case where K is an algebraic function field of one variable over a finite field $\mathbb{F}_q$, $K\overline{\mathbb{F}}_q$ can be also interpreted as an extension of K obtained by adjoining all roots of unity

[9]As for the story of the discovery of Iwasawa theory, there is an article by Iwasawa [**48**] (in Japanese) and there is an interview report [**130**] (especially interesting is the last part of the interview article, which is concerned with the early history of the research of Iwasawa theory).

to K. Fortunately, the analog of the second interpretation of $K\overline{\mathbb{F}}_q$ makes sense for algebraic number fields and this is the way which we take in Iwasawa theory.

For a given number field K, Iwasawa fixed a prime number p and studied an infinite extension K_∞^{cyc} called the cyclotomic $\mathbb{Z}_p$-extension which is obtained by adjoining all p-power roots of unity to K[10].

Let us introduce the Iwasawa main conjecture, which is an analog of the above theorem on the determinant expression of $\zeta^*(K, s)$ for the case when K is a number field. For simplicity, we assume that $p > 2$, $K = \mathbb{Q}(\mu_p)$, $K_\infty^{\mathrm{cyc}} = \mathbb{Q}(\mu_{p^\infty})$ for the rest of this subsection. First, as a corollary of Theorem A, the representation

$$V = \left(\varprojlim_n \mathrm{Cl}(\mathbb{Q}(\mu_{p^n}))[p^\infty] \right) \otimes_{\mathbb{Z}_p} \mathbb{Q}_p$$

is a finite dimensional vector space over $\mathbb{Q}_p$ (see Theorem 3.4 for a more general statement). This result is regarded as an Iwasawa theoretic analog of the finiteness of ideal class groups.

The Galois group $\mathrm{Gal}(\mathbb{Q}(\mu_p)/\mathbb{Q})$ acts on V and we can decompose V as

$$V = \bigoplus_{i=0}^{p-2} V^{\omega^i}$$

where V^{ω^i} is the part of V on which $\mathrm{Gal}(\mathbb{Q}(\mu_p)/\mathbb{Q})$ acts by ω^i (see Section 1.2 for the definition of Teichmüller character ω). Each component V^{ω^i} has a $\mathbb{Q}_p$-linear action of $\mathrm{Gal}(K_\infty^{\mathrm{cyc}}/K) \cong 1 + p\mathbb{Z}_p$. Recall that $\chi_{\mathrm{cyc}} : \mathrm{Gal}(K_\infty^{\mathrm{cyc}}/K) \xrightarrow{\sim} 1 + p\mathbb{Z}_p$ is the p-adic cyclotomic character (see Proposition 2.1). Let $\gamma \in \mathrm{Gal}(K_\infty^{\mathrm{cyc}}/K)$ be the topological generator such that we have $\chi_{\mathrm{cyc}}(\gamma) = 1 + p$. Now we introduce the Iwasawa main conjecture as follows:

THEOREM C (Iwasawa main conjecture/Mazur–Wiles). [11] *For each odd integer satisfying $0 < i < p-1$, we have the following equality of the sets of zeros (as well as the coincidence of multiplicity) of two functions:*

$$\{\textit{zeros of } \zeta_p^{(i)}\}$$
$$= \{\textit{zeros of the characteristic polynomial of the action of } \gamma \textit{ on } V^{\omega^i}\}.$$

The equality in the above theorem was first conjectured by Iwasawa and the equality was proved by Iwasawa under the assumption of Vandiver's conjecture (Conjecture 3.25)[12]. Later, the conjecture was solved by Mazur–Wiles in the 1980s. Note that the left-hand side of the equality of the Iwasawa main conjecture is an "analytic object" consisting of the zeros of a p-adic analytically interpolated function and the right-hand side is an "algebraic object" consisting of the zeros

[10] We might think of studying the field $K(\mu_\infty)$ obtained by adjoining all roots of unity to K. However, the class group of $K(\mu_\infty)$ is too big and we can not obtain delicate Iwasawa invariants over $K(\mu_\infty)$ (see [**46**], [**67**] for example).

[11] The Iwasawa main conjecture for ideal class groups has the $(-)$-version (see Theorem 3.66) and the $(+)$-version (see Theorem 3.67). The two versions are equivalent to each other. The actual theorem corresponds to the $(-)$ version of the Iwasawa main conjecture, but it is stated in a different expression as Theorem 3.66 without using the notion of characteristic ideal (see Remark 2.45 (3)). We note that the theorem of Ferrero–Washington (Theorem 3.60) assures that this statement is equivalent to Theorem 3.66.

[12] See around Corollary 4.5.4 in [**13**].

of the characteristic polynomial of a Galois representation. The Iwasawa main conjecture plays the most important role among a lot of theorems in Iwasawa theory since the Iwasawa main conjecture establishes a bridge between two completely different objects: an analytic object and an algebraic object. However, we also remark that even after the solution of the Iwasawa main conjecture, there remain some important problems left in both the analytic side and the algebraic side (see Section 3.5).

Finally, we should not forget that generalizations of the (classical) Iwasawa main conjecture as above are actively studied, which is the theme of this book.

1.4. Wiles's proof and Iwasawa theory

As already discussed in 1.1, from a historical point of view, the study of Iwasawa theory is deeply motivated by Fermat's last theorem. In this section, we recall another important impact of Iwasawa theory which was observed in the dramatic solution of Fermat's last theorem by Wiles in mid 1990s. More precisely, in late 1980s, Ribet [**96**] reduced Fermat's last theorem to Shimura–Taniyama conjecture. Then, in his paper [**125**] and his joint paper [**116**] with Taylor published in 1995, Wiles proved that the Shimura–Taniyama conjecture holds true under certain conditions, which implied Fermat's last theorem combined with above-mentioned result by Ribet.

As explained in the earlier part of the section, Wiles already had worked on Iwasawa theory before this work. Hence, we find ideas from Iwasawa theory in many parts of his proof of Fermat's last theorem. Here, we point out two concrete examples with which we can say that Iwasawa theory is present in the proof of Fermat's last theorem.

(1) Recall that the analytic class number formula which is stated very roughly as

(1.5) the size of the ideal class group of $\mathbb{Q}(\mu_p)$
$$= \text{the special value of the zeta function of } \mathbb{Q}(\mu_p)$$

was the central interest of Iwasawa theory for ideal class groups (see (3.69) and (3.73) for precise statements of the analytic class number formula). It is known that this analytic class formula is generalized as the "Tamagawa number conjecture of Bloch–Kato" [13]:

(1.6) the size of the Selmer group for V
$$= \text{the special value of the zeta function of } V$$

It is known that, when V is equal to the Galois representation $(\varprojlim_n \mu_{p^n}) \otimes_{\mathbb{Z}_p} \mathbb{Q}_p$ over $\mathbb{Q}(\mu_p)$, the Selmer group for V over $\mathbb{Q}(\mu_p)$ is an extension of the ideal class group of $\mathbb{Q}(\mu_p)$ by the group of units of $\mathbb{Q}(\mu_p)$. Thus, we can regard the "Tamagawa number conjecture of Bloch–Kato" (1.6) as a generalization of the analytic class number formula (1.5). As an important breakthrough, Wiles discovered that the Shimura–Taniyama conjecture is reduced to the the Tamagawa number conjecture

[13] Below is only a vague expression of the conjecture. We need to prepare a lot of notations to state the precise statement of the Tamagawa number conjecture of Bloch–Kato. See [**5**] for example.

of Bloch–Kato for $V = \mathrm{ad}^0(V_f) = \mathrm{End}(V_f)^{\mathrm{tr}=0}$ where V_f is the rank-two Galois representation[14] associated to a cuspidal eigenform f. Taylor–Wiles [**116**] and Wiles [**125**] solved the Tamagawa number conjecture of Bloch–Kato for $V = \mathrm{ad}^0(V_f)$. Thus the work of Wiles depends on Iwasawa theory deeply from a conceptual point of view.

(2) We also remark that the method by Wiles in the proof of the Tamagawa number conjectures of Bloch–Kato for $V = \mathrm{ad}^0(V_f)$ has a common idea as the method by Rubin when he gave another proof of the Iwasawa main conjecture of the ideal class group (theorem of Mazur–Wiles). Rubin used the Euler system of cyclotomic units consisting of elements of the form

$$\{1 - \zeta_{p^n \ell_1 \cdots \ell_t} \in \mathbb{Q}(\mu_{p^n \ell_1 \cdots \ell_t}) \,|\, p, \ell_1, \ldots, \ell_t \text{ are distinct prime numbers}\},$$

which forms a certain (twisted) norm compatible system with respect to norm maps for the extensions $\mathbb{Q}(\mu_{p^n \ell_1 \cdots \ell_t})/\mathbb{Q}(\mu_{p^n \ell_1 \cdots \ell_{t-1}})$. In the first announcement of the result by Wiles, his method of proof was a generalization of the method of Rubin. In fact, Wiles insisted that he constructed an Euler system for $V = \mathrm{ad}^0(V_f)$ and he proved the Shimura–Taniyama conjecture by applying the Euler system method. However, short time after his announcement, it turned out that there was a gap in the construction of his Euler system for $V = \mathrm{ad}^0(V_f)$. Wiles tried hard to correct the construction of the Euler system, but he failed in the correction[15]. Finally, he took a detour avoiding the part which depended on the expected Euler system and succeeded to go through with a new technique (which is called a method of "Taylor–Wiles system") developed in a joint paper with Taylor [**116**]. Though Taylor–Wiles system is a setting quite different from Euler systems, there is a common basic technical point that it is also a system parametrized by square-free natural numbers $\ell_1 \cdots \ell_t$. Thus the work of Wiles is deeply related to Iwasawa theory from a technical point of view.

In the above example, we discussed the importance of Iwasawa theory through the relation with Fermat's last theorem. More recently, Iwasawa theory contributed also to the solution of the Catalan conjecture (see [**105**] for example). Thus, Iwasawa theory will continue to strongly influence and have applications in various aspects of Number theory.

---

[14]See Theorem A.1 for the definition of the Galois representation V_f associated to f.

[15]No one has since succeeded in the correction of this original proof of Wiles. If we can fix the proof, we will obtain another important proof of the Shimura–Taniyama conjecture and Fermat's last theorem.

CHAPTER 2

$\mathbb{Z}_p$-extension and Iwasawa algebra

From now on, p means a fixed prime number. To avoid too much complication, we assume that $p \geq 3$ throughout the book, unless otherwise mentioned specifically[1]. Throughout the book, we fix the algebraic closure $\overline{\mathbb{Q}}$ of the field of rationals $\mathbb{Q}$ as a subfield of the field of complex numbers $\mathbb{C}$. We also fix an embedding $\overline{\mathbb{Q}} \hookrightarrow \overline{\mathbb{Q}}_p$ of $\overline{\mathbb{Q}}$ into the algebraic closure $\overline{\mathbb{Q}}_p$ of the field of p-adic numbers $\mathbb{Q}_p$ (see (5.2) for a more precise situation and the reason to fix these embeddings). As is well-known, $\overline{\mathbb{Q}}_p$ is not complete. We denote the completion of $\overline{\mathbb{Q}}_p$ by $\mathbb{C}_p$.

2.1. Basics on $\mathbb{Z}_p$-extensions

In this section, we give various properties of $\mathbb{Z}_p$-extensions of finite algebraic number fields.

To follow some of the arguments of this section, the reader is required to be familiar with the Galois theory for infinite extensions using Krull topology as well as local and global class field theory. Also, it will be important to be familiar with the additive group structure and the multiplicative group structure of a p-adic field. As for these basic facts, we refer the reader to Appendix of **[117]** where we can find a compact quick review.

We also need to know fundamental algebraic number theory concerning the ideal class group and the group of units of an algebraic number field, the structure of the multiplicative group of a p-adic field as well as the language of adèles and idèles. As for these subjects, we refer the reader to **[57]** for example.

We denote the group of m-th roots of unity in $\overline{\mathbb{Q}}$ by μ_m and we put $\mu_{p^\infty} = \varinjlim_n \mu_{p^n}$.

---

[1] The prime $p = 2$ often requires an exceptional way of description or extra complexity in arithmetic. Here, we list up some possible complexities of the prime $p = 2$ which happens in Iwasawa theory: (i) For $p \geq 3$, the torsion subgroup of $\mathbb{Z}_p^\times$ is a cyclic group of order $p-1$. While, the torsion subgroup of $\mathbb{Z}_p^\times$ is a cyclic group of order p when $p = 2$. Because of this, the structure of the group $\mathbb{Z}_p^\times$, $\mathbb{Q}_p^\times$ is described in a different manner depending on $p \geq 3$ or $p = 2$, which requires us an extra effort when we work with $p = 2$. (ii) After Chapter 4 of the book, we consider a p-primary torsion module A with continuous action of the absolute Galois group G_K of a number field K and we associate a group called "Selmer group for A" to A, which plays a principal role in generalized Iwasawa theory. This "Selmer group for A" is defined to be a subgroup of the Galois cohomology $H^1(K, A)$ whose elements are required to satisfy a local condition in $H^1(K_v, A)$ for each prime v of K. When $p \geq 3$, we have $H^1(K_v, A) = 0$ for every non archimedean place v and the study of Selmer group becomes easier for $p \geq 3$. (iii) When we discuss the "period" and the "algebraicity" of L-value, we want a $\pm$-decomposition of cohomologies according to the action of the complex conjugate. Since 2 is a unit in $\mathbb{Z}_{(p)}$ for $p \geq 3$, the localization at p allows us to define such a $\pm$-decomposition.

PROPOSITION 2.1. *For each natural number n, p^n-th cyclotomic extension $\mathbb{Q}(\mu_{p^n})/\mathbb{Q}$ is a Galois extension. The following map which was obtained by choosing a primitive p^n-th root of unity $\zeta_{p^n} \in \overline{\mathbb{Q}}$:*

$$\chi_{\mathrm{cyc},n} : \mathrm{Gal}(\mathbb{Q}(\mu_{p^n})/\mathbb{Q}) \longrightarrow (\mathbb{Z}/p^n\mathbb{Z})^\times, \ \zeta_{p^n}^g = \zeta_{p^n}^{\chi_{\mathrm{cyc},n}(g)} \tag{2.1}$$

is an isomorphism of groups. Further, the isomorphism $\chi_{\mathrm{cyc},n}$ is canonical[2] *in the sense that the map $\chi_{\mathrm{cyc},n}$ does not change even if we replace ζ_{p^n} by another primitive p^n-th root of unity ζ'_{p^n}.*

PROOF. Let us explain the organization of the proof. In the first half of the proof, we will prove that the map (2.1) is an injective and canonical map which is a group homomorphism. In the latter half of the proof, we will show that $\chi_{\mathrm{cyc},n}$ is also surjective. We remark that the argument in the first half holds even if $\mathbb{Q}$ is replaced by a general commutative field K whose characteristic is prime to p. Hence, we prove the same statement as the first half of the proof for such a general field K. This will help us to understand that checking of the surjectivity is the deepest part of the proof.

Let us choose and fix a primitive p^n-th root of unity ζ_{p^n} inside the algebraic closure of K. The minimal polynomial $f(x) \in K[x]$ of ζ_{p^n} over K is an irreducible factor of $x^{p^n} - 1$. Any conjugate of ζ_{p^n} over K is written as $\zeta_{p^n}^i$ for some integer i and it is clear that $\zeta_{p^n}^i \in K(\mu_{p^n})/K$. Hence, $K(\mu_{p^n})/K$ is a normal extension. Since the characteristic of K is prime to p, the polynomial $f(X)$ has no double roots in $K(\mu_{p^n})$. Hence, $K(\mu_{p^n})/K$ is a Galois extension. Now, we recall the map $\chi_{\mathrm{cyc},n}$ given in (2.1) and prove that the map is an injective homomorphism. For any $g \in \mathrm{Gal}(K(\mu_{p^n})/K)$, there exists a unique $i = i(g) \in (\mathbb{Z}/p^n\mathbb{Z})^\times$ such that $g(\zeta_{p^n}) = \zeta_{p^n}^i$ because the characteristic of K is prime to p. By the uniqueness of i, the map $\chi_{\mathrm{cyc},n}$ defined by $\chi_{\mathrm{cyc},n}(g) = i$ is well-defined. Let us take two elements $g, g' \in \mathrm{Gal}(K(\mu_{p^n})/K)$. If we have $\chi_{\mathrm{cyc},n}(g) = i$, $\chi_{\mathrm{cyc},n}(g') = i'$, then $\zeta_{p^n}^{\chi_{\mathrm{cyc},n}(gg')}$ is calculated as follows.

$$\zeta_{p^n}^{\chi_{\mathrm{cyc},n}(gg')} = (gg')(\zeta_{p^n}) = g(g'(\zeta_{p^n})) = g(\zeta_{p^n}^{i'}) = (g(\zeta_{p^n}))^{i'} = (\zeta_{p^n}^i)^{i'} = \zeta_{p^n}^{ii'}.$$

Thus $\chi_{\mathrm{cyc},n}$ is a group homomorphism. Suppose that the image of an element $g \in \mathrm{Gal}(K(\mu_{p^n})/K)$ by $\chi_{\mathrm{cyc},n}$ is trivial.

Since g acts trivially on ζ_{p^n} by the definition of $\chi_{\mathrm{cyc},n}$ and $K(\zeta_{p^n})$ is generated by ζ_{p^n} over K, g acts trivially on $K(\zeta_{p^n})$. Thus $\chi_{\mathrm{cyc},n}$ is injective.

Let us choose now another primitive p^n-th root of unity ζ'_{p^n} and define another character $\chi'_{\mathrm{cyc},n}$ as follows:

$$\chi'_{\mathrm{cyc},n} : \mathrm{Gal}(\mathbb{Q}(\mu_{p^n})/\mathbb{Q}) \longrightarrow (\mathbb{Z}/p^n\mathbb{Z})^\times, \ (\zeta'_{p^n})^g = (\zeta'_{p^n})^{\chi'_{\mathrm{cyc},n}(g)}.$$

Since there exists $j \in (\mathbb{Z}/p^n\mathbb{Z})^\times$ such that $\zeta'_{p^n} = \zeta_{p^n}^j$, we have

$$(\zeta'_{p^n})^{\chi'_{\mathrm{cyc},n}(g)} = g(\zeta_{p^n}^j) = (g(\zeta_{p^n}))^j = (\zeta_{p^n}^j)^{\chi_{\mathrm{cyc},n}(g)} = (\zeta'_{p^n})^{\chi_{\mathrm{cyc},n}(g)}.$$

Thus we prove that $\chi_{\mathrm{cyc},n}$ is a canonical character which does not depend on the choice of the primitive p^n-th root of unity ζ_{p^n}.

---

[2]The fact that χ_{cyc} is canonical and independent of an artificial choice of a primitive p^n-th root of unity ζ_{p^n} is very important when we describe the interpolation property of various p-adic L-functions (see Theorem 3.30 for example).

In the rest of the proof, we restrict ourselves to the original situation of $K = \mathbb{Q}$ and we will prove the surjectivity of $\chi_{\mathrm{cyc},n}$ for any natural number n. Note that the minimal polynomial $f(X)$ of ζ_{p^n} over $\mathbb{Q}$ divides

$$\Phi_{p^n}(X) = \frac{X^{p^n} - 1}{X^{p^{n-1}} - 1} = (X^{p^{n-1}})^{p-1} + (X^{p^{n-1}})^{p-2} + \cdots + (X^{p^{n-1}}) + 1$$

in the ring $\mathbb{Q}[X]$. Now, we set $\Psi_{p^n}(Y) = \Phi_{p^n}(X)|_{X=Y+1}$ by change of variable. $q = p^{n-1}$. By putting $q = p^n$, we have

$$\begin{aligned}\Psi_{p^n}(Y) &\equiv (Y^q + 1)^{p-1} + (Y^q + 1)^{q-2} + \cdots + (Y^q + 1) + 1\\ &\equiv \frac{(Y^q + 1)^p - 1}{(Y^q + 1) - 1} \equiv \frac{Y^{pq}}{Y^q} \equiv Y^{q(p-1)} \bmod p.\end{aligned}$$

Since $\Psi_{p^n}(Y)|_{Y=0} = p$, the polynomial $\Psi_{p^n}(Y)$ is an Eisenstein polynomial in the variable Y. This implies $\Psi_{p^n}(Y) \in \mathbb{Q}[Y]$ is irreducible and thus $\Phi_{p^n}(X) \in \mathbb{Q}[X]$ is irreducible. By this ireducibility, we have proved that $\Phi_{p^n}(X)$ itself is the minimal polynomial of ζ_{p^n} over $\mathbb{Q}$ [3]. Note that we obtained an inequality $[\mathbb{Q}(\mu_{p^n}) : \mathbb{Q}] \geq p^{n-1}(p-1)$ and this implies the surjectivity of $\chi_{\mathrm{cyc},n} : \mathrm{Gal}(\mathbb{Q}(\mu_{p^n})/\mathbb{Q}) \longrightarrow (\mathbb{Z}/p^n\mathbb{Z})^\times$. Since the injectivity of $\chi_{\mathrm{cyc},n}$ is already proved in a more general setting, we have obtained the isomorphism $\mathrm{Gal}(\mathbb{Q}(\mu_{p^n})/\mathbb{Q}) \xrightarrow{\sim} (\mathbb{Z}/p^n\mathbb{Z})^\times$. □

By taking the projective limit of (2.1), we have

(2.2) $$\chi_{\mathrm{cyc}} : \mathrm{Gal}(\mathbb{Q}(\mu_{p^\infty})/\mathbb{Q}) \xrightarrow{\sim} \mathbb{Z}_p^\times$$

REMARK 2.2. The above proof of the surjectivity of $\chi_{\mathrm{cyc},n}$ is a method to use the ramification at p. By applying this method for the ℓ^n-th cyclotomic extension $\mathbb{Q}(\mu_{\ell^n})/\mathbb{Q}$ with any prime ℓ and any power ℓ^n of ℓ, we will be able to prove

(2.3) $$\mathrm{Gal}(\mathbb{Q}(\mu_m)/\mathbb{Q}) \cong (\mathbb{Z}/m\mathbb{Z})^\times \text{ for any natural number } m.$$

To complete the proof of (2.3) by this "ramification method", we need to show that $\mathbb{Q}(\mu_m) \cap \mathbb{Q}(\mu_{m'}) = \mathbb{Q}$ when $(m, m') = 1$. (See [**117**, Prop. 2.4] for the proof of this fact.) Another proof of Proposition 2.1 and (2.3) by a "global method" using Frobenius maps at unramified places of $\mathrm{Gal}(\mathbb{Q}(\mu_m)/\mathbb{Q})$ is quite well-known. (See [**58**, Thm 5.4] for example.)

DEFINITION 2.3. Let us recall the isomorphism $\mathrm{Gal}(\mathbb{Q}(\mu_{p^\infty})/\mathbb{Q}) \cong (\mathbb{Z}/p\mathbb{Z})^\times \times (1 + p\mathbb{Z}_p)$ obtained by (2.2). By the Galois theory for infinite extensions, the invariant subfield $\mathbb{Q}(\mu_{p^\infty})^{(\mathbb{Z}/p\mathbb{Z})^\times}$ for the action of the closed subgroup $(\mathbb{Z}/p\mathbb{Z})^\times \subset \mathrm{Gal}(\mathbb{Q}(\mu_{p^\infty})/\mathbb{Q})$ denoted by $\mathbb{Q}_\infty$ is a Galois extension of $\mathbb{Q}$ and $\mathrm{Gal}(\mathbb{Q}_\infty/\mathbb{Q})$ is isomorphic to $\mathrm{Gal}(\mathbb{Q}(\mu_{p^\infty})/\mathbb{Q})/(\mathbb{Z}/p\mathbb{Z})^\times = 1 + p\mathbb{Z}_p \cong \mathbb{Z}_p$. We call the extension $\mathbb{Q}_\infty/\mathbb{Q}$ the **cyclotomic $\mathbb{Z}_p$-extension** of $\mathbb{Q}$. For a general finite algebraic number field K [4], we call the composite field $K_\infty^{\mathrm{cyc}} = K\mathbb{Q}_\infty$ in $\overline{\mathbb{Q}}$ the **cyclotomic $\mathbb{Z}_p$-extension** of K[5].

[3] We sometimes call $\Phi_{p^n}(X)$ the **cyclotomic polynomial** of degree p^n.

[4] The term "algebraic number field" sometimes means a finite extension of $\mathbb{Q}$. However, since algebraic number fields with infinite degree over $\mathbb{Q}$ often appear in Iwasawa theory. In this book, we will call a finite extension of $\mathbb{Q}$ a finite algebraic number field, to be precise.

[5] $\mathrm{Gal}(K_\infty^{\mathrm{cyc}}/K)$ is naturally identified with an open subgroup of $\mathrm{Gal}(\mathbb{Q}_\infty/\mathbb{Q}) \cong \mathbb{Z}_p$ with index $[\mathbb{Q}_\infty \cap K : \mathbb{Q}]$. Especially, we have $\mathrm{Gal}(K_\infty^{\mathrm{cyc}}/K) \cong \mathbb{Z}_p$ since any open subgroup of $\mathbb{Z}_p$ is isomorphic to $\mathbb{Z}_p$.

DEFINITION 2.4. In general, a Galois extension of a finite algebraic number field K_∞/K such that $\mathrm{Gal}(K_\infty/K)$ is isomorphic to $\mathbb{Z}_p$ as a topological group is called a $\mathbb{Z}_p$-extension of K.

We also prepare the following terminology which we will play important roles later to describe the interpolation property of p-adic L-functions.

DEFINITION 2.5. Let us denote the Galois group $\mathrm{Gal}(\mathbb{Q}_\infty/\mathbb{Q})$ of the cyclotomic $\mathbb{Z}_p$-extension $\mathbb{Q}_\infty/\mathbb{Q}$ of $\mathbb{Q}$ by Γ_{cyc}. For a generally finite algebraic number field K, we denote $\mathrm{Gal}(K_\infty^{\mathrm{cyc}}/K)$ by $\Gamma_{\mathrm{cyc},K}$. There is a natural injection $\Gamma_{\mathrm{cyc},K} \hookrightarrow \Gamma_{\mathrm{cyc}}$ and this injection is an isomorphism if $\mathbb{Q}_\infty \cap K = \mathbb{Q}$. Since we have

$$\mathrm{Gal}(\mathbb{Q}(\mu_{p^\infty})/\mathbb{Q}) = \mathrm{Gal}(\mathbb{Q}(\mu_p)/\mathbb{Q}) \times \Gamma_{\mathrm{cyc}},$$

the canonical isomorphism $\chi_{\mathrm{cyc}} : \mathrm{Gal}(\mathbb{Q}(\mu_{p^\infty})/\mathbb{Q}) \xrightarrow{\sim} \mathbb{Z}_p^\times$ is decomposed as the product of the canonical isomorphisms $\kappa_{\mathrm{cyc}} : \Gamma_{\mathrm{cyc}} \xrightarrow{\sim} 1 + p\mathbb{Z}_p$ and $\omega : \mathrm{Gal}(\mathbb{Q}(\mu_p)/\mathbb{Q}) \xrightarrow{\sim} (\mathbb{Z}/p\mathbb{Z})^\times$. We call χ_{cyc} and κ_{cyc} a **p-adic cyclotomic character** and we call ω the Teichmüller character.

The existence of the cyclotomic $\mathbb{Z}_p$-extension shows that any finite algebraic number field K has at least one $\mathbb{Z}_p$-extension. Now, for a given K, we might wonder how many $\mathbb{Z}_p$-extensions does K have. It is easy to see that, if K has two different $\mathbb{Z}_p$-extensions, K has infinitely many different $\mathbb{Z}_p$-extensions. Thus, the "number of $\mathbb{Z}_p$-extensions" does not make sense literally. However, if K' is a Galois extension of K such that $\mathrm{Gal}(K'/K) \cong \mathbb{Z}_p^d$ and K''/K is an $\mathbb{Z}_p$-extension which is not contained in K', we have $\mathrm{Gal}(K'K''/K) \cong \mathbb{Z}_p^{d+1}$ and the "$\mathbb{Z}_p$-rank of the Galois group of the composite of all $\mathbb{Z}_p$-extensions over K" makes sense as we will see below.

Let $r_1(K)$ be the number of inclusion maps of K into $\mathbb{R}$ and $2r_2(K)$ the number of inclusion maps of K into $\mathbb{C}$ which do not factor through $\mathbb{R}$[6]. We remark that the equality $r_1(K) + 2r_2(K) = [K : \mathbb{Q}]$ holds. We denote by $\mathfrak{F}(K)$ (resp. $\mathfrak{R}(K)$) be the set of nonarchimedean places of K (the set of real archimedean places of K).

Now, we have the following results.

THEOREM 2.6. (1) *Let K_∞/K be an arbitrary $\mathbb{Z}_p$-extension. The primes of K which are ramified in the extension K_∞/K divide (p)* [7]*. Also, there is at least one prime of K which is ramified in the extension K_∞/K.*

(2) *When we denote by $\widetilde{K}_\infty$ the composite of all $\mathbb{Z}_p$-extensions of K inside the algebraic closure $\overline{\mathbb{Q}}$, $\mathrm{Gal}(\widetilde{K}_\infty/K)$ is a free $\mathbb{Z}_p$-module of finite rank and the inequality*

$$r_2(K) + 1 \leq \mathrm{rank}_{\mathbb{Z}_p} \mathrm{Gal}(\widetilde{K}_\infty/K) \leq [K : \mathbb{Q}].$$

holds.

PROOF. Let us prove the first half of assertion (1). Let λ be a prime of K over (ℓ) for a prime number ℓ different from p and let K_λ be an ℓ-adic field obtained by the completion of K with respect to the λ-adic topology. The statement "The extension of global fields K_∞/K is unramified at the prime λ" which we want to prove is equivalent to the statement "The extension of ℓ-adic fields $K_\infty K_\lambda/K_\lambda$ is

[6] If we have such an embedding $K \hookrightarrow \mathbb{C}$, we also have its complex conjugate. Thus the number of such embeddings must be a even number.

[7] Often, we express the situation by the expression "K_∞/K is unramified outside (p)".

unramified". We will prove the latter statement. Since $\mathrm{Gal}(K_\infty K_\lambda/K_\lambda)$ is naturally identified with a closed subgroup of $\mathrm{Gal}(K_\infty/K) \cong \mathbb{Z}_p$, $\mathrm{Gal}(K_\infty K_\lambda/K_\lambda)$ is isomorphic to $\mathbb{Z}_p$ or $\mathrm{Gal}(K_\infty K_\lambda/K_\lambda)$ is rivial. When $\mathrm{Gal}(K_\infty K_\lambda/K_\lambda)$ is trivial[8], there is nothing to prove. Hence, we will assume that $\mathrm{Gal}(K_\infty K_\lambda/K_\lambda)$ is isomorphic to $\mathbb{Z}_p$ and we show the following general statement holds true:

> For any ℓ-adic field k such that $\ell \neq p$ and for a Galois extension k'/k such that $\mathrm{Gal}(k'/k) \cong \mathbb{Z}_p$, k'/k is unramified.

We will prove this by contradiction. Suppose that the extension k'/k is ramified. Since every nontrivial closed subgroup of $\mathbb{Z}_p$ is isomorphic to $\mathbb{Z}_p$, the inertia subgroup of $\mathrm{Gal}(k'/k)$ is isomorphic to $\mathbb{Z}_p$. Recall that, for the maximal abelian extension k^{ab} of an ℓ-adic field k, the inertia subgroup of $\mathrm{Gal}(k^{\mathrm{ab}}/k)$ is isomorphic to the multiplicative group $\mathcal{O}_k^\times$ which consists of the units of the ring of integers $\mathcal{O}_k$. On the other hand, we have an isomorphism:

$$\mathcal{O}_k^\times \cong \mu(k) \times \mathbb{Z}_\ell^{[k:\mathbb{Q}_\ell]}$$

where $\mu(k)$ is a finite group which consists of all roots of unity in k. Since there is no nontrivial homomorphism from $\mathbb{Z}_\ell$ to $\mathbb{Z}_p$ when $\ell \neq p$, the inertia subgroup of $\mathrm{Gal}(k^{\mathrm{ab}}/k)$ has no subquotient which is isomorphic to $\mathbb{Z}_p$. Since the inertia subgroup of $\mathrm{Gal}(k'/k)$ is a subquotient of the inertia subgroup of $\mathrm{Gal}(k^{\mathrm{ab}}/k)$, this contradicts to the assumption. We have thus proved that $K_\infty K_\lambda/K_\lambda$ is unramified at any nonarichimedean prime $\lambda \nmid p$.

Since the group $\mathbb{Z}_p$ does not have a closed subgroup isomorphic to $\mathbb{Z}/2\mathbb{Z}$, the inertia subgroup at an archimedean prime is always trivial. Thus we complete the proof of the first half of the assertion (1).

We shall prove the latter half of assertion (1). Let us denote by $K^{\mathrm{ur,ab}}$ the maximal unramified abelian extension of K. By the unramified global class field theory, we have an isomorphism:

$$\mathrm{Gal}(K^{\mathrm{ur,ab}}/K) \cong \mathrm{Cl}(K). \tag{2.4}$$

Since $\mathrm{Cl}(K)$ is a finite group, $\mathrm{Gal}(K^{\mathrm{ur,ab}}/K)$ must be also a finite group. If the extension K_∞/K is unramified at all primes, we have $K_\infty \subset K^{\mathrm{ur,ab}}$ and this contradicts to the finiteness of $\mathrm{Gal}(K^{\mathrm{ur,ab}}/K)$ explained above. Thus we have proven the latter half of the assertion (1) by contradiction.

Next, we shall prove the assertion (2). Let us consider the injective map:

$$K^\times \hookrightarrow \bigoplus_{\mathfrak{p}|(p)} K_{\mathfrak{p}}^\times \oplus \bigoplus_{\substack{\lambda \in \mathfrak{F}(K) \\ \lambda \nmid (p)}} \mathbb{Z} \oplus \bigoplus_{\lambda \in \mathfrak{R}(K)} \mathbb{Z}/(2)$$

Here, the map to the first component is a natural embedding where $K_{\mathfrak{p}}^\times$ is the completion of K at each $\mathfrak{p}$ dividing (p), the map to the second component is a direct sum of additive discrete valuation maps $\mathrm{ord}_\lambda : K^\times \longrightarrow \mathbb{Z}$ which are normalized as $\mathrm{ord}_\lambda(\lambda) = 1$ at each nonarchimedean prime $\lambda \nmid (p)$ and the map to the third component is assigning the signature $\{\pm 1\} \cong \mathbb{Z}/(2)$ for each real embedding of K.

[8]This case happens when the prime λ is totally decomposed in the extension K_∞/K.

Let us denote by $K^{\mathrm{ab},(p)}$ the maximal abelian extension of K which is unramified outside (p). By the global class field theory, we have the following exact sequence

$$\begin{aligned}(2.5)\quad \{1\} &\longrightarrow \overline{K^\times} \\ &\longrightarrow \bigoplus_{\mathfrak{p}|(p)} K_{\mathfrak{p}}^\times \oplus \bigoplus_{\substack{\lambda\in\mathfrak{F}(K)\\ \lambda\nmid(p)}} \mathbb{Z} \oplus \bigoplus_{\lambda\in\mathfrak{R}(K)} \mathbb{Z}/(2) \\ &\longrightarrow \mathrm{Gal}(K^{\mathrm{ab},(p)}/K) \longrightarrow \{1\}\end{aligned}$$

where $\overline{K^\times}$ is the topological completion of $K^\times$ in the injective map explained before (2.5) and the first map of (2.5) is the extension of this injective map to $\overline{K^\times}$.

When we denote the ring of integers of the p-adic field $K_{\mathfrak{p}}$ by $\mathcal{O}_{\mathfrak{p}}$, the group $\prod_{\mathfrak{p}|(p)}\mathcal{O}_{\mathfrak{p}}^\times \oplus \bigoplus_{\lambda\in\mathfrak{R}(K)}\mathbb{Z}/(2)$ is naturally regarded as a subgroup of the middle term of (2.5) by assigning 0 at the components $\mathbb{Z}$ associated to nonarchimedean primes $\lambda \nmid (p)$. It is easy to see that

$$(2.6)\qquad K^\times \cap \left(\prod_{\mathfrak{p}|(p)} \mathcal{O}_{\mathfrak{p}}^\times \oplus \bigoplus_{\lambda\in\mathfrak{R}(K)} \mathbb{Z}/(2)\right) = \mathfrak{r}_K^\times$$

holds inside the middle term of (2.5) where $\mathfrak{r}_K$ is the ring of integers of K. By a diagram chasing argument combining (2.4), (2.5) and (2.6), we obtain the following exact sequence:

$$\begin{aligned}(2.7)\quad \{1\} &\longrightarrow p\text{-part}\left(\left(\prod_{\mathfrak{p}|(p)} \mathcal{O}_{\mathfrak{p}}^\times\right)\Big/\overline{\mathfrak{r}_K^\times}\right) \\ &\longrightarrow p\text{-part}\left(\mathrm{Gal}(K^{\mathrm{ab},(p)}/K)\right) \longrightarrow p\text{-part}\left(\mathrm{Cl}(K)\right) \longrightarrow \{1\}\end{aligned}$$

where $\overline{\mathfrak{r}_K^\times}$ means the closure of the image of $\mathfrak{r}_K^\times$ embedded diagonally into $\prod_{\mathfrak{p}|(p)}\mathcal{O}_{\mathfrak{p}}^\times$ and "p-part" means to take the (pro-)p-part of (pro)finite abelian groups in the sequence. Recall that

$$(2.8)\qquad \mathrm{rank}_{\mathbb{Z}_p}\, p\text{-part}\left(\prod_{\mathfrak{p}|(p)} \mathcal{O}_{\mathfrak{p}}^\times\right) = [K:\mathbb{Q}] = r_1(K) + 2r_2(K).$$

Since we have the following equality by Dirichlet's unit theorem

$$\mathrm{rank}_{\mathbb{Z}}\, \mathfrak{r}_K^\times = r_1(K) + r_2(K) - 1,$$

we have

$$(2.9)\qquad \mathrm{rank}_{\mathbb{Z}_p}\, p\text{-part}\left(\overline{\mathfrak{r}_K^\times}\right) \le r_1(K) + r_2(K) - 1.$$

We obtain the following inequality by combining (2.7), (2.8), and (2.9):

$$(2.10)\qquad r_2(K) + 1 \le \mathrm{rank}_{\mathbb{Z}_p}\, p\text{-part}\left(\left(\prod_{\mathfrak{p}|(p)} \mathcal{O}_{\mathfrak{p}}^\times\right)\Big/\overline{\mathfrak{r}_K^\times}\right) \le [K:\mathbb{Q}].$$

Since the ideal class group is finite, (2.7) and (2.10) imply

$$r_2(K) + 1 \le \mathrm{rank}_{\mathbb{Z}_p} p\text{-part}\left(\mathrm{Gal}(K^{\mathrm{ab},(p)}/K)\right) \le [K:\mathbb{Q}].$$

Since we have $\widetilde{K}_\infty \subset K^{\mathrm{ab},(p)}$ as shown in the assertion (1), we have

$$\mathrm{rank}_{\mathbb{Z}_p}\, p\text{-part}\left(\mathrm{Gal}(K^{\mathrm{ab},(p)}/K)\right) = \mathrm{rank}_{\mathbb{Z}_p}\mathrm{Gal}(\widetilde{K}_\infty/K).$$

This completes the proof. □

The following conjecture is widely believed.

CONJECTURE 2.7 (Leopoldt conjeture). *Let K be a finite algebraic number field, p a prime number. Then the following equivalent statements hold true*[9].

(i) $r_2(K)+1 = \mathrm{rank}_{\mathbb{Z}_p}\mathrm{Gal}(\widetilde{K}_\infty/K)$.

(ii) The natural map $\mathfrak{r}_K^\times \otimes_{\mathbb{Z}} \mathbb{Z}_p \longrightarrow \left(\overline{\mathfrak{r}_K^\times}\right) \subset \prod_{\mathfrak{p}|(p)} \mathcal{O}_{\mathfrak{p}}^\times$ *is injective.*

(iii) $\mathrm{rank}_{\mathbb{Z}_p}$ *p-part* $\left(\overline{\mathfrak{r}_K^\times}\right) = r_1(K)+r_2(K)-1$.

For a finite algebraic number field K, we put

$$\delta_{K,p} = \mathrm{rank}_{\mathbb{Z}_p}\mathrm{Gal}(\widetilde{K}_\infty/K) - r_2(K) - 1$$

and we call $\delta_{K,p}$ the Leopoldt defect of K. As for the Leopoldt conjecture, the following results are known.

REMARK 2.8. (1) When K is a finite abelian extension of $\mathbb{Q}$ or when K is a finite abelian extension of an imaginary quadratic field, the Leopoldt conjecture is true for any prime number p. The proof is based on the theory of linear independence of p-adic logarithms of algebraic numbers proved by Brumer[**8**], which is the p-adic analogue of Baker's theory on linear independence of logarithms of algebraic numbers in transcendental number theory. For the proof of the Leopoldt conjecture in this case, we refer the reader to [**81**, Chap. X §3] or [**117**, §5.5].

(2) Suppose that K is a totally real algebraic number field with $d_K = [K : \mathbb{Q}] < \infty$. Then, the following statements are equivalent to each other[10]:

(i) The Leopoldt conjecture for (K,p) holds true.

(ii) An invariant R_p called p-adic regulator which is defined by

$$R_p = \det(\log_p \sigma_i(\epsilon_j))_{1\le i,j\le d_K-1}$$

is nonzero. Here, $\sigma_1,\cdots,\sigma_{d_K-1}$ is a set of d_K-1 embeddings which are chosen in an arbitrary manner [11] from the set of d_K-1 different embeddings $K \hookrightarrow \mathbb{C}_p$, and ϵ_j runs through the set of fundamental units of $\mathfrak{r}_K^\times$ $\{\epsilon_1,\cdots,\epsilon_{d_K-1}\}$.

(iii) The denominator of the p-adic L-function $L_p(K,s) \in \mathrm{Frac}(\mathbb{Z}_p[[\Gamma_{\mathrm{cyc}}]])$ of K [12] is divisible by $\gamma-1$ where γ is a topological generator of Γ_{cyc}.

[9]The fact that statements (i), (ii), and (iii) are equivalent to each other is checked by the arguments in the proof of Theorem 2.6 (2).

[10]The equivalence between the statement (i) and the statement (ii) below is due to the equivalence between the statement (i) and the statement (ii) of Conjecture 2.7. The proof of the equivalence between the statement (ii) and the statement (iii) is due to a result on the residue of the p-adic L-function $\zeta_p(K,s)$ for K proved by Colmez [**18**].

[11]The condition that R_p should be nonzero does not depend on the choice of $k-1$ embeddings σ_i.

[12]This p-adic L-function $L_p(K,s)$ is the same as case $F=K$, $\psi=\mathbf{1}$ of the p-adic L-function $L_p(F,\psi)$ which we introduce in Theorem 3.96 of this book later.

We refer the reader to [**81**, Chap. X, §3] for a fairy exhaustive list on other related results for the Leopoldt conjecture as well as equivalent reformulations of the conjecture and some insights on the conjecture.

2.2. Definition of Iwasawa algebra and its various aspects

In Iwasawa theory, the algebra called Iwasawa algebra plays an essential role. There are several different ways to define and introduce the Iwasawa algebra. All of them reflect various aspects of the theory and we sometimes switch between different definitions of Iwasawa theory. Since it is very important to understand a lot of different aspects of Iwasawa theory, we will give four subsections for four different aspects of the Iwasawa algebra.

As for the background to follow the arguments of this section, the reader is recommended to be familiar with some basics on the commutative algebra theory such as the notion of the projective limit, the completion of a module by an ideal, the regular local ring and the Krull-dimension of a commutative algebra. For example, we might be required to have the level of the knowledge found in the text [**2**].

Let us denote by Γ a profinite topological group which is isomorphic to the multiplicative group $1 + p\mathbb{Z}_p \subset \mathbb{Z}_p^\times$. We denote its unique cyclic quotient Γ/Γ^{p^n} of order p^n by Γ_n. We remark that we have $\Gamma \cong \varprojlim_n \Gamma_n$. Throughout the book, we will denote by $\mathcal{O}$ the ring of integers of a finite extension of $\mathbb{Q}_p$ and we denote by ϖ a fixed uniformizer of the discrete valuation ring $\mathcal{O}$.

DEFINITION 2.9. We call the complete group algebra $\mathcal{O}[[\Gamma]] = \varprojlim_n \mathcal{O}[\Gamma_n]$ an **Iwasawa algebra**, and we denote it by $\Lambda_{\mathcal{O}}$. When we have $\mathcal{O} = \mathbb{Z}_p$, we sometimes denote $\Lambda_{\mathcal{O}}$ by Λ.

REMARK 2.10. Let us denote by $\widetilde{\Gamma}_K$ the Galois group of the composite of all $\mathbb{Z}_p$ extensions of K studied in Theorem2.6. Then, in a similar manner as above, we can define a more general Iwasawa algebra $\mathcal{O}[[\widetilde{\Gamma}_K]] = \varprojlim_n \mathcal{O}[\widetilde{\Gamma}_K/\widetilde{\Gamma}_K^{p^n}]$. In fact, when G is a topological group isomorphic to a finite product of $\mathbb{Z}_p$, $\mathcal{O}[[G]]$ is defined similarly and analogous statements of the most of properties which we will prove for $\mathcal{O}[[\Gamma]]$ in this section hold true. We remark that topological groups isomorphic to a finite product of $\mathbb{Z}_p$ appear as a torus subgroup of the group of $\mathbb{Z}_p$-points of reductive groups (the torus subgroup of $GL_n(\mathbb{Z}_p)$ for example). In Hida theory, which we explain later in the book, the generalized Iwasawa algebra $\mathcal{O}[[G]]$ plays an important role (see §2.2.4 and §7.2.1).

2.2.1. Iwasawa algebra as the ring of power series. After Iwasawa published a series of articles on the study of the $\mathbb{Z}_p$-extension from late 1950s, Serre immediately realized the importance of this theory and he proved the following theorem on the identification of the Iwasawa algebra with the ring of one-variable formal power series over $\mathbb{Z}_p$ in his article [**106**][13].

THEOREM 2.11. *Let us fix a topological generator* γ[14]. *Then there exists a unique (noncanonical) isomorphism of topological* $\mathcal{O}$*-algebra* $\Lambda_{\mathcal{O}} \cong \mathcal{O}[[T]]$ *which sends* $\gamma \in \Lambda_{\mathcal{O}}$ *to* $1 + T \in \mathcal{O}[[T]]$.

[13] In the article [**106**], Serre explains the proof of the Iwasawa class number formula as well as the explanation on the characteristic ideal.

[14] A set of elements $g_1, \cdots, g_k$ of a topological group G is called topological generators if the subgroup of G which they generate is a dense subgroup of G.

PROOF. For each natural number n, we denote by $\overline{\gamma} \in \mathcal{O}[\Gamma_n]$ the image of γ via the natural surjection $\mathcal{O}[[\Gamma]] \twoheadrightarrow \mathcal{O}[\Gamma_n]$. Also, we define $\omega_n(T) \in \mathcal{O}[T]$ to be $\omega_n(T) = (T+1)^{p^n} - 1$ for each natural number n. We claim that there exists an $\mathcal{O}$-algebra isomorphism

$$\mathcal{O}[\Gamma_n] \xrightarrow{\sim} \mathcal{O}[T]/(\omega_n(T)) \tag{2.11}$$

which sends $\overline{\gamma}_n$ to $1+T$. By putting $T' = 1+T$, we have an $\mathcal{O}$-algebra isomorphism $\mathcal{O}[T]/(\omega_n(T)) \cong \mathcal{O}[T']/((T')^{p^n} - 1)$. Since we have the presentation of the group $\Gamma_n = \langle \overline{\gamma}_n \,|\, (\overline{\gamma}_n)^{p^n} = \mathbf{1} \rangle$, we have an $\mathcal{O}$-algebra isomorphism $\mathcal{O}[\Gamma_n] \cong \mathcal{O}[T']/((T')^{p^n} - 1)$. We thus verified (2.11). Since the isomorphism (2.11) is compatible with natural projective systems with respect to n on the both sides, the projective limit of (2.11) gives us

(2.12)
$$\mathcal{O}[[\Gamma]] \xrightarrow{\sim} \varprojlim_{n} \mathcal{O}[T]/(\omega_n(T)) = \varprojlim_{n} \mathcal{O}[T]/(\omega_n(T)) = \varprojlim_{m,n} \mathcal{O}[T]/(\varpi^m, \omega_n(T)).$$

On the other hand, by the definition of the ring of formal power seiris, we have

$$\mathcal{O}[[T]] = \varprojlim_{n'} \mathcal{O}[T]/(T^{n'}) = \varprojlim_{m',n'} \mathcal{O}[T]/(\varpi^{m'}, T^{n'}). \tag{2.13}$$

For each pair (m', n'), we have $(\varpi^m, \omega_n(T)) \subset (\varpi^{m'}, T^{n'})$ when m and n are large enough. For each pair (m, n), we have $(\varpi^m, \omega_n(T)) \supset (\varpi^{m'}, T^{n'})$ when m' and n' are large enough. Hence, we obtain

$$\varprojlim_{m,n} \mathcal{O}[T]/(\varpi^m, \omega_n(T)) \cong \varprojlim_{m',n'} \mathcal{O}[T]/(\varpi^{m'}, T^{n'}). \tag{2.14}$$

By combining (2.12), (2.13), and (2.14), we complete the proof. □

Though the definition of the Iwasawa algebra $\mathcal{O}[[\Gamma]]$ looks a bit abstract, Theorem 2.11 allows us to identify $\mathcal{O}[[\Gamma]]$ with a more concrete algebra $\mathcal{O}[[T]]$, which is technically easier to work with. On the other hand, this identification is a noncanonical identification which depends on the choice of a topological generator $\gamma \in \Gamma$. In this book, we would like to avoid noncanonical identification as much as possible. For example, we do not use a presentation over a power series algebra $\mathcal{O}[[\Gamma]]$ in any of principal statements such as the Iwasawa main conjecture, the theorem of the existence of p-adic L-function, etc.

DEFINITION 2.12. A monic polynomial $P(T) = T^l + a_{l-1}T^{l-1} + \cdots + a_1T + a_0 \in \mathcal{O}[T]$ is called a **distinguished polynomial** if $a_0, \cdots, a_{l-1}$ is $\mathcal{O}$ are contained in the maximal ideal (ϖ) of $\mathcal{O}$.

Since $\mathcal{O}$ is a complete regular local ring of Krull dimension 1 with the maximal ideal (ϖ), we obtain the following corollary of Theorem 2.11.

COROLLARY 2.13. *The ring $\Lambda_{\mathcal{O}} \cong \mathcal{O}[[T]]$ is a complete regular local ring of Krull dimension 1 with the maximal ideal (ϖ, T). All the prime ideals of $\mathcal{O}[[T]]$ except (0) and the maximal ideal are principal prime ideals. Further, a principal prime ideal of $\mathcal{O}[[T]]$ is either (ϖ) or an ideal $(P(T))$ generated by an irreducible distinguished polynomial $P(T) \in \mathcal{O}[[T]]$.*

The proof of Corollary 2.13 is omitted. However we remark that the proof of the final statement is due to the (p-adic) Weierstrass preparation theorem, which will be given below.

REMARK 2.14. In the case of $G \cong \mathbb{Z}_p^d$, we can also prove that the generalized Iwasawa algebra $\mathcal{O}[[G]]$ is noncanonically identified with the the ring of d-variable formal power series $\mathcal{O}[[T_1, \cdots, T_d]]$ over $\mathcal{O}$. This fact is proved by a similar argument as the proof of Theorem 2.11 by an inductive argument with respect to d if we note that there is a decomposition of the group $G \cong G' \times G''$ such that we have $G' \cong \mathbb{Z}_p^{d-1}$, $G'' \cong \mathbb{Z}_p$ which gives us $\mathcal{O}[[G]] \cong \mathcal{O}[[G']][[G'']] \cong \mathcal{O}[[G']]\widehat{\otimes}_{\mathcal{O}}\mathcal{O}[[G'']]$[15].

THEOREM 2.15 (p-adic Weierstrass preparation theorem). *For a given power series $F(T) = \sum_{i=0}^{\infty} a_i T^i \in \mathcal{O}[[T]]$, there exist a nonnegative integer m, distinguished polynomial $P(T)$ and $U(T) \in \mathcal{O}[[T]]^\times$ which are all uniquely determined from $F(T)$ such that we have $F(T) = \varpi^m P(T)U(T)$. Further, the integer m is characterized as the largest integer m such that $\frac{1}{\varpi^m}F(T) \in \mathcal{O}[[T]]$ holds true and the degree of $P(T)$ is characterized to be the smallest nonnegative integer k such that $\frac{a_k}{\varpi^m} \in \mathcal{O}^\times$ holds true.*

We give some lemmas for the preparation of the proof of Theorem 2.15.

LEMMA 2.16. *Let $\mathcal{R}$ be a complete local ring. Then a power series $F(T) = a_0 + a_1 T + \cdots + a_k T^k + \cdots \in \mathcal{R}[[T]]$ is a unit in $\mathcal{R}[[T]]$ if and only if the constant term $a_0 \in \mathcal{R}$ is a unit in $\mathcal{R}$. Further, if $a_0 \in \mathcal{R}^\times$ holds, the inverse element of $F(T)^{-1}$ of $F(T)$ is given by*

$$F(T)^{-1} = \frac{1}{a_0}\sum_{i=0}^{\infty}(-1)^i Q(T)^i$$

where we put $Q(T) = \frac{1}{a_0}(a_1 T + \cdots + a_k T^k + \cdots)$.

PROOF OF LEMMA 2.16. Suppose that $a_0 \notin \mathcal{R}^\times$. Since $\mathcal{R}$ is a local ring, a_0 is contained in the maximal ideal $\mathfrak{M}_{\mathcal{R}}$ of $\mathcal{R}$. Note that, for each natural number k, $\mathcal{R}[T]/(T^k)$ is a local ring whose maximal ideal is generated by $\mathfrak{M}_{\mathcal{R}}$ and T. Hence, the image of $F(T)$ in $\mathcal{R}[T]/(T^k)$ is contained in the maximal ideal for each natural number k. Since we have $\mathcal{R}[[T]] = \varprojlim_k \mathcal{R}[T]/(T^k)$, $F(T)$ is contained in the maximal ideal of $\mathcal{R}[[T]]$ and thus $F(T)$ is not a unit of $\mathcal{R}[[T]]$. This proves the necessity in the first assertion of the condition $a_0 \in \mathcal{R}^\times$.

Now, let us assume that $a_0 \in \mathcal{R}^\times$. Since the element $Q(T)$ defined in the statement of the lemma is contained in the maximal ideal of $\mathcal{R}[[T]]$, the series $\sum_{i=0}^{\infty}(-1)^i Q(T)^i$ converges in $\mathcal{R}[[T]]$ and we have $(1 + Q(T)) \cdot (\sum_{i=0}^{\infty}(-1)^i Q(T)^i) = 1$. Since we have $F(T) = a_0(1 + Q(T))$ by definition, we conclude that $\frac{1}{a_0}\sum_{i=0}^{\infty}(-1)^i Q(T)^i$ is the inverse of $F(T)$. This completes the proof of the sufficiency in the first assertion of the condition $a_0 \in \mathcal{R}^\times$ and the proof of the second assertion at the same time. □

LEMMA 2.17. *Let $\mathcal{R}$ be a complete local Noetherian ring with the maximal ideal $\mathfrak{M}_{\mathcal{R}}$ and $F(T) = a_0 + a_1 T + \cdots + a_i T^i + \cdots$ a formal power series in $\mathcal{R}[[T]]$. Assume that there exists a positive integer $l \geq 1$ such that $a_0, a_1, \cdots, a_{l-1} \in \mathfrak{M}_{\mathcal{R}}$ and $a_l \in \mathcal{R}^\times$. Then there exist a unique monic polynomial $P(T) \in \mathcal{R}[T]$ of degree l whose coefficients in degree $\leq l-1$ are all contained in $\mathfrak{M}_{\mathcal{R}}$ and a unique unit $U(T) \in \mathcal{R}[[T]]^\times$ such that we have $F(T) = P(T)U(T)$.*

[15] The symbol $\widehat{\otimes}_{\mathcal{O}}$ means a formal tensor product over $\mathcal{O}$. For example, we mean $\mathcal{O}[[G']]\widehat{\otimes}_{\mathcal{O}}\mathcal{O}[[G'']] = \varprojlim_{n,n'} \mathcal{O}[G'/(G')^{p^n}]\otimes_{\mathcal{O}}\mathcal{O}[G''/(G'')^{p^{n'}}]$ here.

PROOF OF LEMMA 2.17. For each natural number n, we denote the mod $\mathfrak{M}_{\mathcal{R}}^n$ reduction map $\mathcal{R}[[T]] \twoheadrightarrow (\mathcal{R}/\mathfrak{M}_{\mathcal{R}}^n)[[T]]$ by $r_{(n)}$. It suffices to find a projective system $\{U_{(n)}(T) \in (\mathcal{R}/\mathfrak{M}_{\mathcal{R}}^n)[[T]]\}_{n\in\mathbb{N}}$ satisfying the following conditions:

(i) For each natural number n, $U_{(n)}(T)$ is a unit in $(\mathcal{R}/\mathfrak{M}_{\mathcal{R}}^n)[[T]]$.

(ii) For each natural number n, $r_{(n)}(F(T))U_{(n)}(T)^{-1} \in (\mathcal{R}/\mathfrak{M}_{\mathcal{R}}^n)[[T]]$ is a monic polynomial of degree l (the coefficient in degree l is 1 and the coefficients in degree $\geq l+1$ are all 0) whose coefficients in degree $\leq l-1$ are all contained in the maximal ideal of $\mathcal{R}/\mathfrak{M}_{\mathcal{R}}^n$.

Since we assume that $\mathcal{R}$ is Noetherian, we have

$$\bigcap_{n=1}^{\infty} \mathfrak{M}^n = 0$$

and

$$\mathcal{R}[[T]] = \varprojlim_n (\mathcal{R}/\mathfrak{M}^n)[[T]].$$

Thus, once we have such a projective system

$$\{U_{(n)}(T) \in (\mathcal{R}/\mathfrak{M}_{\mathcal{R}}^n)[[T]]\}_{n\in\mathbb{N}},$$

we define $U(T) \in \mathcal{R}[[T]]^\times$ to be $U(T) = \varprojlim_n U_{(n)}(T)$. Then, if we put $P(T) = F(T)U(T)^{-1}$, $P(T)$ is a monic polynomial of degree l whose coefficients in degree $\leq l-1$ are all contained in $\mathfrak{M}_{\mathcal{R}}$.

For the rest of the proof, we will construct a projective system $\{U_{(n)}(T) \in (\mathcal{R}/\mathfrak{M}_{\mathcal{R}}^n)[[T]]\}_{n\in\mathbb{N}}$ satisfying the conditions (i) and (ii) by an inductive manner with respect to i. Let us start from the case $n=1$. We have

$$r_{(1)}(F(T)) = \overline{a}_l T^l + \overline{a}_{l+1}T^{l+1} + \cdots,$$

where we denote the image of the coefficients $a_i \in \mathcal{R}$ by $\overline{a}_i \in \mathcal{R}/\mathfrak{M}$. If we set $U_{(1)}(T) = T^{-l}r_{(1)}(F(T))$, the condition (i) holds true thanks to Lemma 2.16 and the condition (ii) holds true by the definition of $U_{(1)}(T)$. Next, we will construct $U_{(k+1)}(T)$ assuming that the desired $U_{(1)}(T)$ exists up to $n=k$. Let us take a lifting $\widetilde{U_{(k)}(T)} \in (\mathcal{R}/\mathfrak{M}^{k+1})[[T]]$ of $U_{(k)}(T) \in (\mathcal{R}/\mathfrak{M}^k)[[T]]$. Since the constant term of $\widetilde{U_{(k)}(T)}$ is a unit in $\mathcal{R}/\mathfrak{M}^{k+1}$, we have $\widetilde{U_{(k)}(T)} \in (\mathcal{R}/\mathfrak{M}^{k+1})[[T]]^\times$ thanks to Lemma 2.16. Consider the expansion:

$$r_{(k+1)}(F(T))\widetilde{U_{(k)}(T)}^{-1} = b_0 + b_1T + \cdots + b_iT^i + \cdots. \tag{2.15}$$

Since the image of b_l in $\mathcal{R}/\mathfrak{M}^k$ is equal to 1, we have $b_l \in (\mathcal{R}/\mathfrak{M}^{k+1})^\times$. By modifying $\widetilde{U_{(k)}(T)}$ with multiplication of b_l if necessary, we can assume that $b_l = 1$ in (2.15) without loss of generality. It remains to show that we can take $\widetilde{U_{(k)}(T)}$ so that $b_{l+1} = b_{l+2} = \cdots = 0$. Suppose that there is a natural number $l' > l$ such that $b_{l'} \neq 0$ and take $l' > l$ to be the smallest one among those. Then, $\widetilde{U_{(k)}(T)}' = \widetilde{U_{(k)}(T)}(1 - b_{l'}T^{l'-l}) \in (\mathcal{R}/\mathfrak{M}^{k+1})[[T]]^\times$ is also a lift of $U_{(k)}(T)$. By calculation, $r_{(k+1)}(F(T))\big(\widetilde{U_{(k)}(T)}'\big)^{-1}$ has zero as the coefficient at degree l' and the coefficients at degree $< l'$ coincide as those of $r_{(k+1)}(F(T))\widetilde{U_{(k)}(T)}^{-1}$. If we have $l'' > l'$ such that the coefficient at degree l'' of $r_{(k+1)}(F(T))\big(\widetilde{U_{(k)}(T)}'\big)^{-1}$ is nonzero, we can repeat the same modification to $\widetilde{U_{(k)}(T)}'$. Since an infinite product of the form $\prod_{i\geq l}(1-\alpha_i T^i)$ converges in $(\mathcal{R}/\mathfrak{M}^{k+1})[[T]]$, we modify $\widetilde{U_{(k)}(T)}$ by the multiplication of such an infinite product to obtain $U_{(k+1)}(T)$ of the desired projective system.

Once we constructed the desired projective system satisfying the conditions (i) and (ii), this will give $U(T)$ and $P(T)$ as explained earlier. Since the uniqueness assertion of the statement in the assertion is clear from our construction, this completes the proof. □

Now we can prove Theorem 2.15 using Lemma 2.17.

PROOF OF THEOREM 2.15. For a given power series $F(T)=\sum_{i=0}^{\infty}a_iT^i\in\mathcal{O}[[T]]$, we take the largest power ϖ^m of the uniformizer ϖ which divides $F(T)$ and we set $G(T)=\frac{1}{\varpi^m}F(T)\in\mathcal{O}[[T]]$. Let us give the expansion $G(T)=a'_0+a'_1T+\cdots+a'_iT^i+\cdots\in\mathcal{O}[[T]]$ of $G(T)$. By the definition of $G(T)$, there exists a natural number l such that $a'_0,a'_1,\cdots,a'_{l-1}$ are all divisible by ϖ and $a'_l\in\mathcal{O}^\times$. We complete the proof by applying Lemma 2.17 to $G(T)$ with $\mathcal{R}=\mathcal{O}$. □

2.2.2. Iwasawa algebra as the ring of measures.

DEFINITION 2.18. A **measure** μ with values in $\mathcal{O}$ is an $\mathcal{O}$-linear function

$$\mu:\ \{\mathcal{O}\text{-valued locally constant functions on }\Gamma\}\longrightarrow\mathcal{O}$$

where we regard the first set naturally as an $\mathcal{O}$-module. We denote by $\mathrm{Meas}(\Gamma;\mathcal{O})$ the $\mathcal{O}$-module which consists of $\mathcal{O}$-valued measure on Γ. Further, for a locally constant $\mathcal{O}$-valued function $u=u(g)$ on Γ, we sometimes denote $\mu(u)$ by $\int_\Gamma u(g)d\mu$.

Let $\mu_\Gamma\in\mathrm{Meas}(\Gamma;\mathcal{O})$ be a Haar measure which is normalized by $\mu_\Gamma(\Gamma)=1$. The measure μ_Γ is characterized by having the property $\mu(g\Gamma^{p^n})=\frac{1}{p^n}$ for any $g\in\Gamma$ and $n\in\mathbb{N}$. Recall that, for $\mu,\mu'\in\mathrm{Meas}(\Gamma;\mathcal{O})$, we define a convolution product $\mu*\mu'\in\mathrm{Meas}(\Gamma;\mathcal{O})$ to be a unique measure whose integral for any locally constant $\mathcal{O}$-valued function $u(g)$ on Γ is given by

$$\int_\Gamma u(g)d(\mu*\mu'):=\int_{\Gamma\times\Gamma}u(g_1g_2)d\mu\times d\mu'.$$

The $\mathcal{O}$-module $\mathrm{Meas}(\Gamma;\mathcal{O})$ can be equipped with a commutative $\mathcal{O}$-algebra structure by defining the product by convolution and by setting a multiplicative unit of $\mathrm{Meas}(\Gamma;\mathcal{O})$ to be the Dirac measure μ_{1_Γ} at the unit 1_Γ of the group Γ[16].

LEMMA 2.19. *We have a natural isomorphism* $\Lambda_{\mathcal{O}}\xrightarrow{\sim}\mathrm{Meas}(\Gamma;\mathcal{O})$ *of* $\mathcal{O}$-*algebra.*

PROOF. Each of $\Lambda_{\mathcal{O}}$ and $\mathrm{Meas}(\Gamma;\mathcal{O})$ is presented as a projective limit of finite $\mathcal{O}$-algebra. We have $\Lambda_{\mathcal{O}}:=\varprojlim_n\mathcal{O}[\Gamma_n]$ by definition. We have $\mathrm{Meas}(\Gamma;\mathcal{O})=\varprojlim_n\mathrm{Meas}(\Gamma_n;\mathcal{O})$ where we define the finite $\mathcal{O}$-algebra $\mathrm{Meas}(\Gamma_n;\mathcal{O})$ similarly as in Definition 2.18. Hence, we construct a projective system natural isomorphism $\mathcal{O}[\Gamma_n]\xrightarrow{\sim}\mathrm{Meas}(\Gamma_n;\mathcal{O})$ of finite $\mathcal{O}$-algebra, we obtain the desired isomorphism $\Lambda_{\mathcal{O}}\xrightarrow{\sim}\mathrm{Meas}(\Gamma;\mathcal{O})$ by taking the projective limit.

---

[16]For each $g\in\Gamma$, we define the Dirac measure μ_g at g to be a measure such that we have

$$\int_\Gamma\chi_U d\mu_g=\begin{cases}1 & \text{when } g\in U,\\ 0 & \text{when } g\notin U,\end{cases}$$

for any compact open subset $U\in\Gamma$ where χ_U is the characteristic function of the set U.

Since Γ_n is finite, any measure $\overline{\mu} \in \mathrm{Meas}(\Gamma_n; \mathcal{O})$ is an $\mathcal{O}$-linear combination $\overline{\mu} = \sum_{\overline{g} \in \Gamma_n} a_g \mu_{\overline{g}}$ of Dirac measures $\mu_{\overline{g}}$. In order to calculate the product of two measures, we discuss the product of Dirac measures $\mu_{\overline{g}_1}, \mu_{\overline{g}_2} \in \mathrm{Meas}(\Gamma_n; \mathcal{O})$. For any locally constant $\mathcal{O}$-valued function $u(\overline{g})$ on Γ_n, we have $\displaystyle\int_{\Gamma_n} u(\overline{g}) d(\mu_{\overline{g}_1} * \mu_{\overline{g}_2}) = u(\overline{g}_1 \overline{g}_2)$ by the definition of the Dirac measure. The second term is equal to $\displaystyle\int_{\Gamma_n} u(\overline{g}) d\mu_{\overline{g}_1 \overline{g}_2}$ again by the definition of the Dirac measure. We thus verified

$$\mu_{\overline{g}_1} * \mu_{\overline{g}_2} = \mu_{\overline{g}_1 \overline{g}_2}$$

for any $\overline{g}_1, \overline{g}_2 \in \Gamma_n$. Hence, the $\mathcal{O}$-linear isomorphism

$$\mathcal{O}[\Gamma_n] \xrightarrow{\sim} \mathrm{Meas}(\Gamma_n; \mathcal{O}), \ \sum_{\overline{g} \in \Gamma_n} a_{\overline{g}} \cdot [\overline{g}] \mapsto \sum_{\overline{g} \in \Gamma_n} a_{\overline{g}} \mu_{\overline{g}} \tag{2.16}$$

is compatible with the product defined on each side. This completes the proof. □

Let $\eta : \Gamma \longrightarrow \overline{\mathbb{Q}}_p^\times$ be a continuous character. Then η is naturally extended to a ring homomorphism $\Lambda_{\mathcal{O}} = \mathcal{O}[[\Gamma]] \xrightarrow{\eta} \overline{\mathbb{Q}}_p$. On the other hand, it makes sense to integrate η by a given measure $\mu \in \mathrm{Meas}(\Gamma; \mathcal{O})$. Note that there exists a finite extension of $\mathbb{Q}_p$ which contsins all values of η. We denote by $\mathcal{O}'$ the ring of integers of a finite extension of $\mathbb{Q}_p$ which contains $\mathcal{O}$ and all values of η. We have a natural injection map $\mathrm{Meas}(\Gamma; \mathcal{O}) \hookrightarrow \mathrm{Meas}(\Gamma; \mathcal{O}')$ by extending the coefficients. Hence, it makes sense to integrate any locally constant $\mathcal{O}'$-valued function on Γ by $\mu \in \mathrm{Meas}(\Gamma; \mathcal{O})$. Since η is a continuous $\mathcal{O}'$-valued function on Γ, it also makes sense to integrate η by the measure $\mu \in \mathrm{Meas}(\Gamma; \mathcal{O})$. We denote the integral by $\displaystyle\int_\Gamma \eta(g) d\mu$.

The above argument and the argument in the proof of the previous lemma gives us the following proposition.

PROPOSITION 2.20. *Let $\eta : \Gamma \longrightarrow \overline{\mathbb{Q}}_p^\times$ be a continuous character. Then we have the following commutative diagram:*

$$\begin{array}{ccc} \mathrm{Meas}(\Gamma; \mathcal{O}) & \xrightarrow{\mu \mapsto \int_\Gamma \eta(g) d\mu} & \overline{\mathbb{Q}}_p \\ \| & & \| \\ \Lambda_{\mathcal{O}} & \xrightarrow[x \mapsto \eta(x)]{} & \overline{\mathbb{Q}}_p. \end{array} \tag{2.17}$$

Here, the left vertical identification is the canonical isomorphism obtained in Lemma 2.19.

2.2.3. Iwasawa algebra as a p-adic analogue of the ring of holomorphic functions. We denote by $\mathrm{C}_{\mathrm{cont}}(\Gamma, \overline{\mathbb{Q}}_p^\times)$ the set of continuous characters $\Gamma \longrightarrow \overline{\mathbb{Q}}_p^\times$. Let $| \ |_p$ be a p-adic absolute value on $\overline{\mathbb{Q}}_p$ normalized as $|p|_p = \frac{1}{p}$. We denote a p-adic open disc in $\overline{\mathbb{Q}}_p$ of radius 1 centered at $1 \in \overline{\mathbb{Q}}_p$ by

$$U(1, 1; \overline{\mathbb{Q}}_p) = \{x \in \overline{\mathbb{Q}}_p \mid |x - 1|_p < 1\}.$$

Each time we choose a topological generator γ of Γ, we have the following non-canonical bijection map depending on the choice of γ:

$$\mathrm{C}_{\mathrm{cont}}(\Gamma, \overline{\mathbb{Q}}_p^\times) \xrightarrow{\sim} U(1, 1; \overline{\mathbb{Q}}_p), \ \eta \mapsto \eta(\gamma). \tag{2.18}$$

In fact, the inverse of this isomorphism is the map which assigns a unique character η_x such that $\eta_x(\gamma) = x$ to any $x \in U(1, 1; \overline{\mathbb{Q}}_p)$.

Let us take a character $\eta \in \mathrm{C}_{\mathrm{cont}}(\Gamma, \overline{\mathbb{Q}}_p^\times)$. Then, for each $f \in \Lambda_{\mathcal{O}}$, $\eta(f) \in \overline{\mathbb{Q}}_p$ makes sense since the character η is naturally extended to a ring homomorphism $\Lambda_{\mathcal{O}} = \mathcal{O}[[\Gamma]] \xrightarrow{\eta} \overline{\mathbb{Q}}_p$ as explained just after Lemma 2.19. Now, let us set $f(\eta) := \eta(f)$ assuming that f remains fixed and the characters η vary. In this way, we regard an element $f \in \Lambda_{\mathcal{O}}$ as a function on the set $\mathrm{C}_{\mathrm{cont}}(\Gamma, \overline{\mathbb{Q}}_p^\times)$.

Let us fix a topological generator $\gamma \in \Gamma$ and denote by $F \in \mathcal{O}[[T]]$ the image of $f \in \Lambda_{\mathcal{O}}$ via the noncanonical isomorphism $\Lambda_{\mathcal{O}} \xrightarrow{\sim} \mathcal{O}[[T]]$ sending γ to $1 + T$. By definition, we have the following commutative diagram:

$$\begin{array}{ccc} \mathcal{O}[[T]] & \xrightarrow{F \mapsto F(x-1)} & \overline{\mathbb{Q}}_p \\ \| & & \| \\ \Lambda_{\mathcal{O}} & \xrightarrow[f \mapsto f(\eta_x)]{} & \overline{\mathbb{Q}}_p. \end{array} \tag{2.19}$$

The following theorem is an analogue of the identity theorem in the theory of holomorphic functions when we regard $\Lambda_{\mathcal{O}}$ as a p-adic analogue of the ring of holomorphic functions on an open disc of radius 1 in $\mathbb{C}$.

PROPOSITION 2.21 (Identity Theorem). *If $f \in \Lambda_{\mathcal{O}}$ vanishes at infinitely many different $\eta \in \mathrm{C}_{\mathrm{cont}}(\Gamma, \overline{\mathbb{Q}}_p^\times)$, we have $f = 0$.*

PROOF. Let us fix a topological generator $\gamma \in \Gamma$ and denote by $F \in \mathcal{O}[[T]]$ the image of $f \in \Lambda_{\mathcal{O}}$ via the noncanonical isomorphism $\Lambda_{\mathcal{O}} \xrightarrow{\sim} \mathcal{O}[[T]]$ sending γ to $1+T$. By the commutative diagram (2.19), $f(\eta_x) = 0$ is equivalent to $F(x-1) = 0$ for each $x \in U(1, 1; \overline{\mathbb{Q}}_p)$. So the condition of the proposition implies that $F(x-1) = 0$ holds for infinitely many different $x \in U(1, 1; \overline{\mathbb{Q}}_p)$. By the p-adic Weierstrass preparation theorem (Theorem 2.15), any $F(T) \in \mathcal{O}[[T]]$ has only finitely many zeros as far as $F(T) \neq 0$. Thus, we must have $F(T) = 0$ when $F(T)$ has infinitely many zero. This completes the proof. □

In the case where $\Gamma = \Gamma_{\mathrm{cyc}}$, we will introduce a set of special points in $\mathrm{C}_{\mathrm{cont}}(\Gamma_{\mathrm{cyc}}, \overline{\mathbb{Q}}_p^\times)$ as follows, which plays important roles throughout the book.

DEFINITION 2.22. Let w be an integer. A character $\eta \in \mathrm{C}_{\mathrm{cont}}(\Gamma_{\mathrm{cyc}}, \overline{\mathbb{Q}})$ is called an **arithmetic character** of weight w if there exists an open subgroup $U \subset \Gamma_{\mathrm{cyc}}$ (depending on η) and we have $\eta|_U = \kappa_{\mathrm{cyc}}^w$ (Recall that κ_{cyc} was introduced in Definition 2.5).

Note that, for each $w \in \mathbb{Z}$, there exist countably infinite arithmetic characters of weight w. Later, we discuss various p-adic L-functions which satisfy a suitable interpolation property for their evaluations at arithmetic characters of specific weights. Hence, the identity theorem stated above allow us to characterize the p-adic L-functions uniquely from their interpolation properties.

The notion of arithmetic characters is generalized to higher rank group $G \cong \mathbb{Z}_p^d$ or to their Iwasawa algebra $\mathcal{O}[[G]]$. This generalization plays an essential role in Part III (Chapters 7 and 8) where we discuss "Iwasawa theory of p-adic Galois deformations".

2.2.4. Iwasawa algebra as deformation ring or Hecke algebra. Iwasawa algebra appears in various aspects of p-adic theory in number theory and arithmetic geometry. Below, we review aspects of Iwasawa algebra as "Deformation algebra of Galois representation" and "ordinary p-adic Hecke algebra", which will be important in Part III.

(1) Thanks to the theory of deformation for Galois representations initiated by Mazur (see §7.3.2 of the second volume of this book), when we are given an irreducible representation $\overline{\rho}: G_K \longrightarrow \mathrm{Aut}_{\mathbb{F}_q}(\overline{T})$ of rank d over a finite field $\mathbb{F}_q$ of characteristic p of the absolute Galois group G_K of a finite algebraic number field K, the set of deformations of $\overline{\rho}$ satisfying certain deformation conditions is represented by a universal deformation ring $\mathcal{R}$ on which we have a universal Galois deformation $\mathcal{T} \cong \mathcal{R}^{\oplus d}$.. (See §7.3.2 of the second volume of this book, Mazur's original article [**72**] as well as some related articles in [**133**].) When $d = 1$, it is not difficult to see that a generalized Iwasawa algebra $\mathcal{O}[[\widetilde{\Gamma}_K]]$ which appeared in Remark 2.10 gives an example of the universal deformation ring under certain conditions on K. Even for general d, the universal deformation ring is known to be isomorphic to an Iwasawa algebra of several variables when the deformation is "unobstructed".

(2) Let us consider the space of automorphic forms on a certain reductive group G over a number field K with specified weight and level. Since automorphic forms on G are functions on the group of adèles-valued points $G(\mathbb{A}_K)$ of G which satisfy certain invariance conditions with respect to certain open subgroup of $G(\mathbb{A}_K)$, the space of automorphic forms on G has an action of the Hecke algebra, which is generated by appropriate classes of double cosets of $G(\mathbb{A}_K)$ by an open subgroup of $G(\mathbb{A}_K)$.

In Part II (from Chapter 4 to Chapter 6) and Part III (from 7 to Chapter 8) of this book, the case of $G = \mathrm{GL}(2)_K$ will appear as an important example. However, since the case of $G = \mathrm{GL}(1)_K$ is more basic than the case of $G = \mathrm{GL}(2)_K$, we start from this case. In fact, an automorphic form on $G = \mathrm{GL}(1)_K$ is nothing but an algebraic Hecke character introduced in §A.2 and we can say that Part I corresponds to an Iwasawa theory for automorphic forms on $G = \mathrm{GL}(1)_K$.

The space of automorphic forms on $\mathrm{GL}(1)_K$ with specified weight and level is independent of the weight and has the action of the Hecke algebra which is isomorphic to the group ring $\mathcal{R}[\mathrm{Cl}(K, \mathfrak{N})]$ where $\mathcal{R}$ is a certain coefficient algebra, $\mathfrak{N} \subset \mathfrak{r}_K$ is the specified level and $\mathrm{Cl}(K, \mathfrak{N})$ is a narrow ideal class group of conductor $\mathfrak{N}$ (see [**38**] for more detail). Recall that, for each natural number n, the Iwasawa algebra was isomorphic to the limit of group rings $\mathbb{Z}_p[\Gamma_n]$ where Γ_n is canonically isomorphic to the p-Sylow subgroup of the narrow ideal class group of $\mathbb{Q}$ with conductor p^n. Thus the Iwasawa algebra is also regarded as the projective limit of the Hecke algebra for the space of automorphic forms on $\mathrm{GL}(1)_K$ with level p^n.

The main point of this book is to discuss the generalization and the reconstruction of Iwasawa theory through the idea of deformations. In Hida deformation which is explained in §7.2 of the second volume of this book, we consider the projective limit of the Hecke algebra of level p^n with coefficients in $\mathbb{Z}_p$. Hence, it will be also important to interpret Iwasawa algebra as Hecke algebra of the setting of Hida deformation.

2.3. Iwasawa modules

In Iwasawa theory, it is important to study compact modules over an Iwasawa algebra (we often call such a module an **Iwasawa module**) and we study various invariants associated to Iwasawa modules. In this section, we prepare some fundamental results on Iwasawa modules.

In most of textbooks on Iwasawa theory, we often restrict ourselves to Iwasawa modules over Iwasawa algebras $\Lambda_{\mathcal{O}}$ of one variable. However, in Part III of this book, we will work with Iwasawa algebras of several variables. Also, some of results for Iwasawa modules are naturally generalized to modules over quite general algebras. If we state the result and prove it in a setting as general as possible, it will help us to understand the principles and the hypothesis with which the proof works. Hence, we first state and prove the result in a setting as general as possible in §2.3.2. Then, in §2.3.3, we specialize these results to the case of Iwasawa modules to deduce and state well-known standard results on Iwasawa modules.

In this section, the reader might be required a bit stronger background on the commutative algebra compared to §2.2. For example, the reader might be required the level of the knowledge found in the text [**71**] to follow the proof.

2.3.1. Basics on Iwasawa modules.

LEMMA 2.23. *Let $\mathcal{M}$ be a compact $\mathcal{O}$-module with continuous Γ-action. Then M is naturally endowed with a structure of compact $\Lambda_{\mathcal{O}}$-module.*

PROOF. Let us denote by $\mathcal{M}_{\Gamma^{p^n}}$ the largest Γ^{p^n}-coinvariant quotient of $\mathcal{M}$. That is, $\mathcal{M}_{\Gamma^{p^n}}$ is the largest quotient of $\mathcal{M}$ on which Γ^{p^n} acts trivially. Since a finite group $\Gamma_n = \Gamma/\Gamma^{p^n}$ acts on the $\mathcal{O}$-module $\mathcal{M}_{\Gamma^{p^n}}$, $\mathcal{M}_{\Gamma^{p^n}}$ has a structure of $\mathcal{O}[\Gamma_n]$-module. Hence, $\varprojlim_n \mathcal{M}_{\Gamma^{p^n}}$ is naturally endowed with a structure of compact-module over $\Lambda_{\mathcal{O}} = \varprojlim_n \mathcal{O}[\Gamma_n]$. To complete the proof, it is sufficient to show that the natural map $\mathcal{M} \to \varprojlim_n \mathcal{M}_{\Gamma^{p^n}}$ is an isomorphism. By using a topological generator γ of Γ, the kernel of the map $\mathcal{M} \to \varprojlim_n \mathcal{M}_{\Gamma^{p^n}}$ is isomorphic to $(\gamma^{p^n}-1)\mathcal{M}$ for each natural number n. Hence, the kernel of the map $\mathcal{M} \to \varprojlim_n \mathcal{M}_{\Gamma^{p^n}}$ is isomorphic to $\bigcap_{n\geq 1}(\gamma^{p^n} - 1)\mathcal{M}$. By the continuity of the action of Γ on $\mathcal{M}$, we have $\bigcap_{n\geq 1}(\gamma^{p^n} - 1)\mathcal{M}$, which implies the injectivity of the map $\mathcal{M} \to \varprojlim_n \mathcal{M}_{\Gamma^{p^n}}$.

In order to prove the surjectivity, let us take an element $\varprojlim_n m_n \in \varprojlim_n \mathcal{M}_{\Gamma^{p^n}}$. For each natural number n, an element $m_n \in \mathcal{M}_{\Gamma^{p^n}}$ is a closed subset of $\mathcal{M}_{\Gamma^{p^n}}$. Hence, the inverse image $Z_n \subset \mathcal{M}$ of m_n via $\mathcal{M} \longrightarrow \mathcal{M}_{\Gamma^{p^n}}$ is a nonempty closed subset of $\mathcal{M}$ and we have $Z_{n+1} \subset Z_n$. Since $\mathcal{M}$ is compact, we obtain $\bigcap_{n\geq 1} Z_n \neq \emptyset$, which implies the surjectivity of the map $\mathcal{M} \to \varprojlim_n \mathcal{M}_{\Gamma^{p^n}}$. □

REMARK 2.24. Let A be a discrete $\mathcal{O}$-module with continuous Γ-action. Then the Pontrjagin dual $\mathcal{M} := A^\vee$ of A is a compact $\mathcal{O}$-module with continuous Γ-action. Hence, $\mathcal{M}$ has a structure of compact $\Lambda_{\mathcal{O}}$-module. Thus $A = (A^\vee)^\vee$ has a structure of discrete $\Lambda_{\mathcal{O}}$-module.

In general, a topological $\mathcal{O}$-module $\mathcal{M}$ with continuous Γ-action might not have a structure of $\Lambda_{\mathcal{O}}$-module if $\mathcal{M}$ is neither compact nor discrete. For example, $\mathcal{O}[\Gamma]$ is a $\Lambda_{\mathcal{O}}$-module is a $\mathcal{O}$-module with continuous Γ-action, but it doe not have a natural structure of $\Lambda_{\mathcal{O}}$-module.

THEOREM 2.25 (Topological Nakayama's lemma for Iwasawa modules). *Let $\mathcal{M}$ be a compact $\Lambda_{\mathcal{O}}$-module and we denote the maximal ideal of $\Lambda_{\mathcal{O}}$ by $\mathfrak{M}$. Then the following statements hold:*

(1) *If $\mathcal{M}/\mathfrak{M}\mathcal{M} = 0$, we have $\mathcal{M} = 0$.*
(2) *If $\mathcal{M}/\mathfrak{M}\mathcal{M}$ is of finite cardinality, $\mathcal{M}$ is a finitely generated $\Lambda_{\mathcal{O}}$-module.*

COROLLARY 2.26. *Let $\mathcal{M}$ be a compact $\Lambda_{\mathcal{O}}$-module. If there exists a non unit $x \in \Lambda_{\mathcal{O}}$ such that $\mathcal{M}/(x)\mathcal{M}$ is a finitely generated $\Lambda_{\mathcal{O}}/(x)$-module, $\mathcal{M}$ is a finitely generated $\Lambda_{\mathcal{O}}$-module.*

PROOF OF COROLLARY 2.26. Since we assume $x \notin (\Lambda_{\mathcal{O}})^{\times}$, we have $(x) \subset \mathfrak{M}$. Hence, $\mathcal{M}/\mathfrak{M}\mathcal{M}$ is a finitely generated $\Lambda_{\mathcal{O}}/\mathfrak{M}$-module. Since $\Lambda_{\mathcal{O}}/\mathfrak{M}$ is a finite field, the module $\mathcal{M}/\mathfrak{M}\mathcal{M}$ is of finite cardinality. Thus, Theorem 2.25 (2) implies that $\mathcal{M}$ is a finitely generated $\Lambda_{\mathcal{O}}$-module. □

For generalized Iwasawa theory for Galois deformations which we discuss in this book, general rings, such as Iwasawa algebras of several variables, are important. Hence, we will state Theorem 2.27 below which is a generalized and refined version of Theorem 2.25 above. Then we will prove Theorem 2.27, which will prove Theorem 2.25 at the same time.

THEOREM 2.27 (Generalized topological Nakayama's lemma). *Let $\mathcal{R}$ be a complete local Noetherian ring with finite residue field*[17], *$\mathcal{M}$ a compact $\mathcal{R}$-module. Let us denote the maximal ideal of $\mathcal{R}$ by $\mathfrak{M}$.*

Assume that there exist elements $x_1, \cdots, x_s \in \mathcal{M}$ such that the $\mathcal{R}/\mathfrak{M}$-module $\mathcal{M}/\mathfrak{M}\mathcal{M}$ is generated by the images $\overline{x}_1, \cdots, \overline{x}_s \in \mathcal{M}/\mathfrak{M}\mathcal{M}$ of $x_1, \cdots, x_s \in \mathcal{M}$. Then, $\mathcal{M}$ is generated by $x_1, \cdots, x_s$. Especially, if $\mathcal{M}/\mathfrak{M}\mathcal{M}$ is a finitely generated $\mathcal{R}/\mathfrak{M}$-module, $\mathcal{M}$ is a finitely generated $\mathcal{R}$-module.

PROOF OF THEOREM 2.27 (AND OF THEOREM 2.25). Let us take $x_1, \cdots, x_s \in \mathcal{M}$ which satisfy the assumption of Theorem 2.27 and consider the closed subgroup $\sum_{i=1}^{s} \mathcal{R}x_i \subset \mathcal{M}$. Let us take an arbitrary element $m \in \mathcal{M}$. It suffices to show that $m \in \sum_{i=1}^{s} \mathcal{R}x_i$. Since $\mathcal{M}/\mathfrak{M}\mathcal{M}$ is generated by the images $\overline{x}_1, \cdots, \overline{x}_s \in \mathcal{M}/\mathfrak{M}\mathcal{M}$ of $x_1, \cdots, x_s$, there exists an element $m_{(1)} \in \sum_{i=1}^{s} \mathcal{R}x_i$ such that $m - m_{(1)} \in \mathfrak{M}\mathcal{M}$. By an induction argument, we will show that, for any natural number n, there exists an element $m_{(n)} \in \sum_{i=1}^{s} \mathcal{R}x_i$ such that $m - m_{(n)} \in \mathfrak{M}^n\mathcal{M}$. Suppose that this holds true up to $n = k$ and take an element $m_{(k)} \in \sum_{i=1}^{s} \mathcal{R}x_i$ satisfying the desired condition. By assumption, there exist $r \in \mathfrak{M}^k$ and $m' \in \mathcal{M}$ such that $m - m_{(k)} = rm'$. My applying the assumption of Theorem 2.27 to m', we find an element $m'_{(1)} \in \sum_{i=1}^{s} \mathcal{R}x_i$ such that $m' - m'_{(1)} \in \mathfrak{M}\mathcal{M}$. By putting $m_{(k+1)} = m_{(k)} + rm'_{(1)}$, we obtain

$$m - m_{(k+1)} = r(m' - m'_{(1)}) \in \mathfrak{M}^k\mathfrak{M}\mathcal{M} = \mathfrak{M}^{k+1}\mathcal{M}.$$

[17] The finiteness of the residue field is necessary in order that $\mathcal{R}$ is a compact group whose base is given by an $\mathfrak{M}$-adic topology.

This shows the existence of a desired element $m_{(n)}$ for any natural number n. Since $\{m_{(n)}\}_{\{n\geq 1}$ forms a Cauchy sequence in a compact subset $\sum_{i=1}^{s}\mathcal{R}x_i$ of $\mathcal{M}$ with respect to $\mathfrak{M}$-adic topology, we have the limit $\lim_{n\to\infty} m_{(n)}$ in $\sum_{i=1}^{s}\mathcal{R}x_i$ which coincides with m. Thus m must is written as an $\mathcal{R}$-linear combination of $x_1, \cdots, x_s \in \mathcal{M}$. This completes the proof of Theorem 2.27. □

COROLLARY 2.28. *Let $\mathcal{R}$ be a complete local Noetherian ring with finite residue field, $\mathcal{M}$ a compact $\mathcal{R}$-module. Let us denote the maximal ideal of $\mathcal{R}$ by $\mathfrak{M}$. Then, if there exists a nonunit $x \in \mathcal{R}$ such that $\mathcal{M}/(x)\mathcal{M}$ is generated by s elements over $\mathcal{R}/(x)$, $\mathcal{M}$ is generated by s elements over $\mathcal{R}$.*

PROOF. Since $x \notin \mathcal{R}^\times$, we have $(x) \subset \mathfrak{M}$ and $\mathcal{M}/\mathfrak{M}\mathcal{M}$ is generated by s elements over $\mathcal{R}/\mathfrak{M}$. By Theorem 2.27, $\mathcal{M}$ is generated by s elements over $\mathcal{R}$. □

2.3.2. Structure theorem of modules over normal Noetherian domains. Later in §2.3.3, we will state the structure theorem of Iwasawa modules (Theorem 2.39). Theorem 2.39 itself is well-known in the usual setting of Iwasawa modules of one variable. However, in a more general setting of Iwasawa modules of several variables studied in the latter part of the book, some theorems analogous to Theorem 2.39 hold true. As some of the specific natures of Iwasawa algebra, such as completeness, are not essential in the proof of the structure theorem, we will state and prove a general structure theorem (Theorem 2.36) for finitely generated modules over a general normal Noetherian domain $\mathcal{R}$. Then, in §2.3.3, we prove that the structure theorem of Iwasawa modules (Theorem 2.39) follows from Theorem 2.34 and a general structure theorem (Theorem 2.36).

For a general normal Noetherian domain $\mathcal{R}$, we denote the set of prime ideals of height one in $\mathcal{R}$ by $P^1(\mathcal{R})$. By the assumption that $\mathcal{R}$ is normal, the localization $\mathcal{R}_\mathfrak{p}$ at each $\mathfrak{p} \in P^1(\mathcal{R})$ is a discrete valuation ring. For a finitely generated torsion $\mathcal{R}$-module $\mathcal{M}$ and for a prime ideal $\mathfrak{p} \in P^1(\mathcal{R})$, the length of the localization $\mathcal{M}_\mathfrak{p}$ as an $\mathcal{R}_\mathfrak{p}$-module is finite. We denote the length by $l(\mathfrak{p}, \mathcal{M})$.

DEFINITION 2.29. Let $\mathcal{R}$ be a normal Noetherian domain and $\mathcal{M}$ a finitely generated $\mathcal{R}$-module.

(1) If we have $l(\mathfrak{p}, \mathcal{M}) = 0$ for any $\mathfrak{p} \in P^1(\mathcal{R})$, we call $\mathcal{M}$ a **pseudo-null $\mathcal{R}$-module**[18].
(2) The largest pseudo-null $\mathcal{R}$-submodule of $\mathcal{M}$ is called **the maximal pseudo-null $\mathcal{R}$-submodule** of $\mathcal{M}$ and we denote it by $\mathcal{M}_{\text{null}}$.[19]
(3) Let $\mathcal{N}$ be also a finitely generated $\mathcal{R}$-module and $f : \mathcal{M} \longrightarrow \mathcal{N}$ an $\mathcal{R}$-linear homomorphism. We call f a **pseudo-isomorphism** if the kernel and the cokernel of f are both pseudo-null $\mathcal{R}$modules.

REMARK 2.30. (1) Let $\mathcal{R}$ be a normal Noetherian domain. For finitely generated $\mathcal{R}$-modules $\mathcal{M}$ and $\mathcal{N}$, the existence of a pseudo-isomorphism $f : \mathcal{M} \longrightarrow \mathcal{N}$ does not necessarily give an equivalence relation between $\mathcal{M}$ and $\mathcal{N}$. For example, in the case, $\mathcal{R} = \Lambda$, $\mathcal{M} = (p, T)$, $\mathcal{N} = \mathcal{R}$, a natural inclusion map $\mathcal{M} \longrightarrow \mathcal{N}$ is a pseudo-isomorphism, but we have no pseudo-isomorphism $\mathcal{N} \longrightarrow \mathcal{M}$ in other direction.

[18]Note that a pseudo-null $\mathcal{R}$-module is a torsion $\mathcal{R}$-module.

[19]We leave the reader to check that such a module $\mathcal{M}_{\text{null}}$ exists uniquely and $\mathcal{M}_{\text{null}}$ is also characterized to be the smallest $\mathcal{R}$-submodule $\mathcal{N}$ of $\mathcal{M}$ such that $\mathcal{M}/\mathcal{N}$ contains no nonzero pseudo-null $\mathcal{R}$-submodule.

(2) Let $\mathcal{O}$ be the ring of integers of a p-adic field, $\mathcal{R} = \Lambda_{\mathcal{O}}$. Then, a finitely generated $\mathcal{R}$-module $\mathcal{M}$ is a pseudo-null $\mathcal{R}$-module if and only if we have $\#\mathcal{M} < \infty$.

As discussed in Remark 2.30 (1), between two finitely generated $\mathcal{R}$-modules, the existence of a pseudo-isomorphism does not give an equivalence relation. However, between two finitely generated torsion $\mathcal{R}$-modules, the existence of a pseudo-isomorphism does not give an equivalence relation. That is, if there exists a pseudo-isomorphism $\mathcal{M} \longrightarrow \mathcal{N}$ between finitely generated torsion $\mathcal{R}$-modules $\mathcal{M}$ and $\mathcal{N}$, there also exists a pseudo-isomorphism $\mathcal{N} \longrightarrow \mathcal{M}$ [20]. For finitely generated torsion $\mathcal{R}$-modules $\mathcal{M}$ and $\mathcal{N}$, we will denote by $\mathcal{M} \sim \mathcal{N}$ the equivalence relation given by the existence of a pseudo-isomorphism.

DEFINITION 2.31. Let $\mathcal{R}$ be a ring, $\mathcal{N}$ a finitely generated $\mathcal{R}$-module. Let us denote the functor $\mathrm{Hom}_{\mathcal{R}}(\ \mathcal{R})$ taking the $\mathcal{R}$-linear dual by $(\ )^*$ for short. Then, if the canonical map $\mathcal{N} \longrightarrow \mathcal{N}^{**}$ associating the double $\mathcal{R}$-linear dual is isomorphism, $\mathcal{N}$ is called a **reflexive $\mathcal{R}$-module** .

As is seen immediately from the definition, any finitely generated free $\mathcal{R}$-module is a reflexive $\mathcal{R}$-module. Also, when $\mathcal{R}$ is a domain, any reflexive $\mathcal{R}$-module has no nontrivial torsion element.

PROPOSITION 2.32. *Let $\mathcal{R}$ be a normal Noetherian domain. Let $\mathcal{M}$ and $\mathcal{N}$ be finitely generated torsion-free $\mathcal{R}$-modules. Then the following statements hold.*

(1) $\mathcal{M}^* = \bigcap_{\mathfrak{p}\in P^1(\mathcal{R})}(\mathcal{M}^*)_{\mathfrak{p}}$.
(2) $\mathcal{N}^{**} = \bigcap_{\mathfrak{p}\in P^1(\mathcal{R})}\mathcal{N}_{\mathfrak{p}}$.

PROOF. Let us prove the assertion (1). Since $\mathcal{M}^*$ is a torsion-free $\mathcal{R}$-module, we have $\mathcal{M}^* \subset (\mathcal{M}^*)_{\mathfrak{p}}$ for each $\mathfrak{p} \in P^1(\mathcal{R})$. Hence, we have $\mathcal{M}^* \subset \bigcap_{\mathfrak{p}\in P^1(\mathcal{R})}(\mathcal{M}^*)_{\mathfrak{p}}$. Now, Let us take an element $f \in \bigcap_{\mathfrak{p}\in P^1(\mathcal{R})}(\mathcal{M}^*)_{\mathfrak{p}}$. Since $\mathcal{R}$ is normal, we have $\mathcal{R} = \bigcap_{\mathfrak{p}\in P^1(\mathcal{R})}\mathcal{R}_{\mathfrak{p}}$. Hence, we have $f(m) \in \mathcal{R}$ for any $m \in \mathcal{M}$, which means that $f \in \mathcal{M}^* = \mathrm{Hom}_{\mathcal{R}}(\mathcal{M}, \mathcal{R})$. This completes the proof of the opposite inclusion $\mathcal{M}^* \supset (\mathcal{M}^*)_{\mathfrak{p}}$.

Let us prove the assertion (2). By applying the result of the assertion (1) to $\mathcal{M} = \mathcal{N}^*$, we obtain $\mathcal{N}^{**} = \bigcap_{\mathfrak{p}\in P^1(\mathcal{R})}(\mathcal{N}^{**})_{\mathfrak{p}}$.

Since taking the localization functor and the dual functor commutes to each other, $(\mathcal{N}^{**})_{\mathfrak{p}}$ is canonically isomorphic to the double $\mathcal{R}_{\mathfrak{p}}$-linear dual of $\mathcal{N}_{\mathfrak{p}}$. Note that any finitely generated torsion-free $\mathcal{R}_{\mathfrak{p}}$-module is free since $\mathcal{R}_{\mathfrak{p}}$ is a discreta valuation ring. Hence, $(\mathcal{N}^{**})_{\mathfrak{p}}$ is canonically isomorphic to $\mathcal{N}_{\mathfrak{p}}$, which completes the proof. □

COROLLARY 2.33. *Let $\mathcal{R}$ be a normal Noetherian domain. Then, for any nontrivial finitely generated torsion-free $\mathcal{R}$-module $\mathcal{M}$, the $\mathcal{R}$-linear dual $\mathcal{M}^*$ is a nontrivial reflexive $\mathcal{R}$-module.*

[20] We do not prove this facts here, but this can be checked by the argument of the proof of Theorem 2.36 (1).

PROOF. By Proposition 2.32 (2), a finitely generated torsion-free $\mathcal{R}$-module $\mathcal{N}$ is reflexive if and only if

$$\mathcal{N} = \bigcap_{\mathfrak{p} \in P^1(\mathcal{R})} \mathcal{N}_{\mathfrak{p}} \tag{2.20}$$

holds. On the other hand, $\mathcal{N} = \mathcal{M}^*$ satisfies (2.20) by Proposition 2.32 (1). This completes the proof. □

THEOREM 2.34. *Let $\mathcal{R}$ be a complete local regular Noetherian domain whose Krull dimension is equal or smaller than 2. Then, any reflexive $\mathcal{R}$-module $\mathcal{N}$ is a free $\mathcal{R}$-module.*

PROOF. When the Krull dimension of $\mathcal{R}$ is 0 (resp. 1), $\mathcal{R}$ is a field (resp. a discrete valuation ring). Since the conclusion is clear in these cases, in the rest of the proof, we will discuss the case where the Krull dimension of $\mathcal{R}$ is 2. By the assumption that $\mathcal{R}$ is a complete local regular Noetherian domain of Krull dimension 2, there exists an element $x \in \mathcal{R}$ such that $\mathcal{R}/(x)$ becomes a discrete valuation ring. Since we have an isomorphism

$$\mathcal{N}^{**}/(x)\mathcal{N}^{**} = \mathrm{Hom}_{\mathcal{R}}(\mathcal{N}^*, \mathcal{R}) \otimes_{\mathcal{R}} \mathcal{R}/(x) \cong \mathrm{Hom}_{\mathcal{R}}(\mathcal{N}^*, \mathcal{R}/(x)),$$

$\mathcal{N}^{**}/(x)\mathcal{N}^{**}$ is a finitely generated torsion-free $\mathcal{R}/(x)$-module. By the assumption that $\mathcal{R}/(x)$ is a discrete valuation ring, $\mathcal{N}^{**}/(x)\mathcal{N}^{**}$ must be a free $\mathcal{R}/(x)$-module of finite rank. Since $\mathcal{N}$ is a reflexive $\mathcal{R}$-module, $\mathcal{N}/(x)\mathcal{N}$ is also a free $\mathcal{R}/(x)$ of finite rank. Now, let us give a presentation of the $\mathcal{R}$-module $\mathcal{N}$ with its minimal set of generators as follows:

$$0 \longrightarrow \mathrm{Ker}(q) \longrightarrow \mathcal{R}^{\oplus s} \xrightarrow{q} \mathcal{N} \longrightarrow 0. \tag{2.21}$$

Note that $\mathcal{N}$ is a torsion-free $\mathcal{R}$-module because $\mathcal{N}$ is a reflexive $\mathcal{R}$-module. Hence, we obtain the following exact sequence by applying $\otimes_{\mathcal{R}}\mathcal{R}/(x)$ to (2.21):

$$0 \longrightarrow \mathrm{Ker}(q)/(x)\mathrm{Ker}(q) \longrightarrow (\mathcal{R}/(x))^{\oplus s} \xrightarrow{q} \mathcal{N}/(x)\mathcal{N} \longrightarrow 0.$$

By the minimality of the number of generators s and by the fact that $\mathcal{N}/(x)\mathcal{N}$ is a free $\mathcal{R}/(x)$-module, Corollary 2.28 of the generalized topological Nakayama's lemma (Theorem 2.27) implies $\mathrm{Ker}(q)/(x)\mathrm{Ker}(q) = 0$. Again, by applying Corollary 2.28 to $\mathrm{Ker}(q)$, we obtain $\mathrm{Ker}(q) = 0$. Hence, (2.21) implies that $\mathcal{N}$ is a free $\mathcal{R}$-module of finite rank. □

REMARK 2.35. When $\mathcal{R}$ is a complete local regular Noetherian domain whose Krull-dimension is equal to or greater than 3, a finitely generated reflexive $\mathcal{R}$-module $\mathcal{N}$ is not necessarily a free $\mathcal{R}$-module. For example, let us consider a complete local regular Noetherian domain $\mathcal{R} = \mathbb{Z}_p[[T_1, T_2]]$ and let us consider a projective resolution of $\mathcal{R}$-module $\mathbb{F}_p$ as follows:

$$0 \longrightarrow \mathbb{Z}_p[[T_1,T_2]] \xrightarrow{a_1} \mathbb{Z}_p[[T_1,T_2]]^{\oplus 3} \xrightarrow{a_2} \mathbb{Z}_p[[T_1,T_2]]^{\oplus 3} \xrightarrow{a_3} \mathbb{Z}_p[[T_1,T_2]] \longrightarrow \mathbb{F}_p \longrightarrow 0,$$

where the map a_1 is given by $x \mapsto (xT_2, xp, -xT_1)$, the map a_2 is given by $(x_1, x_2, x_3) \mapsto (x_1T_1 + x_3T_2, -x_1p + x_2T_2, -x_2T_1 - x_3p)$, the map a_3 is given by $(y_1, y_2, y_3) \mapsto y_1p + y_2T_1 + y_3T_2$. Put $\mathcal{N} = \mathrm{Im}(a_2)$. By the definition of $\mathcal{N}$, we can decompose the above long exact sequence to two short exact sequences as follows:

$$0 \longrightarrow \mathbb{Z}_p[[T_1,T_2]] \xrightarrow{a_1} \mathbb{Z}_p[[T_1,T_2]]^{\oplus 3} \xrightarrow{a_2} \mathcal{N} \longrightarrow 0, \tag{2.22}$$

$$0 \longrightarrow \mathcal{N} \longrightarrow \mathbb{Z}_p[[T_1,T_2]]^{\oplus 3} \xrightarrow{a_3} \mathbb{Z}_p[[T_1,T_2]] \longrightarrow \mathbb{F}_p \longrightarrow 0. \tag{2.23}$$

By applying the functor $\mathrm{Hom}_{\mathbb{Z}_p[[T_1,T_2]]}(\ \mathbb{Z}_p[[T_1,T_2]])$ to (2.22), we obtain

$$\begin{aligned}(2.24)\quad 0 \longrightarrow \mathcal{N}^* \xrightarrow{a_2^*} \mathbb{Z}_p[[T_1,T_2]]^{\oplus 3} \xrightarrow{a_1^*} \mathbb{Z}_p[[T_1,T_2]] \\ \longrightarrow \mathrm{Ext}^1_{\mathbb{Z}_p[[T_1,T_2]]}(\mathcal{N},\mathbb{Z}_p[[T_1,T_2]]) \longrightarrow 0.\end{aligned}$$

By definition, the $\mathbb{Z}_p[[T_1,T_2]]$-linear dual map a_1^* of a_1 is identified with a_3. Hence, by comparing (2.23) and (2.24), we obtain $\mathcal{N} \cong \mathcal{N}^*$ and

$$\mathrm{Ext}^1_{\mathbb{Z}_p[[T_1,T_2]]}(\mathcal{N},\mathbb{Z}_p[[T_1,T_2]]) \cong \mathbb{F}_p.$$

The former fact $\mathcal{N} \cong \mathcal{N}^*$ says that $\mathcal{N}$ is a reflexive $\mathbb{Z}_p[[T_1,T_2]]$-module. The latter fact says that $\mathcal{N}$ is not a free $\mathbb{Z}_p[[T_1,T_2]]$-module.

THEOREM 2.36 (General structure theorem). *Let $\mathcal{R}$ be a normal Noetherian domain. Then the following statements hold:*

(1) *Let $\mathcal{M}$ be a finitely generated $\mathcal{R}$-module and let us denote the maximal torsion $\mathcal{R}$-submodule of $\mathcal{M}$ by $\mathcal{M}_{\mathrm{tor}}$. Then there exists a pseudo-isomorphism $a : \mathcal{M} \longrightarrow (\mathcal{M}/\mathcal{M}_{\mathrm{tor}}) \oplus \mathcal{M}_{\mathrm{tor}}$ such that the restriction $a|_{\mathcal{M}_{\mathrm{tor}}}$ to $\mathcal{M}_{\mathrm{tor}}$ is a multiplication by certain constant on $\mathcal{M}_{\mathrm{tor}}$.*
(2) *For any finitely generated torsion-free $\mathcal{R}$-module $\mathcal{M}$, there exists a pseudo-isomorphism $b : \mathcal{M} \longrightarrow \mathcal{N}$ such that $\mathcal{N}$ is a reflexive $\mathcal{R}$-module.*
(3) *For any finitely generated torsion $\mathcal{R}$-module $\mathcal{M}$, there exists a pseudo-isomorphism $c : \mathcal{M} \longrightarrow \bigoplus_{i=1}^u \mathcal{R}/\mathfrak{q}_i^{r_i}$ where $\mathfrak{q}_1, \cdots, \mathfrak{q}_u$ are height-one prime ideals of $\mathcal{R}$ and $r_1, \cdots, r_u$ are natural numbers. Here, the components $\mathcal{R}/\mathfrak{q}_i^{r_i}$ are unique modulo permutation and we allow the case $i \neq j$ and $\mathfrak{q}_i = \mathfrak{q}_j$.*

PROOF. First, we shall prove the assertion (1). Let us consider the quotient map $\mathcal{M} \twoheadrightarrow \mathcal{M}/\mathcal{M}_{\mathrm{tor}}$ which we denote by a_1. We want to construct an $\mathcal{R}$-linear map $a_2 :\ \mathcal{M} \longrightarrow \mathcal{M}_{\mathrm{tor}}$ so that

$$a = a_1 \oplus a_2 :\ \mathcal{M} \longrightarrow (\mathcal{M}/\mathcal{M}_{\mathrm{tor}}) \oplus \mathcal{M}_{\mathrm{tor}},\ x \mapsto a_1(x) \oplus a_2(x)$$

satisfies the same property as the desired map. Let us consider $\{\mathfrak{p}_1, \cdots, \mathfrak{p}_t\} = \{\mathfrak{p} \in P^1(\mathcal{R}) \mid (\mathcal{M}_{\mathrm{tor}})_{\mathfrak{p}} \neq 0\}$ and define a multiplicative system $S \subset \mathcal{R}$ to be $S = \mathcal{R} - \cup_{i=1}^t \mathfrak{p}_i$. If the set $\{\mathfrak{p}_1, \cdots, \mathfrak{p}_t\}$ is empty, $\mathcal{M}_{\mathrm{tor}}$ is a pseudo-null $\mathcal{R}$-module. In this case, The map a satisfies the desired properties if we put $a_2 = 0$. Hence, from now on, we will assume that $t \geq 1$, that is, $\mathcal{M}_{\mathrm{tor}}$ is not a pseudo-null $\mathcal{R}$-module. Since $\mathcal{R}$ is a normal Noetherian domain and the prime ideals $\mathfrak{p}_1, \cdots, \mathfrak{p}_t$ are of height one, the localized ring $S^{-1}\mathcal{R}$ is a Dedekind domain. Since the Dedekind domain $S^{-1}\mathcal{R}$ has only finitely many prime ideals, $S^{-1}\mathcal{R}$ is a PID. By using the structure theorem of finitely generated modules over a PID, we have a decomposition:

$$S^{-1}\mathcal{M} = S^{-1}\mathcal{M}/S^{-1}\mathcal{M}_{\mathrm{tor}} \oplus S^{-1}\mathcal{M}_{\mathrm{tor}}.$$

We denote the projection map $S^{-1}\mathcal{M} \longrightarrow S^{-1}\mathcal{M}_{\mathrm{tor}}$ to the second component by q_{tor}. By the well-known property of the localization $S^{-1}\mathrm{Hom}_{\mathcal{R}}(\mathcal{M},\mathcal{M}_{\mathrm{tor}}) = \mathrm{Hom}_{S^{-1}\mathcal{R}}(S^{-1}\mathcal{M}, S^{-1}\mathcal{M}_{\mathrm{tor}})$, there exist an $\mathcal{R}$-linear map $a_2' : \mathcal{M} \longrightarrow \mathcal{M}_{\mathrm{tor}}$ and an element $s \in S$ such that $sq_{\mathrm{tor}} = a_2'$. On the other hand, since we have $q_{\mathrm{tor}}|_{S^{-1}\mathcal{M}_{\mathrm{tor}}} = \mathrm{Id}_{S^{-1}\mathcal{M}_{\mathrm{tor}}}$, there exist $s' \in S$ such that $s'q_{\mathrm{tor}}|_{\mathcal{M}_{\mathrm{tor}}} = s'\mathrm{Id}_{\mathcal{M}_{\mathrm{tor}}}$. Hence, by putting $a_2 = s'a_2'$, the map $a = a_1 \oplus a_2$ gives us a desired homomorphism. This completes the proof of the assertion (1).

Next, we shall prove the assertion (2). Since $\mathcal{M}$ is a finitely generated torsion-free $\mathcal{R}$-module, we have a canonical injection $b : \mathcal{M} \hookrightarrow \mathcal{N}$ where $\mathcal{N} = \mathcal{M}^{**}$. By Corollary 2.33, $\mathcal{N}$ is a nontrivial reflexive $\mathcal{R}$-module. Also, for any $\mathfrak{p} \in P^1(\mathcal{R})$, the localization $\mathcal{M}_\mathfrak{p}$ of $\mathcal{M}$ at $\mathfrak{p}$ is a free $\mathcal{R}_\mathfrak{p}$-module of finite rank. Hence, the localization $b_\mathfrak{p} : \mathcal{M}_\mathfrak{p} \longrightarrow \mathcal{N}_\mathfrak{p}$ of the map b at $\mathfrak{p}$ is a canonical isomorphism. Thus $b : \mathcal{M} \longrightarrow \mathcal{N}$ gives a desired pseudo-isomorphism. This completes the proof of the assertion (2).

Finally, we shall prove the assertion (3). As in the proof of the assertion (1), we consider the set $\{\mathfrak{p}_1, \cdots, \mathfrak{p}_t\} = \{\mathfrak{p} \in P^1(\mathcal{R}) \mid (\mathcal{M}_{\mathrm{tor}})_\mathfrak{p} \neq 0\}$. For each $\mathfrak{p}_i \in \{\mathfrak{p}_1, \cdots, \mathfrak{p}_t\}$, there exist natural numbers $r_{i,j}$ such that $\mathcal{M}_{\mathfrak{p}_i} \cong \oplus_{j=1}^{s_i} \mathcal{R}_{\mathfrak{p}_i}/\mathfrak{p}^{r_{i,j}}$ since $\mathcal{R}_{\mathfrak{p}_i}$ is a discrete valuation ring. We label all these natural numbers $r_{1,1}, r_{1,2}, r_{1,3}, \ldots, r_{t,s_t-2}, r_{t,s_t-1}, r_{t,s_t}$ by $r_1, r_2, \ldots, r_u$ and we denote the ideal in $\{\mathfrak{p}_1, \cdots, \mathfrak{p}_t\}$ corresponding to each r_i by $\mathfrak{q}_i$. Then we obtain $S^{-1}\mathcal{M} \cong S^{-1}\left(\oplus_{i=1}^{u} \mathcal{R}/\mathfrak{q}_i^{r_i}\right)$. By the same argument as the proof of (1), we obtain the desired pseudo-isomorphism $c : \mathcal{M} \longrightarrow \bigoplus_{i=1}^{u} \mathcal{R}/\mathfrak{q}_i^{r_i}$. This completes the proof of the assertion (3). □

Let us define the following invariants associated to a module over a normal Noetherian domain.

DEFINITION 2.37. Let $\mathcal{R}$ be a normal Noetherian domain and $\mathcal{M}$ a finitely generated torsion $\mathcal{R}$-module. Then a **divisor ideal** $\mathrm{Div}_\mathcal{R}(\mathcal{M})$ of $\mathcal{M}$ is defined by

$$\mathrm{Div}_\mathcal{R}(\mathcal{M}) = \prod_{\mathfrak{p} \in P^1(\mathcal{R})} \mathfrak{p}^{l(\mathfrak{p}, \mathcal{M})}.$$

DEFINITION 2.38. Let $\mathcal{R}$ be a normal Noetherian domain and $\mathcal{M}$ a finitely generated torsion $\mathcal{R}$-module. Then a **characteristic ideal** $\mathrm{char}_\mathcal{R}(\mathcal{M})$ of $\mathcal{M}$ is defined by

$$\mathrm{char}_\mathcal{R}(\mathcal{M}) = \{x \in \mathcal{R} \mid \mathrm{ord}_\mathfrak{p}\,(x) \geq l(\mathfrak{p}, \mathcal{M})\ \forall \mathfrak{p} \in P^1(\mathcal{R})\},$$

where, for each $\mathfrak{p} \in P^1(\mathcal{R})$, $\mathrm{ord}_\mathfrak{p}$ is an additive discrete valuation map which is normalized so that the value at a uniformizer of $\mathcal{R}_\mathfrak{p}$ is 1.

In the end of this subsection, we refer the reader to [**7**, Chap.7 §4], [**81**, Chap.V §1] for a further detail of some of contents in this subsection.

2.3.3. Structure theorem of Iwasawa modules and Iwasawa invariants. If we restrict ourselves to the situation of the Iwasawa algebra $\mathcal{R} = \Lambda_\mathcal{O}$, we have a bit more precise version of the general structure theorem given at §2.3.2 given for normal Noetherian domains $\mathcal{R}$. In this subsection, we formulate such a more precise structure theorem over Iwasawa algebras. Then we define and discuss the characteristic ideal and Iwasawa invariants associated to a finitely generated torsion $\Lambda_\mathcal{O}$-module.

THEOREM 2.39 (Structure theorem of Iwasawa modules). *Let $\mathcal{M}$ be a finitely generated $\mathcal{O}[[T]]$-module. Then we have the following 4-term exact sequence of $\mathcal{O}[[T]]$-modules:*

$$(2.25)\quad 0 \longrightarrow Z \longrightarrow \mathcal{M}$$
$$\longrightarrow \mathcal{O}[[T]]^{\oplus r} \oplus \bigoplus_{i=1}^{s} \mathcal{O}[[T]]/(F_i(T)^{n_i}) \oplus \bigoplus_{j=1}^{t} \mathcal{O}[[T]]/(\varpi^{m_j}) \longrightarrow Z' \longrightarrow 0,$$

where s, t are nonnegative integers, Z, Z' are finite abelian groups, $F_i(T)$ are irreducible distinguished polynomials (Note that we allow to have different $i \neq j$ such that $F_i(T) = F_j(T)$). Given a finitely generated $\mathcal{O}[[T]]$-module $\mathcal{M}$, the nonnegative integer $r = r(\mathcal{M})$ depends only on $\mathcal{M}$ and does not depend on the choice of a 4-term exact sequence as above. We call the invariant $r = r(\mathcal{M})$ the ***rank*** *of $\mathcal{M}$.*

We shall deduce Theorem 2.39 from Theorem 2.34 and Theorem 2.36 proved in §2.3.2.

PROOF OF THEOREM 2.39. By applying Theorem 2.36 to $\mathcal{R} = \mathcal{O}[[T]]$, there exists a pseudo-isomorphism

$$\mathcal{M} \longrightarrow \mathcal{N} \oplus \bigoplus_{i=1}^{u} \mathcal{O}[[T]]/\mathfrak{q}_i^{r_i}, \tag{2.26}$$

where $\mathcal{N}$ is a reflexive $\mathcal{O}[[T]]$-module, $\mathfrak{q}_1, \cdots, \mathfrak{q}_u$ are height-one prime ideals of $\mathcal{O}[[T]]$ and $q_1, \cdots, q_u$ are natural numbers. Here, the components $\mathcal{O}[[T]]/\mathfrak{q}_i^{r_i}$ are unique modulo permutation and we allow the case $i \neq j$ and $\mathfrak{q}_i = \mathfrak{q}_j$. By Remark 2.30 (2), the kernel and the cokernel of (2.26) are finite groups. Since $\mathcal{O}[[T]]$ is a complete local regular Noetherian domain of Krull dimension 2, the reflexive $\mathcal{O}[[T]]$-module $\mathcal{N}$ is a finitely generated free $\mathcal{O}[[T]]$-module by Theorem 2.34. Note that, any height-one prime ideal $\mathfrak{q}_i$ of $\mathcal{O}[[T]]$ is principal since $\mathcal{O}[[T]]$ is a local regular Noetherian domain. By the p-adic Weierstrass preparation theorem (Theoreom 2.15), a generator x_i of the principal ideal $\mathfrak{p}_i$ is equal to either a uniformizer ϖ of $\mathcal{O}$ or an irreducible distinguished polynomial $F_i(T)$ of $\mathcal{O}[T]$ modulo multiplication by an element of $\mathcal{O}[[T]]^\times$. We complete the proof of Theorem 2.39. □

COROLLARY 2.40. *Let $\mathcal{M}$ be a finitely generated $\mathcal{O}[[T]]$-module and $x \in \mathcal{O}[[T]]$ an irreducible element. Then, if $\mathcal{M}/x\mathcal{M}$ is a finitely generated torsion $\mathcal{O}[[T]]/(x)$-module, $\mathcal{M}$ is a finitely generated torsion $\mathcal{O}[[T]]$-module.*

PROOF. Let us decompose the 4-term exact sequence (2.25) into two short exact sequences

$$0 \longrightarrow Z \longrightarrow \mathcal{M} \longrightarrow \mathcal{M}/Z \longrightarrow 0$$

and

$$0 \longrightarrow \mathcal{M}/Z \longrightarrow \mathcal{O}[[T]]^{\oplus r} \oplus \bigoplus_{i=1}^{s} \mathcal{O}[[T]]/(F_i(T)^{n_i}) \oplus \bigoplus_{j=1}^{t} \mathcal{O}[[T]]/(\varpi^{m_j}) \longrightarrow Z' \longrightarrow 0.$$

Taking a tensor product with $\mathcal{O}[[T]]/(x)$ of the former exact sequence, we obtain the following exact sequence:

$$Z/xZ \longrightarrow \mathcal{M}/x\mathcal{M} \longrightarrow (\mathcal{M}/Z)/x(\mathcal{M}/Z) \longrightarrow 0.$$

Since Z/xZ is a finite group, $\mathcal{M}/x\mathcal{M}$ is a finitely generated torsion $\mathcal{O}[[T]]/(x)$-module if and only if $(\mathcal{M}/Z)/x(\mathcal{M}/Z)$ is a finitely generated torsion $\mathcal{O}[[T]]/(x)$-module. Taking a tensor product with $\mathcal{O}[[T]]/(x)$ of the latter exact sequence, we obtain the following exact sequence:

$$\begin{aligned}(2.27)\quad &\mathrm{Tor}_{\mathcal{O}[[T]]}(\mathcal{O}[[T]]/(x), Z') \longrightarrow (\mathcal{M}/Z)/x(\mathcal{M}/Z)\\ &\longrightarrow (\mathcal{O}[[T]]/(x))^{\oplus r} \oplus \bigoplus_{i=1}^{s} \mathcal{O}[[T]]/(x, F_i(T)^{n_i}) \oplus \bigoplus_{j=1}^{t} \mathcal{O}[[T]]/(x, \varpi^{m_j})\\ &\longrightarrow Z'/xZ' \longrightarrow 0.\end{aligned}$$

Since Z' is a pseudo-null module, $\mathrm{Tor}_{\mathcal{O}[[T]]}(\mathcal{O}[[T]]/(x), Z')$ is also a pseudo-null $\mathcal{O}[[T]]$-module. Hence, $\mathcal{M}/(x)\mathcal{M}$ is a finitely generated torsion $\mathcal{O}[[T]]/(x)$-module if and only if $(\mathcal{M}/Z)/(x)(\mathcal{M}/Z)$ is a finitely generated torsion $\mathcal{O}[[T]]/(x)$-module. In particular, if $\mathcal{M}/(x)\mathcal{M}$ is a finitely generated torsion $\mathcal{O}[[T]]/(x)$-module, r must be zero in (2.27). This completes the proof. □

DEFINITION 2.41. Let $\mathcal{M}$ be a finitely generated $\mathcal{O}[[T]]$-module. Then the following type of module, which appeared in the exact sequence (2.25),

$$\mathcal{O}[[T]]^{\oplus r} \oplus \bigoplus_{i=1}^{s} \mathcal{O}[[T]]/(F_i(T)^{n_i}) \oplus \bigoplus_{j=1}^{t} \mathcal{O}[[T]]/(\varpi^{m_j})$$

is called an **elementary Iwasawa module** (associated to $\mathcal{M}$). A fundamental module associated to $\mathcal{M}$ is uniquely determined from $\mathcal{M}$ modulo $\mathcal{O}[[T]]$-linear isomorphism (see the remark below).

REMARK 2.42. There seems to be a different version of the definition of the fundamental module with which we do not assume that the distinguished polynomials $F_i(T)$'s which appear in Theorem 2.39 are irreducible. If we take this version of the definition of the fundamental module, a fundamental module associated to $\mathcal{M}$ is necessarily unique modulo $\mathcal{O}[[T]]$-linear isomorphism. For example, let us consider $\mathcal{M} = \mathcal{O}[[T]]/(F(T)G(T))$ where $F(T), G(T) \in \mathcal{O}[[T]]$ are distinguished polynomials which do not have a common irreducible factor. We have the following exact sequence:

$$0 \longrightarrow \mathcal{M} \longrightarrow \mathcal{O}[[T]]/(F(T)) \bigoplus \mathcal{O}[[T]]/G(T) \longrightarrow \mathcal{O}[[T]]/(F(T), G(T)) \longrightarrow 0$$

where $\mathcal{O}[[T]]/(F(T), G(T))$ is a finite abelian group. Hence, $\mathcal{O}[[T]]/(F(T)) \bigoplus \mathcal{O}[[T]]/G(T)$ is the elementary Iwasawa module in the sense of Definition 2.41. On the other hand, the module $\mathcal{M}$ admits another exact sequence:

$$0 \longrightarrow \mathcal{M} \longrightarrow \mathcal{O}[[T]]/(F(T)G(T)) \longrightarrow 0$$

where $F(T)G(T)$ is a reducible distinguished polynomial. Hence, $\mathcal{O}[[T]]/(F(T)G(T))$ can be also an elementary Iwasawa module associated to $\mathcal{M}$ if we take a different version of the definition of the fundamental module stated above.

DEFINITION 2.43. Let $\mathcal{M}$ be a finitely generated $\mathcal{O}[[T]]$-module.

(1) The **characteristic ideal** $\mathrm{char}_{\mathcal{O}[[T]]}(\mathcal{M})$ associated to $\mathcal{M}$ is defined to be

$$\mathrm{char}_{\mathcal{O}[[T]]}(\mathcal{M}) = \begin{cases} \left(\prod_{i=1}^{s} F_i(T)^{n_i} \prod_{j=1}^{t} \varpi^{m_j}\right) & \text{when } r = 0, \\ 0 & \text{when } r \neq 0, \end{cases}$$

where $F_i(T)$, n_i and m_j are given as in Theorem 2.39 and we have $\mathrm{char}_{\mathcal{O}[[T]]}(\mathcal{M}) = \mathcal{R}$ when $\mathcal{M} = 0$.

(2) We call $\lambda(\mathcal{M}) = \sum_{i=1}^{s} n_i \deg F_i(T)$ the **Iwasawa λ-invariant** for $\mathcal{M}$, $\mu(\mathcal{M}) = \sum_{j=1}^{t} m_j$ the **Iwasawa μ-invariant** for $\mathcal{M}$.

REMARK 2.44. In the structure theorem of Iwasawa modules (Theorem 2.39), the definitions of Elementary Iwasawa module (Definition 2.41), the characteristic ideal and Iwasawa invariants (Definition 2.43), it is not essential to regard an Iwasawa module $\mathcal{M}$ as an $\mathcal{O}[[T]]$-module through a non canonical ring isomorphism $\Lambda_{\mathcal{O}} \cong \mathcal{O}[[T]]$.

Let $\mathcal{M}$ be a finitely generated torsion $\Lambda_{\mathcal{O}}$-module.

(1) Let us choose a topological generator γ of Γ. This defines an isomorphism $u : \Lambda_{\mathcal{O}} \xrightarrow{\sim} \mathcal{O}[[T]]$ of local complete $\mathcal{O}$-algebras characterized by $\gamma \mapsto 1+T$. By regarding $\mathcal{M}$ as an $\mathcal{O}[[T]]$-module via the map u^{-1}, it makes sense to consider $\mathrm{char}_{\mathcal{O}[[T]]}(\mathcal{M})$. Then, the characteristic ideal $\mathrm{char}_{\Lambda_{\mathcal{O}}}(\mathcal{M})$ is given by the image of $\mathrm{char}_{\mathcal{O}[[T]]}(\mathcal{M})$ by u. Though each of u and u^{-1} depends on the choice of a topological generator γ of Γ, the ideal $\mathrm{char}_{\Lambda_{\mathcal{O}}}(\mathcal{M})$ thus obtained is canonical and it does not depend of γ.

(2) To $\mathcal{M}$, we associate three different ideals: a divisor ideal $\mathrm{Div}_{\Lambda_{\mathcal{O}}}(\mathcal{M})$ (Definition2.37), a characteristic ideal $\mathrm{char}_{\Lambda_{\mathcal{O}}}(\mathcal{M})$ in the sense of Definition 2.38), and a characteristic ideal $\mathrm{char}_{\Lambda_{\mathcal{O}}}(\mathcal{M})$ in the sense of Definition 2.43 (see also the above assertion (1)). It is known that all these ideals coincide.

(3) The rank $r = r(\mathcal{M})$ and the invariants $\lambda(\mathcal{M})$, $\mu(\mathcal{M})$ can be defined without using an isomorphism $u : \Lambda_{\mathcal{O}} \xrightarrow{\sim} \mathcal{O}[[T]]$. In fact, $r(\mathcal{M})$ is the dimension of the finite dimensional Frac(Frac($\Lambda_{\mathcal{O}}$))-vector space $\mathcal{M} \otimes_{\Lambda_{\mathcal{O}}} \mathrm{Frac}(\Lambda_{\mathcal{O}})$. When $r(\mathcal{M}) = 0$, $\lambda(\mathcal{M})$ is the dimension of the finite dimensional Frac($\mathcal{O}$)-vector space $\mathcal{M} \otimes_{\mathcal{O}} \mathrm{Frac}(\mathcal{O})$ and $\mu(\mathcal{M})$ is $\mathrm{length}_{(\Lambda_{\mathcal{O}})_{(\varpi)}} \mathcal{M}_{(\varpi)}$ where $\mathcal{M}_{(\varpi)}$ is the localization of $\mathcal{M}$ at the prime ideal $(\varpi) \subset \Lambda_{\mathcal{O}}$.

We list up some facts and some properties on characteristic ideals and Iwasawa invariants.

REMARK 2.45. (1) When $\mathcal{R}$ is a Dedekind domain and $\mathcal{M}$ is a finitely generated torsion $\mathcal{R}$-module, we have an isomorphism $\mathcal{M} \cong \oplus_{i=1}^{s} \mathcal{R}/\mathfrak{p}_i^{l_i}\mathcal{R}$ where $\mathfrak{p}_1, \ldots, \mathfrak{p}_s$ are maximal ideals of $\mathcal{R}$ which are not necessarily distinct to each other. Hence, we have $\mathrm{char}_{\mathcal{R}}(\mathcal{M}) = \prod_{i=1}^{s} \mathfrak{p}_i^{l_i}$. In particular, when $\mathcal{M}$ is a finitely generated torsion $\mathbb{Z}$-module, we have $\mathrm{char}_{\mathbb{Z}}(\mathcal{M}) = (\#\mathcal{M})\mathbb{Z}$. As the above example shows, we can say that the characteristic ideal measures the size of the module.

(2) When $\mathcal{M}$ is a finitely generated torsion $\Lambda_{\mathcal{O}}$-module, the following facts are important.
   (a) $\mathcal{M}$ is finitely generated as an $\mathcal{O}$-module if and only if $\mu(\mathcal{M}) = 0$ holds.
   (b) $\mathcal{M}$ is a finite abelian group if and only if $\lambda(\mathcal{M}) = \mu(\mathcal{M}) = 0$ holds.

(3) Let $\mathcal{M}$ be a finitely generated torsion $\Lambda_{\mathcal{O}}$-module. The polynomial part $\prod_{i=1}^{s} F_i(T)^{n_i}$ of the characteristic ideal

$$\mathrm{char}_{\mathcal{O}[[T]]}(\mathcal{M}) = \left(\prod_{i=1}^{s} F_i(T)^{n_i} \prod_{j=1}^{t} \varpi^{m_j}\right)$$

can be regarded as a determinant, or in other words, as the characteristic polynomial of a certain operator acting a certain space. Let us fix a topological generator γ of Γ which induces an isomorphism $\Lambda_{\mathcal{O}} \xrightarrow{\sim} \mathcal{O}[[T]]$, $\gamma \mapsto 1+T$. Since Γ acts continuously on $V = \mathcal{M} \otimes_{\mathcal{O}} \mathrm{Frac}(\mathcal{O})$, the characteristic polynomial $\det(xE_\lambda - (\gamma-1); V) \in \mathcal{O}[x]$ of the action of $\gamma - 1$ gives

$$\prod_{i=1}^{s} F_i(T)^{n_i} = \det(xE_\lambda - (\gamma-1); V)|_{x=T}$$

where E_λ is an identity matrix of size $\lambda = \lambda(\mathcal{M})$[21]

(4) Let $\mathcal{R}$ be a commutative algebra and $\mathcal{M}$ a finitely generated torsion $\mathcal{R}$-module. Below, we introduce the Fitting ideal which is another important ideal related to the structure of the module $\mathcal{M}$. We have a presentation of $\mathcal{M}$ as follows:

$$P_1 \longrightarrow P_0 \longrightarrow \mathcal{M} \longrightarrow 0 \tag{2.28}$$

where P_0 is a free $\mathcal{R}$-module of finite rank and P_1 is a free $\mathcal{R}$-module. We put $d = \mathrm{rank}_{\mathcal{R}}\mathcal{M}$. For each integer i satisfying $0 \leq i \leq d$, we consider the $\mathcal{R}$-linear homomorphism induced by the $d-i$-th exterior product of (2.28) as follows:

$$\overset{d-i}{\wedge} P_1 \longrightarrow \overset{d-i}{\wedge} P_0 \cong \mathcal{R}^{\oplus \frac{d!}{(d-i)!i!}} \tag{2.29}$$

The ideal generated by the sums of the elements in $\frac{d!}{(d-i)!i!}$ components of the image of (2.29) does not depend on the presentation (2.28)[22]. We call this ideal the **i-th Fitting ideal** of $\mathcal{M}$ and we denote this ideal by $\mathrm{Fitt}_i(\mathcal{M})$.

Often, we call the 0-th Fitting ideal of $\mathcal{M}$ simply the Fitting ideal of $\mathcal{M}$ and we denote it by $\mathrm{Fitt}_{\mathcal{R}}(\mathcal{M})$. When we have $\mathcal{R} = \Lambda_{\mathcal{O}}$ and $\mathcal{R}$-module $\mathcal{M}$ is an elementary Iwasawa module, we can check that

$$\mathrm{char}_{\Lambda_{\mathcal{O}}}(\mathcal{M}) = \mathrm{Fitt}_{\Lambda_{\mathcal{O}}}(\mathcal{M}).$$

However, $\mathrm{char}_{\Lambda_{\mathcal{O}}}(\mathcal{M})$ does not coincide with $\mathrm{Fitt}_{\Lambda_{\mathcal{O}}}(\mathcal{M})$ in general. For example, for a pseudo-null module $\mathcal{M} = \Lambda_{\mathcal{O}}/\mathfrak{M}$, we have $\mathrm{char}_{\Lambda_{\mathcal{O}}}(\mathcal{M}) = (1)$ and $\mathrm{Fitt}_{\Lambda_{\mathcal{O}}}(\mathcal{M}) = \mathfrak{M}$. In general, we have $\mathrm{char}_{\Lambda_{\mathcal{O}}}(\mathcal{M}) = \mathrm{Fitt}_{\Lambda_{\mathcal{O}}}(\mathcal{M})^{**}$ for a finitely generated torsion $\Lambda_{\mathcal{O}}$-module $\mathcal{M}$ where ** means the double $\Lambda_{\mathcal{O}}$-linear dual (see [**84**, Prop. 9.6] for the proof).

There are two different methods for

2.3.4. Specialization of Iwasawa modules. In this subsection, we prove Theorem 2.46, which is used essentially in the proof of Theorem 3.1. Though the proof of Theorem 2.46 is a bit technical, it will help us to check and understand the details, such as comments given in Remark 3.5. Since the technical detail of the proof does not affect our understanding of later sections of the book, those who do not wish to know the proof of Theorem 2.46 can skip it by admitting the result of Theorem 2.46.

---

[21] In this book, we do not use a presentation as a determinant of the characteristic ideal. But we can formulate the cyclotomic Iwasawa main conjectures using such presentations if we wish (see Chapter 1 for such a presentation of the Iwasawa main conjecture).

[22] We refer the reader to the book [**26**, Chap. 20] for the proof of this fact as well as some fundamental properties of the Fitting ideal.

Throughout §2.3.4, we denote the cardinality of the residue field of the ring of integers of $\mathcal{O}$ our fixed p-adic field by q and we denote the absolute ramification degree of $\mathrm{Frac}(\mathcal{O})$ by e.

THEOREM 2.46. *Let $\mathcal{M}$ be a finitely generated torsion $\mathcal{O}[[T]]$-module. Assume that the intersection between the characteristic ideal $\mathrm{char}_{\mathcal{O}[[T]]}(\mathcal{M})$ and the principal ideal $(\omega_n(T))$ is trivial for every natural number n*[23]. *Then, there exists $\nu = \nu(\mathcal{M}) \in \mathbb{Z}$ such that we have*

$$\# \left(\mathcal{M}/\omega_n(T)\mathcal{M}\right) = q^{\lambda en + \mu p^n + \nu}, \tag{2.30}$$

for sufficiently large n. Here $\lambda = \lambda(\mathcal{M})$ (resp. $\mu = \mu(\mathcal{M})$) is the Iwasawa λ-invariant (resp. the Iwasawa μ-invariant) for the module $\mathcal{M}$.

As for the estimate of "sufficiently large n", it will become clear if we follow the proof of Theorem 2.46. We shall prepare some lemmas which are necessary for the proof of the above theorem.

LEMMA 2.47. (1) *For a pseudo-null $\mathcal{O}[[T]]$-module Z, we have $Z/\omega_n(T)Z \cong Z$ for sufficiently large n.*

(2) *Let $\mathcal{M}$ be a a finitely generated torsion $\mathcal{O}[[T]]$-module. Take a natural number n such that $(\omega_n(T), \mathrm{char}_{\mathcal{O}[[T]]}\mathcal{M}) = 1$. Then, $\mathcal{M}/\omega_n(T)\mathcal{M}$ is a finite group and we have*

$$\begin{aligned}&\# \left(\mathcal{M}/\omega_n(T)\mathcal{M}\right)\\ &= \# \left(Z/\omega_n(T)Z\right) \prod_{i=1}^{s} \# \left(\mathcal{O}[[T]]/(F_i(T)^{n_i}, \omega_n(T))\right) \prod_{j=1}^{t} \# \left(\mathcal{O}[[T]]/(\varpi^{m_j}, \omega_n(T))\right),\end{aligned}$$

where Z, $F_i(T)$, m_i, and n_j are as given in the fundamental exact sequence:

$$\begin{aligned}&0 \longrightarrow Z \longrightarrow \mathcal{M}\\ &\longrightarrow \bigoplus_{i=1}^{s} \mathcal{O}[[T]]/(F_i(T)^{n_i}) \oplus \bigoplus_{j=1}^{t} \mathcal{O}[[T]]/(\varpi^{m_j}) \longrightarrow Z' \longrightarrow 0\end{aligned} \tag{2.31}$$

(see Theorem 2.39 for the fundamental exact sequence).

PROOF. Note that the annihilator ideal of Z contains n-th power of the maximal ideal of $\mathcal{O}[[T]]$ for sufficiently large n. Hence, $\omega_n(T)Z = 0$ for sufficiently large n. This shows the assertion (1). Let us prove the assertion (2). Let us denote by $\mathrm{Fund}(\mathcal{M})$ a fundamental Iwasawa module

$$\bigoplus_{i=1}^{s} \mathcal{O}[[T]]/(F_i(T)^{n_i}) \oplus \bigoplus_{j=1}^{t} \mathcal{O}[[T]]/(\varpi^{m_j})$$

which appeared in the sequence (2.31). Let us consider the following commutative diagram for each natural number n:

$$\begin{array}{ccccccccc} 0 & \longrightarrow & Z & \longrightarrow & \mathcal{M} & \longrightarrow & \mathrm{Fund}(\mathcal{M}) & \longrightarrow & Z' & \longrightarrow & 0 \\ & & \downarrow{\scriptstyle a_n} & & \downarrow{\scriptstyle b_n} & & \downarrow{\scriptstyle c_n} & & \downarrow{\scriptstyle d_n} & & \\ 0 & \longrightarrow & Z & \longrightarrow & \mathcal{M} & \longrightarrow & \mathrm{Fund}(\mathcal{M}) & \longrightarrow & Z' & \longrightarrow & 0 \end{array}$$

[23] By the structure theorem of Iwasawa modules, this assumption is equivalent to assuming $\#\mathcal{M}/\omega_n(T)\mathcal{M} < \infty$ for all n.

where the vertical morphisms a_n, b_n, c_n, d_n are the multiplication by $\omega_n(T)$. Let us set $\mathcal{M}' = \mathcal{M}/Z$ and divide the above commutative diagram into two commutative diagrams as follows:

$$\begin{array}{ccccccccc} 0 & \longrightarrow & Z & \longrightarrow & \mathcal{M} & \longrightarrow & \mathcal{M}' & \longrightarrow & 0 \\ & & \downarrow{\scriptstyle a_n} & & \downarrow{\scriptstyle b_n} & & \downarrow{\scriptstyle b'_n} & & \\ 0 & \longrightarrow & Z & \longrightarrow & \mathcal{M} & \longrightarrow & \mathcal{M}' & \longrightarrow & 0, \end{array} \tag{2.32}$$

$$\begin{array}{ccccccccc} 0 & \longrightarrow & \mathcal{M}' & \longrightarrow & \mathrm{Fund}(\mathcal{M}) & \longrightarrow & Z' & \longrightarrow & 0 \\ & & \downarrow{\scriptstyle b'_n} & & \downarrow{\scriptstyle c_n} & & \downarrow{\scriptstyle d_n} & & \\ 0 & \longrightarrow & \mathcal{M}' & \longrightarrow & \mathrm{Fund}(\mathcal{M}) & \longrightarrow & Z' & \longrightarrow & 0. \end{array} \tag{2.33}$$

The morphism c_n is injective by the assumption that $(\omega_n(T), \mathrm{char}_{\mathcal{O}[[T]]}\mathcal{M}) = 1$. By applying the snake lemma to the diagram (2.33), we have $\mathrm{Ker}(b'_n) = 0$ and the exact sequence:

$$0 \longrightarrow \mathrm{Ker}(d_n) \longrightarrow \mathrm{Coker}(b'_n) \longrightarrow \mathrm{Coker}(c_n) \longrightarrow \mathrm{Coker}(d_n) \longrightarrow 0.$$

The assumption $(\omega_n(T), \mathrm{char}_{\mathcal{O}[[T]]}\mathcal{M}) = 1$ implies that all terms of the above exact sequence are finite. Hence, we obtain the equality $\#\mathrm{Ker}(d_n)\#\mathrm{Coker}(c_n) = \#\mathrm{Coker}(b'_n)\#\mathrm{Coker}(d_n)$ by the exactness of the above sequence. Since Z' is a finite group, we have another equality $\#\mathrm{Ker}(d_n) = \#\mathrm{Coker}(d_n)$. By combining these two equalities, we obtain the equality

$$\#\mathrm{Coker}(b'_n) = \#\mathrm{Coker}(c_n) \tag{2.34}$$

for every natural number n. By applying the snake lemma to the commutative diagram (2.32) and by using the injectivity of the morphism b'_n, we similarly obtain $\#\mathrm{Coker}(b_n) = \#\mathrm{Coker}(a_n)\#\mathrm{Coker}(b'_n)$. By combining this equality with (2.34), we obtain

$$\#\mathrm{Coker}(b_n) = \#\mathrm{Coker}(a_n)\#\mathrm{Coker}(c_n). \tag{2.35}$$

Recall that we have

$$\#\mathrm{Coker}(b_n) = \#\left(\mathcal{M}/\omega_n(T)\mathcal{M}\right), \#\mathrm{Coker}(a_n) = \#\left(Z/\omega_n(T)Z\right)$$

and

$$\#\mathrm{Coker}(c_n) = \prod_{i=1}^{s} \#\left(\mathcal{O}[[T]]/(F_i(T)^{n_i}, \omega_n(T))\right) \prod_{j=1}^{t} \#\left(\mathcal{O}[[T]]/(\varpi^{m_j}, \omega_n(T))\right).$$

This completes the proof. □

LEMMA 2.48. *Let $F(T) \in \mathcal{O}[[T]]$ be a distinguished polynomial of degree l. Assume that we have $(F(T), \omega_n(T)) = 1$ for every n. Then the following statements hold.*

(1) *$\mathcal{O}[[T]]/(F(T), \omega_n(T))$ is a finite abelian group for every n. Further we have*

$$\#(\mathcal{O}'[[T]]/(F(T), \omega_n(T))) = \#(\mathcal{O}[[T]]/(F(T), \omega_n(T)))^{[\mathcal{O}':\mathcal{O}]}$$

for the ring of integers $\mathcal{O}'$ of any finite extension of $\mathrm{Frac}(\mathcal{O})$.

(2) *There exists a constant $\nu(F(T)) \in \mathbb{Z}$ which depends only on $F(T)$ and $\mathcal{O}$ such that the following equality holds for every sufficiently large n:*

$$\# \left(\mathcal{O}[[T]]/(F(T), \omega_n(T))\right) = q^{len+\nu(F(T))}. \tag{2.36}$$

PROOF. For the assertion (1), the first statement follows by the assumption $(F(T), \omega_n(T)) = 1$. The second statement follows from the fact that $\mathcal{O}'$ is flat over $\mathcal{O}$. Let us prove the assertion (2) now. We set e' to be the ramification degree of $\mathrm{Frac}(\mathcal{O}')$ over $\mathbb{Q}_p$ and we set q' to be the cardinality of the residue field of $\mathcal{O}'$. We note that, since we have $[\mathcal{O}' : \mathcal{O}] = \frac{e'}{e}\frac{\log q'}{\log q}$, the following equality holds:

$$(q^{len+\nu(F(T))})^{[\mathcal{O}':\mathcal{O}]} = q'^{le'n+\nu(F(T))\frac{e'}{e}}.$$

Thus, (2.36) holds over $\mathcal{O}$ if and only if (2.36) holds over $\mathcal{O}'$. By replacing $\mathcal{O}$ by the ring of integers of a sufficiently large finite extension if necessary, we may assume that $\mathcal{O}$ contains all roots of $F(T)$. Suppose that $F(T) \in \mathcal{O}[[T]]$ is decomposed as $F(T) = G(T)H(T)$ with two monic polynomials which have no common divisors. Then $G(T)$ and $H(T)$ are distinguished polynomials and the module $Z' = \mathcal{O}[[T]]/(G(T), H(T))$ is a pseudo-null $\mathcal{O}[[T]]$-module. Let us consider the commutative diagram:

$$\begin{array}{ccccccccc}
0 & \longrightarrow & \Lambda_{\mathcal{O}}/(F) & \longrightarrow & \Lambda_{\mathcal{O}}/(G) \oplus \Lambda_{\mathcal{O}}/(H) & \longrightarrow & Z' & \longrightarrow & 0 \\
 & & \big\downarrow \times\omega_n(T) & & \big\downarrow \times\omega_n(T) & & \big\downarrow \times\omega_n(T) & & \\
0 & \longrightarrow & \Lambda_{\mathcal{O}}/(F) & \longrightarrow & \Lambda_{\mathcal{O}}/(G) \oplus \Lambda_{\mathcal{O}}/(H) & \longrightarrow & Z' & \longrightarrow & 0
\end{array}$$

where $\Lambda_{\mathcal{O}}/(F)$, $\Lambda_{\mathcal{O}}/(G)$ and $\Lambda_{\mathcal{O}}/(H)$ denotes $\mathcal{O}[[T]]/(F(T))$, $\mathcal{O}[[T]]/(G(T))$ and $\mathcal{O}[[T]]/(H(T))$ respectively. By a similar argument as the proof of Lemma 2.47 using the snake lemma, we obtain the following equality:

$$\begin{aligned}
&\# \left(\mathcal{O}[[T]]/(F(T), \omega_n(T))\right) \\
&\qquad = \# \left(\mathcal{O}[[T]]/(G(T), \omega_n(T))\right) \times \# \left(\mathcal{O}[[T]]/(H(T), \omega_n(T))\right).
\end{aligned}$$

Thus we have reduced the proof of the second assertion of Lemma 2.48 to the case $F(T) = (T - \alpha)^m$. Let us consider the following diagram:

$$\begin{array}{ccccccccc}
0 & \longrightarrow & \Lambda_{\mathcal{O}}/(T-\alpha)^m & \longrightarrow & \Lambda_{\mathcal{O}}/(T-\alpha)^{m+1} & \longrightarrow & \Lambda_{\mathcal{O}}/(T-\alpha) & \longrightarrow & 0 \\
 & & \big\downarrow \times\omega_n(T) & & \big\downarrow \times\omega_n(T) & & \big\downarrow \times\omega_n(T) & & \\
0 & \longrightarrow & \Lambda_{\mathcal{O}}/(T-\alpha)^m & \longrightarrow & \Lambda_{\mathcal{O}}/(T-\alpha)^{m+1} & \longrightarrow & \Lambda_{\mathcal{O}}/(T-\alpha) & \longrightarrow & 0.
\end{array}$$

By an induction with respect to m using a similar argument as above, we further reduce the proof to the case $F(T) = (T - \alpha)$. Hence, we need to calculate the order of the module

$$\mathcal{O}[[T]]/(T - \alpha, \omega_n(T)) \cong (\mathbb{Z}_p[[T]] \otimes_{\mathbb{Z}_p} \mathcal{O})/(T - \alpha, \omega_n(T)). \tag{2.37}$$

Recall that we have $\omega_n(T) = \prod_{i=0}^{n} \Phi_{p^i}(T+1)$ where $\Phi_{p^i}(T)$ is the minimal polynomial of ζ_{p^i} over $\mathbb{Q}_p$ for each i. We have a natural injection

$$\mathbb{Z}_p[[T]]/(\omega_n(T)) \longrightarrow \prod_{i=0}^{n} \mathbb{Z}_p[\zeta_{p^i}], \quad T \mapsto \{\zeta_{p^i} - 1\}_{0 \le i \le n},$$

where ζ_{p^i} is a fixed p^i-th primitive root of unity for each nonnegative integer i. The cokernel of the above injection is a finite abelian group which we denote by Z_n. By applying the functor $\otimes_{\mathbb{Z}_p}\mathcal{O}$ to the sequence:

$$0 \longrightarrow \mathbb{Z}_p[[T]]/(\omega_n(T)) \longrightarrow \prod_{i=0}^{n} \mathbb{Z}_p[\zeta_{p^i}] \longrightarrow Z_n \longrightarrow 0,$$

we have the following exact sequence:

$$0 \longrightarrow \mathbb{Z}_p[[T]] \otimes_{\mathbb{Z}_p} \mathcal{O}/(\omega_n(T)) \longrightarrow \prod_{i=0}^{n} \mathbb{Z}_p[\zeta_{p^i}] \otimes_{\mathbb{Z}_p} \mathcal{O} \longrightarrow Z_n \otimes_{\mathbb{Z}_p} \mathcal{O} \longrightarrow 0.$$

Note that the multiplication by $\prod_{i=0}^{n}((\zeta_{p^i}-1)-\alpha)$ on the middle term restricted to the first term coincides with the multiplication by $T-\alpha$. Hence, we have the following commutative diagram:

$$\begin{array}{ccccccccc} 0 & \longrightarrow & \mathbb{Z}_p[[T]]_{\mathcal{O}}/(\omega_n(T)) & \longrightarrow & \prod_{i=0}^{n} \mathbb{Z}_p[\zeta_{p^i}]_{\mathcal{O}} & \longrightarrow & (Z_n)_{\mathcal{O}} & \longrightarrow & 0 \\ & & \downarrow{\scriptstyle \times T-\alpha} & & \downarrow{\scriptstyle \times((\zeta_{p^i}-1)-\alpha)_i} & & \downarrow & & \\ 0 & \longrightarrow & \mathbb{Z}_p[[T]]_{\mathcal{O}}/(\omega_n(T)) & \longrightarrow & \prod_{i=0}^{n} \mathbb{Z}_p[\zeta_{p^i}]_{\mathcal{O}} & \longrightarrow & (Z_n)_{\mathcal{O}} & \longrightarrow & 0, \end{array}$$

where $(\ )_{\mathcal{O}}$ denotes $(\ ) \otimes_{\mathbb{Z}_p} \mathcal{O}$. Since $Z_n \otimes_{\mathbb{Z}_p} \mathcal{O}$ is a finite abelian group, the kernel and the cokernel of the right vertical map have the same cardinality. By the snake lemma, we have

(2.38)
$$\#\left(\mathbb{Z}_p[[T]] \otimes_{\mathbb{Z}_p} \mathcal{O}/(T-\alpha, \omega_n(T))\right) = \prod_{i=0}^{n} \#\left(\mathbb{Z}_p[\zeta_{p^i}] \otimes_{\mathbb{Z}_p} \mathcal{O}/((\zeta_{p^i}-1)-\alpha)\right).$$

Since $\mathrm{ord}_p(\zeta_{p^n}-1) = \frac{1}{(p-1)p^{n-1}}$, the following lemma follows by a well-known basic property of p-adic valuation.

LEMMA 2.49. *Let us set $n(\alpha)$ as follows:*

$$n(\alpha) = \begin{cases} \max\left\{ n \in \mathbb{Z}_{\geq 1} \;\middle|\; \frac{1}{(p-1)p^{n-1}} \geq \mathrm{ord}_p(\alpha) \right\} & \text{when } 0 < \mathrm{ord}_p(\alpha) \leq \frac{1}{p-1}, \\ 0 & \text{when } \frac{1}{p-1} < \mathrm{ord}_p(\alpha). \end{cases}$$

Then, for any integer $n > n(\alpha)$, we have

$$\#(\mathbb{Z}_p[\zeta_{p^n}] \otimes_{\mathbb{Z}_p} \mathcal{O})/((\zeta_{p^n}-1)-\alpha) = \#(\mathbb{F}_p \otimes_{\mathbb{Z}_p} \mathcal{O}) = q^e.$$

By combining (2.37), (2.38), and Lemma2.49, the cardinality of $\mathcal{O}[[T]]/(T-\alpha, \omega_n(T))$ is given by

$$q^{e(n-n(\alpha))} \cdot \#\left(\mathbb{Z}_p[[T]] \otimes_{\mathbb{Z}_p} \mathcal{O}/(T-\alpha, \omega_{n(\alpha)}(T))\right)$$

for any $n > n(\alpha)$. This shows that the statement of Lemma 2.48 (2) holds true by setting $\nu(F(T)) = \mathrm{ord}_q\left(\#\left(\mathbb{Z}_p[[T]] \otimes_{\mathbb{Z}_p} \mathcal{O}/(T-\alpha, \omega_{n(\alpha)}(T))\right)\right) - en(\alpha)$. This completes the proof of Lemma 2.48. □

The following lemma concerns the part related to the μ-invariant in Theorem 2.46.

LEMMA 2.50. *For any integer $n \in \mathbb{Z}_{\geq 1}$, we have*

$$\#\left(\mathcal{O}[[T]]/(\varpi, \omega_n(T))\right) = q^{p^n}. \tag{2.39}$$

PROOF. The proof follows immediately by the following isomorpihsms:

$$\mathcal{O}[[T]]/(\varpi, \omega_n(T)) \cong (\mathcal{O}/\varpi)[[T]]/(\omega_n(T)) \cong (\mathcal{O}/\varpi)[[T]]/(T^{p^n}) \cong (\mathcal{O}/\varpi)^{\oplus p^n}.$$

□

Let us get back to the proof of Theorem 2.46 after the above preparations.

PROOF OF THEOREM 2.46. For a finitely generated torsion $\mathcal{O}[[T]]$-module $\mathcal{M}$, we consider the fundamental sequence

$$0 \longrightarrow Z \longrightarrow \mathcal{M} \longrightarrow \bigoplus_{i=1}^{s} \mathcal{O}[[T]]/(F_i(T)^{n_i}) \oplus \bigoplus_{j=1}^{t} \mathcal{O}[[T]]/(\varpi^{m_j}) \longrightarrow Z' \longrightarrow 0$$

obtained in Theorem 2.39. Recall that we defined $\lambda = \lambda(\mathcal{M}) = \sum_{i=1}^{s} n_i \mathrm{deg} F_i(T)$, $\mu = \mu(\mathcal{M}) = \sum_{j=1}^{t} m_j$ in Definition 2.43. We set $\nu(\mathcal{M}) = \sum_{i=1}^{s} n_i \nu(F_i(T)) + \mathrm{ord}_q(\# Z)$ (see Lemma 2.48)), $n(\mathcal{M}) = \mathrm{Max}\,(n'(\mathcal{M}), n''(\mathcal{M}))$ where

$$n'(\mathcal{M}) = \mathrm{Max}\{n(\alpha) \mid \alpha \text{ runs through the roots of } \textstyle\prod_{i=1}^{s} F_i(T)\ \},$$

$$n''(\mathcal{M}) = \mathrm{Min}\{n \in \mathbb{N} \mid \omega_n(T) Z = 0\}.$$

The equality $\#(\mathcal{M}/\omega_n(T)\mathcal{M}) = q^{\lambda e n + \mu p^n + \nu}$ holds for $n > n(\mathcal{M})$ by combining Lemma 2.47, Lemma 2.48, and Lemma 2.50. This completes the proof. □

REMARK 2.51. Let $\mathcal{M}$ be a finitely generated torsion $\mathcal{O}[[T]]$-module, $f(T) \in \mathcal{O}[[T]]$ a polynomial dividing $\omega_{n_0}(T)$ for some natural number n_0. If the characteristic ideal $\mathrm{char}_{\mathcal{O}[[T]]}(\mathcal{M})$ and $(\omega_n(T)/f(T))$ have no common factors for any natural number $n \geq n_0$, by a similar argument as the proof of Theorem 2.46, we can show that there exists an integer ν such that

$$\#(\mathcal{M}/(\omega_n(T)/f(T))\mathcal{M}) = q^{\lambda(\mathcal{M}) e n + \mu(\mathcal{M}) p^n + \nu} \tag{2.40}$$

holds for any sufficiently large n. The statement of Theorem 2.46 is nothhing but the above statement for $f(T) = 1$.

Let us give some examples for the equality (2.30) of Theorem 2.46 whose proofs are left for the reader as an exercise.

(1) Let $\mathcal{M} = \mathbb{Z}_p[[T]]/(T - p^a)$ $(a \geq 1)$. Let us set

$$\lambda(\mathcal{M}) = 1,\ \mu(\mathcal{M}) = 0,\ \nu(\mathcal{M}) = a.$$

Then, by using the formula $\mathrm{ord}_p\binom{p^n}{i} = n - \mathrm{ord}_p(i)$ on the p-adic orders of binomial coefficients, we can check that (2.30) holds for every nonnegative integer n.

(2) Let $\mathcal{M} = \mathbb{Z}_p[[T]]/(T^a - p)$ $(a \geq 1)$. The number $n(\mathcal{M})$ which appeared in the proof of Theorem 2.46 is the largest natural number n satisfying $n \leq 1 + \log_p a - \log_p(p-1)$. Let us set

$$\lambda(\mathcal{M}) = a,\ \mu(\mathcal{M}) = 0,\ \nu(\mathcal{M}) = \mathrm{ord}_p(\#\mathcal{M}/\omega_{n(\mathcal{M})}\mathcal{M}) - an(\mathcal{M}).$$

Then we can check that (2.30) holds for every natural number n satisfying $n > n(\mathcal{M})$. Note that we have $\nu(\mathcal{M}) < 0$ for $a > 1$ and we have $\nu(\mathcal{M}) \to -\infty$ when a tends to ∞.

CHAPTER 3

Cyclotomic Iwasawa theory for ideal class groups

3.1. Algebraic aspect (Selmer group)

In this section (§3.1), we discuss the algebraic side of the Iwasawa theory. These results hold true for any finite algebraic number field K and any $\mathbb{Z}_p$-extension K_∞/K. On the other hand, the results of §3.2 and §3.3 will be on the analytic side of the theory. Often, these results hold true only for limited cases such as cyclotomic $\mathbb{Z}_p$-extensions of totally real finite algebraic number fields K.

In §3.1, most of the arguments of commutative algebra and algebraic number theory are based on the techniques which we already used earlier, except that we often use diagram chase arguments, such as the snake lemma used in §2.3.4. We refer the reader to [**118**] for the basics of homological algebra.

3.1.1. Iwasawa's class number formula. As we remarked earlier in Section 1.1, not many results on class groups are known for a family of infinite numbers of fields. The following theorem, which was proved by Iwasawa in his paper [**47**] published in 1959, shows the interest and importance of $\mathbb{Z}_p$-extensions for the study of ideal class groups.

THEOREM 3.1 (Iwasawa's class number formula). *Let K be a finite algebraic number field, K_∞/K a $\mathbb{Z}_p$-extension. Then there exist nonnegative integers $\lambda = \lambda(K_\infty)$, $\mu = \mu(K_\infty)$ and an (possibly negative) integer $\nu = \nu(K_\infty)$ such that the order of the p-part of the class groups $\mathrm{Cl}(K_n)[p^\infty]$ of n-th intermediate fields $K_n = (K_\infty)^{\Gamma^{p^n}}$ satisfies*[1] *for sufficiently large n:*

$$\#\mathrm{Cl}(K_n)[p^\infty] = p^{\lambda n + \mu p^n + \nu}.$$

The proof of the original paper [**47**] by Iwasawa is based on the method that studies discrete Γ-modules of class groups directly. Here, we take the method that studies Iwasawa modules of compact Γ-modules obtained by the Pontrjagin duals of class groups. We will give some technical preparations on commutative algebra in 2.3.4 and we will prove Theorem 3.1 in §3.1.2. For the rest of this subsection, we introduce another theorem (Theorem 3.4) as a corollary of Theorem 3.1 after some preparation.

Let K be a finite algebraic number field, K_∞ an arbitrary $\mathbb{Z}_p$-extension of K, $\Gamma = \mathrm{Gal}(K_\infty/K)$. We denote by L_∞ the maximal everywhere unramified abelian p-power extension of K_∞.

LEMMA 3.2. *L_∞ is a Galois extension over K.*

[1] K_n is an intermediate field of K_∞/K. Note also that each intermediate field of K_∞/K is equal to one of such K_n's.

PROOF. If L_∞/K is not Galois over K, there must exist an automorphism σ of $\overline{\mathbb{Q}}$ over K such that $\sigma(L_\infty) \neq L_\infty$. Since the composite field $\sigma(L_\infty) \cup L_\infty$ is an abelian and everywhere unramified extension of K_∞, this contradicts to the maximality of L_∞. □

In this book, we denote the compact $\mathbb{Z}_p$-module $\mathrm{Gal}(L_\infty/K_\infty)$ by X_{K_∞}.

LEMMA 3.3. *The module X_{K_∞} is equipped with a natural continuous action of Γ.*

PROOF. Note that we have the following exact sequence by the definition of X_{K_∞}:

$$\{1\} \longrightarrow X_{K_\infty} \longrightarrow \mathrm{Gal}(L_\infty/K) \longrightarrow \Gamma \longrightarrow \{1\}.$$

Let us take an element $g \in \Gamma$ and choose a lift $\widetilde{g} \in \mathrm{Gal}(L_\infty/K)$ of g. We define an action of g on $x \in X_{K_\infty}$ by $g \cdot x := \widetilde{g}x\widetilde{g}^{-1}$. Since X_{K_∞} is a normal subgroup in $\mathrm{Gal}(L_\infty/K)$, we have $\widetilde{g}x\widetilde{g}^{-1} \in X_{K_\infty}$. Let $\widetilde{g}' \in \mathrm{Gal}(L_\infty/K)$ be another lift of g and let us set $y := \widetilde{g}^{-1}\widetilde{g}' \in X_{K_\infty}$. Since X_{K_∞} is an abelian group, we have $\widetilde{g}'x(\widetilde{g}')^{-1} = \widetilde{g}yxy^{-1}\widetilde{g}^{-1} = \widetilde{g}yy^{-1}x\widetilde{g}^{-1} = \widetilde{g}x\widetilde{g}^{-1}$. Thus the action $g \cdot x$ is well-defined independently of a choice of a lift $\widetilde{g} \in \mathrm{Gal}(L_\infty/K)$.

To prove the continuity of the action of Γ on X_{K_∞}, it suffices to prove that, for an arbitrary open subgroup U of X_{K_∞}, there exists an open subgroup U' of Γ such that U' acts trivially on X_{K_∞}/U. Since the fixed field $(L_\infty)^U$ for a given open subgroup $U \subset X_{K_\infty}$ is a finite extension of K_∞, $(L_\infty)^U$ is the field of decomposition of an irreducible polynomial $f(X) \in K_\infty[X]$ of finite degree. Since all coefficients of $f(X)$ is contained in a finite extension of K, the subgroup $U' \subset \Gamma$ consisting of elements acting trivially on the coefficients of $f(X)$ is of finite index in Γ and the action of U' on on X_{K_∞}/U is trivial by construction. This completes the proof. □

Since X_{K_∞} is a compact $\mathbb{Z}_p$-module with continuous Γ-action, X_{K_∞} is naturally regarded as a $\Lambda = \mathbb{Z}_p[[\Gamma]]$-module thanks to Lemma 2.23. The following theorem is a corollary of Iwasawa class number formula.

THEOREM 3.4. *Let K be an arbitrary finite algebraic number field, K_∞/K an arbitrary $\mathbb{Z}_p$-extension. Then X_{K_∞} is a finitely generated torsion Λ-module.*

We will prove Theorem 3.4 in the end of §3.1.2.

REMARK 3.5. We give some remarks on Theorem 3.1. As we will see later in the proof of Theorem 3.1 below, "a sufficiently large n" in the statement of Theorem 3.1 can be made completely explicit from the information of the $\mathbb{Z}_p$-extension K_∞/K and the Iwasawa module X_{K_∞}. The requirement for "a sufficiently large n" comes from the following reasons:

(i) We need that all the primes of K_n which ramify in K_∞/K_n be totally ramified.
(ii) We need that γ^{p^n} act trivially on the largest finite submodule Z of X_{K_∞}.
(iii) We need that the p-adic valuation of $\zeta_{p^n} - 1$ is smaller than the minimum of p-adic valuations of roots of the characteristic polynomial $F(T)$ of X_{K_∞}.

On the other hand, the pseudo-null quotient Z' which appears in the last term of the fundamental sequence (2.25) of Theorem 2.39 does affect at all the ν-invariant nor the requirement of "sufficiently large n" of the statement of the Iwasawa class number formula.

3.1.2. Proof of Iwasawa's class number formula. Let us consider the following conditions on the fixed prime p.

(3.1) There is only one prime of K which is over p.

Under this condition, we discuss the following condition on the fixed $\mathbb{Z}_p$-extension K_∞/K:

(3.2) The only prime of K over p is totally ramified at K_∞/K.

The main idea of the proof of Iwasawa's class number formula is much easier to follow under the conditions (3.1) and (3.2). So we first prove Iwasawa's class number formula under the conditions (3.1) and (3.2). After, we will give a proof for the general situation.

As for the validity of these conditions, we remark that the conditions (3.1) and (3.2) hold true for $K = \mathbb{Q}(\mu_p)$ and its cyclotomic $\mathbb{Z}_p$-extension $K_\infty^{\mathrm{cyc}}/K$. For a finite algebraic number field K and a prime p satisfying (3.1), the cyclotomic $\mathbb{Z}_p$-extension $K_\infty^{\mathrm{cyc}}/K$ satisfies (3.2) if p does not divide the absolute discriminant D_K.

Let us set $K_n = (K_\infty)^{\Gamma^{p^n}}$ for a natural number n. We define L_n to be the maximal everywhere unramified abelian p-power extension of K_n. The field L_n is a subfield of L_∞ by definition and we can show that L_n is a Galois extension of K_n by the same argument as the proof of Lemma 3.2.

PROPOSITION 3.6. *Let us assume the conditions* (3.1) *and* (3.2). *When* γ *is a topological generator of* Γ, *we have the following isomorphism for every* n:

$$X_{K_\infty}/(\gamma^{p^n}-1)X_{K_\infty} \cong \mathrm{Cl}(K_n)[p^\infty].$$

PROOF. We set $G_n = \mathrm{Gal}(L_\infty/K_n)$. Let $[G_n, G_n]$ be the commutator subgroup of G_n, $\overline{[G_n, G_n]}$ the topological closure of $[G_n, G_n]$. Then we claim the following equality:

$$(\gamma^{p^n}-1)X_{K_\infty} = \overline{[G_n, G_n]}. \tag{3.3}$$

In order to check this, we denote by L'_n (resp. L''_n) the fixed field of L_∞ by $(\gamma^{p^n}-1)X_{K_\infty}$ (resp. $\overline{[G_n, G_n]}$). By Galois theory, it suffices to show $L'_n = L''_n$ in order to deduce (3.3). Let us look at the following diagram:

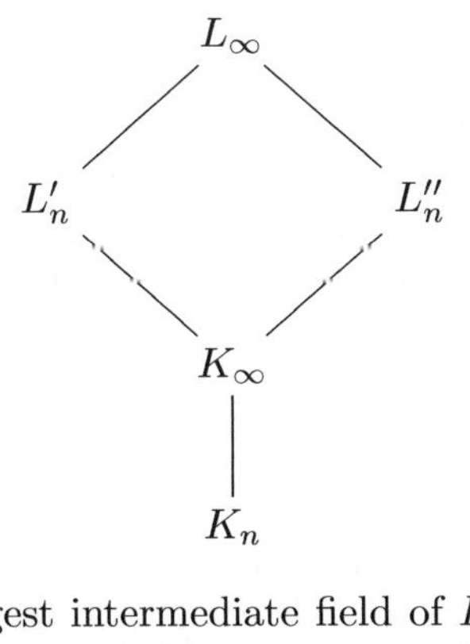

By definition, L''_n is the largest intermediate field of L_∞/K_∞ which is an abelian extension of K_n. On the other hand, L'_n is the largest intermediate field of L_∞/K_∞ such that γ^{p^n} acts trivially on $\mathrm{Gal}(L'_n/K_\infty)$. For an intermediate field L of L_∞/K_∞,

L is an abelian extension of K_n if and only if the conjugate action of γ^{p^n} on $\mathrm{Gal}(L/K_\infty)$ is trivial. Hence, $L'_n = L''_n$ follows and the proof of (3.3) is complete.

Recall that we have only one prime of K which contains (p) by the condition (3.1). We denote this prime by $\mathfrak{p}$. For each natural number n, the extension K_n/K is totally ramified at $\mathfrak{p}$. We denote by I_n the inertia subgroup of G_n at the unique prime of K_n over $\mathfrak{p}$. Since L_∞ is an everywhere unramified extension of K_∞, we have the equality $I_n \bigcap X_{K_\infty} = \{\mathbf{1}\}$ in $\mathrm{Gal}(L_\infty/K)$. On the other hand, the natural map $I_n \longrightarrow G_n/X_{K_\infty}$ must be surjective because the extension K_∞/K is totally ramified at $\mathfrak{p}$. Thus we have a semi-direct product $G_n \cong I_n \ltimes X_{K_\infty}$. Now, the following isomorphisms follow:

$$\begin{aligned}\mathrm{Gal}(L_n/K_n) &\cong G_n/I_n\overline{[G_n, G_n]} \\ &\cong I_n X_{K_\infty}/I_n(\gamma^{p^n} - 1)X_{K_\infty} \cong X_{K_\infty}/(\gamma^{p^n} - 1)X_{K_\infty}.\end{aligned}$$

Here, the first isomorphism is due to the fact that L_n is characterized as the largest intermediate field of L''_n/K_n which is unramified over K_n. The second isomorphism follows from (3.3). We complete the proof noting that we have $\mathrm{Cl}(K_n)[p^\infty] \cong \mathrm{Gal}(L_n/K_n)$ by class field theory. □

Thanks to the above proposition, we can give compact proofs of Theorem 3.4 and Theorem 3.1 under the conditions (3.1) and (3.2).

PROOF OF THEOREM 3.4 UNDER THE CONDITIONS (3.1) AND (3.2). Since the ideal class group $\mathrm{Cl}(K)[p^\infty]$ is a finite abelian p-group, it is regarded as a finitely generated torsion $\mathbb{Z}_p$-module. Hence, $X_{K_\infty}/(\gamma - 1)X_{K_\infty}$ is a finitely generated torsion $\mathbb{Z}_p$-module by Proposition 3.6. By Corollary 2.40, X_{K_∞} must be a finitely generated torsion Λ-module. □

PROOF OF THEOREM 3.1 UNDER CONDITIONS (3.1) AND (3.2). As is shown above, X_{K_∞} is a finitely generated torsion Λ-module under the conditions (3.1) and (3.2). By applying Theorem 2.46 to a finitely generated torsion Λ-module $\mathcal{M} = X_{K_\infty}$ with $\lambda = \lambda(\mathcal{M})$, we show that $\mu = \mu(\mathcal{M})$, there exists an integer $\nu = \nu(\mathcal{M}) \in \mathbb{Z}$ such that

$$\# X_{K_\infty}/(\gamma^{p^n} - 1)X_{K_\infty} = p^{\lambda n + \mu p^n + \nu}$$

holds for sufficiently large n. On the other hand, we have an isomorphism

$$X_{K_\infty}/(\gamma^{p^n} - 1)X_{K_\infty} \cong \mathrm{Cl}(K_n)[p^\infty]$$

for any natural number n by Proposition 3.6. This completes the proof. □

We recall the following fact, which is clear by the proof of Theorem 3.1.

COROLLARY 3.7. *Let $\lambda(K_\infty), \mu(K_\infty)$ be invariants introduced in Theorem 3.1 and let $\lambda(X_{K_\infty}), \mu(X_{K_\infty})$ be invariants introduced in Definition 2.43. Under the conditions (3.1) and (3.2), we have*

$$\lambda(K_\infty) = \lambda(X_{K_\infty}),\ \mu(K_\infty) = \mu(X_{K_\infty}).$$

Now we go into the proof of the general situation.

PROOF OF THEOREM 3.1 IN THE GENERAL SITUATION. **(Step 1)** Let us take a sufficiently large natural number n_0 such that all ramified primes of K_∞/K_{n_0} are totally ramified. We denote by $\mathfrak{p}_1, \cdots, \mathfrak{p}_s$ the primes of K_{n_0} which are ramified in K_∞/K_{n_0}. Note that, if Iwasawa's class number formula for K_∞/K_{n_0} holds,

Iwasawa's class number formula for the original $\mathbb{Z}_p$-extension K_∞/K also holds[2]. Thus, by replacing K by an intermediate field K_{n_0} with a sufficiently large n_0 if necessary, we suppose that the primes $\mathfrak{p}_1, \cdots, \mathfrak{p}_s$ of K which are ramified in K_∞/K are all totally ramified.

(Step 2) By the assumption that all ramified primes are totally ramified, the inertia subgroup $I_{\mathfrak{p}_j} \subset \mathrm{Gal}(L_\infty/K)$ for each ramified prime $\mathfrak{p}_j$ is isomorphic to Γ through the surjection $\mathrm{Gal}(L_\infty/K) \twoheadrightarrow \Gamma$. Let γ be a fixed topological generator of Γ and we denote by $\gamma_j \in I_{\mathfrak{p}_j}$ the pull-back of γ through the isomorphism $I_{\mathfrak{p}_j} \xrightarrow{\sim} \Gamma$. By exchanging a topological generator γ if necessary, we suppose that $\gamma = \gamma_1$ in the following. Let us consider the group

$$Y_{K_\infty} = \overline{\langle (\gamma-1)X_{K_\infty}, \gamma_2\gamma_1^{-1}, \cdots, \gamma_s\gamma_1^{-1} \rangle},$$

which is the topological closure of the subgroup of X_{K_∞} generated by $(\gamma-1)X_{K_\infty}$, $\gamma_2\gamma_1^{-1}, \cdots, \gamma_s\gamma_1^{-1}$. We would like to show that

$$\mathrm{Cl}(K_n)[p^\infty] \cong \mathrm{Gal}(L_n/K_n) \cong X_{K_\infty}/\tfrac{\gamma^{p^n}-1}{\gamma-1}Y_{K_\infty}. \tag{3.4}$$

The first isomorphism of (3.4) is nothing but the global unramified class field theory. Hence, we will concentrate on the proof of the second isomorphism of (3.4) below. As is shown in (3.3), $X_{K_\infty}/(\gamma^{p^n}-1)X_{K_\infty}$ is isomorphic to the maximal abelian quotient $\mathrm{Gal}(L_\infty/K_n)^{\mathrm{ab}}$ of $\mathrm{Gal}(L_\infty/K_n)$[3]. The group $\mathrm{Gal}(L_n/K_n)$ is the quotient of $\mathrm{Gal}(L_\infty/K_n)^{\mathrm{ab}}$ by the minimal closed normal subgroup which contains the inertia subgroups for all ramified primes. For each j, the inertia subgroup of $\mathrm{Gal}(L_\infty/K_n)^{\mathrm{ab}}$ at the unique prime of K_n over the prime $\mathfrak{p}_j$ is generated by $\gamma_j^{p^n} \in \mathrm{Gal}(L_\infty/K)$. Since we have $\mathrm{Gal}(L_\infty/K_n) = X_{K_\infty}\langle\gamma^{p^n}\rangle$, the following isomorphism holds:

$$\begin{aligned}\mathrm{Gal}(L_n/K_n) &\cong \mathrm{Gal}(L_\infty/K)^{\mathrm{ab}}/\overline{\langle\gamma_1^{p^n}, \gamma_2^{p^n}, \cdots, \gamma_s^{p^n}\rangle} \\ &\cong X_{K_\infty}/\overline{\langle(\gamma^{p^n}-1)X_{K_\infty}, \gamma_1^{p^n}, \cdots, \gamma_s^{p^n}\rangle} \cap X_{K_\infty} \\ &\cong X_{K_\infty}/\overline{\langle(\gamma^{p^n}-1)X_{K_\infty}, \gamma_2^{p^n}\gamma_1^{-p^n}, \cdots, \gamma_s^{p^n}\gamma_1^{-p^n}\rangle}.\end{aligned}$$

On the other hand, the identity

$$\gamma_j^{p^n} = (\gamma_j\gamma_1^{-1})(\gamma_1\cdot\gamma_j\gamma_1^{-1}\cdot\gamma_1^{-1})(\gamma_1^2\cdot\gamma_j\gamma_1^{-1}\cdot\gamma_1^{-2})\cdots(\gamma_1^{p^n-1}\cdot\gamma_j\gamma_1^{-1}\cdot\gamma_1^{-(p^n-1)})\gamma_1^{p^n}$$

holds. Hence, we have $\gamma_j^{p^n}\gamma_1^{-p^n} = (\gamma_j\gamma_1^{-1})^{\frac{\gamma^{p^n}-1}{\gamma-1}}$ since Γ acts on X_{K_∞} by conjugation. This implies the isomorphism

$$\begin{aligned}&\overline{\langle(\gamma^{p^n}-1)X_{K_\infty}, \gamma_2^{p^n}\gamma_1^{-p^n}, \cdots, \gamma_s^{p^n}\gamma_1^{-p^n}\rangle} \\ &\qquad\cong \frac{\gamma^{p^n}-1}{\gamma-1}\overline{\langle(\gamma-1)X_{K_\infty}, \gamma_2\gamma_1, \cdots, \gamma_s\gamma_1\rangle}.\end{aligned}$$

This completes the proof of (3.4).

(Step 3) According to what we have proved up to now, we have $\#\mathrm{Cl}(K_n)[p^\infty] = \#(X_{K_\infty}/Y_{K_\infty})\cdot\#(Y_{K_\infty}/\frac{\gamma^{p^n}-1}{\gamma-1}Y_{K_\infty})$ for each n. Let us define a nonnegative integer ν'' by $p^{\nu''} = \#(X_{K_\infty}/Y_{K_\infty}) = \#\mathrm{Cl}(K_0)[p^\infty]$. Now, by applying Theorem 2.46 to

[2] In fact, since $\lambda(K)(n+n_0)+\mu(K)p^{n+n_0}+\mu(K) = \lambda(K_{n_0})n+\mu(K_{n_0})p^n+\mu(K_{n_0})$ holds for sufficiently large n, it follows that $\lambda(K_{n_0}) = \lambda(K)$, $\mu(K_{n_0}) = \mu(K)$, $\nu(K_{n_0}) = \nu(K)+\lambda(K)n_0+p^{n_0}\mu(K)$.

[3] Note that (3.3) is proved without assuming (3.1) or (3.2).

$\mathcal{M} = Y_{K_\infty}$ or by applying Remark 2.51 to $\mathcal{M} = Y_{K_\infty}$ and $f(T) = T$ more precisely, there exists $\nu' \in \mathbb{Z}$ such that we have $\#(Y_{K_\infty}/\frac{\gamma^{p^n}-1}{\gamma-1}Y_{K_\infty}) = p^{\lambda n+\mu p^n+\nu'}$ for all sufficiently large n with $\lambda = \lambda(Y_{K_\infty})$ and $\mu = \mu(Y_{K_\infty})$. By setting $\nu = \nu' + \nu''$, we obtain the desired result and we complete the proof. □

By using the argument of the above proof, we also prove Theorem 3.4.

PROOF OF THEOREM 3.4 IN THE GENERAL SITUATION. For each natural number n, $\mathrm{Cl}(K_n)[p^\infty]$ is a finite abelian group. Take n to be sufficiently large. By Step 3 of the proof of Theorem 3.1, $Y_{K_\infty}/xY_{K_\infty} \cong \mathrm{Cl}(K_n)[p^\infty]$ is a finite abelian group for any irreducible factor x of $\frac{\gamma^{p^n}-1}{\gamma-1} \in \mathbb{Z}_p[[\Gamma]] = \Lambda$. Thus $Y_{K_\infty}/xY_{K_\infty}$ is a torsion $\Lambda/(x)$-module. By Corollary 2.40, Y_{K_∞} is a finitely generated torsion Λ-module. Since Y_{K_∞} is a subgroup of X_{K_∞} with finite index as shown in Step 3 of the proof of Theorem 3.1, X_{K_∞} is also a finitely generated torsion Λ-module. □

REMARK 3.8. The group $X_{K_\infty}/(\gamma^{p^n} - 1)X_{K_\infty}$ can be infinite if we do not assume (3.1) nor (3.2). That is, the characteristic ideal of X_{K_∞} and the principal ideal $(\gamma^{p^n} - 1)$ might have a nontrivial common factor in general. For example, let us consider a CM-field K and let us denote by K^+ its maximal totally real subfield. Greenberg [**32**] shows that, if the number s of the primes of K^+ over p which are decomposed at K/K^+ is nonzero, we have

$$1 \leq \dim_{\mathbb{Q}_p}(X_{K_\infty}/(\gamma - 1)X_{K_\infty}) \otimes_{\mathbb{Z}_p} \mathbb{Q}_p \leq s.$$

In particular, the group $X_{K_\infty}/(\gamma - 1)X_{K_\infty}$ is infinite in this case[4].

3.1.3. Ideal class groups on CM-fields. In this subsection, we shall study a detailed property of the module X_{K_∞} (Theorem 3.16) in a restricted situation where K is a CM-field and K_∞ is the cyclotomic $\mathbb{Z}_p$-extension of K. Since CM-fields are important in later sections of the book, we shall explain definitions and basic properties of CM-fields.

DEFINITION 3.9. Let K be an algebraic number field. If one of the following equivalent conditions (i), (ii) holds true, K is called a **CM-field**[5].

(i) There exists a totally real algebraic number field K^+ such that K is a totally imaginary quadratic extension of K^+.
(ii) Both of the following conditions (a) and (b) hold true.
   (a) Let us denote by $J : \mathbb{C} \longrightarrow \mathbb{C}$ the complex conjugate. For any embedding $\tau : K \hookrightarrow \mathbb{C}$, we have $J(\tau(K)) = \tau(K)$ and the automorphism $J_\tau = \tau^{-1} \circ J \circ \tau$ of the field K does not depend on the choice of τ (When this condition holds true, we might denote J_τ by J by abuse of notation and call it the complex conjugate of K).
   (b) The condition (a) holds true but $J : K \longrightarrow K$ is a nontrivial automorphism of K.

---

[4]This phenomenon is related to the phenomenon of trivial zero which is discussed later in the book.

[5]The word CM is the abbreviation for the complex multiplication. The name is originated from the fact that the ring of endomorpihsms of an abelian variety with complex multiplication is an order of a CM-field.

For a CM-field K, the invariant subfield K^J coincides with K^+. A field K which satisfies only the condition (ii) (a) of the above definition is either a totally real field or a CM-field and such a field is called a **J-field**.

REMARK 3.10. Let us recall some basic facts on J-fields.

(1) K is a totally real field if and only if K is a J-field on which J acts trivially.
(2) The composite KK' of J-fields K, K' in $\overline{\mathbb{Q}}$ is also a J-field.
(3) Let K be a J-field. A subfield $K' \subset K$ which satisfies $K' = J(K')$ is also a J-field.

EXAMPLE 3.11. (1) A cyclotomic field $\mathbb{Q}(\mu_m) = \mathbb{Q}(\zeta_m)$ is a CM-field. In fact, its subfield $\mathbb{Q}(\zeta_m + \zeta_m^{-1})$ is a totally real algebraic number field and $\mathbb{Q}(\mu_m)$ is a totally imaginary quadratic extension of $\mathbb{Q}(\zeta_m + \zeta_m^{-1})$. Since $\mathrm{Gal}(\mathbb{Q}(\mu_m)/\mathbb{Q}) \cong (\mathbb{Z}/m\mathbb{Z})^\times$ is abelian, the complex conjugate J is in the center of $\mathrm{Gal}(\mathbb{Q}(\mu_m)/\mathbb{Q})$. Hence, any intermediate field of $\mathbb{Q}(\mu_m)/\mathbb{Q}$ is a J-field. By Kronecker-Weber's theorem, any abelian extension of $\mathbb{Q}$ is a J-field.

(2) In general, there exist CM-fields which are non abelian over $\mathbb{Q}$. We will give a CM-field which is not even Galois over $\mathbb{Q}$. Let us consider an irreducible cubic polynomial $f(x) = x^3 - 4x + 1 \in \mathbb{Q}[x]$. Since the discriminant $D_f = -4(-4)^3 - 27 = 229$ is positive, $f(x)$ has three distinct real roots. Hence, $\mathbb{Q}[x]/(f(x))$ is a cubic totally real number field. Because $D_f \notin (\mathbb{Q}^\times)^2$, the Galois group of the smallest splitting field of $f(x)$ over $\mathbb{Q}$ is isomorphic to S_3. This implies that $\mathbb{Q}[x]/(f(x))$ is non Galois over $\mathbb{Q}$. Let us set $K^+ = \mathbb{Q}[x]/(f(x))$ and take an arbitrary imaginary quadratic field E. The composite $K = K^+E$ gives an example of a CM-field which is non Galois over $\mathbb{Q}$. As an another example, we consider a quadratic extension $K = K^+(\sqrt{\sqrt{2}-3})$ of a real quadratic field $K^+ = \mathbb{Q}(\sqrt{2})$. Since K is a totally imaginary extension of K^+, K is a CM-field. However, in this case, K is not obtained as a composite K^+E with a certain imaginary quadratic field E.

For an algebraic number field K, we denote by μ_K the group of roots of unity contained in K. We recall some basic results on the structure of the group of units of a CM-field and the principalization (capitulation) of ideals of a CM-field.

PROPOSITION 3.12. *Let K be a CM-field. Then the group $\mathfrak{r}_{K^+}^\times \mu_K$ is a subgroup of $\mathfrak{r}_K^\times$ whose index is less than* 2.

PROOF OF PROPOSITION 3.12. Let us recall the following well-known result.

LEMMA 3.13. *Let $x \in \overline{\mathbb{Q}}$ be an algebraic integer. If we have $|\tau(x)|_\infty = 1$ for any embedding $\tau : \overline{\mathbb{Q}} \hookrightarrow \mathbb{C}$, x is a root of unity.*

This lemma is a well-known fact proved by Kronecker. The proof is left to the reader as an exercise but we give a brief sketch of the proof. In fact, if x is an algebraic integer of degree d which satisfies the assumption of the lemma, the coefficients of the minimal polynomials of x^n are rational integers with a universal bound of the absolute values which are independent of n. Hence, there exists an integer m such that x coincides with x^m by the pigeonhole principle. This implies that x is a root of unity.

Now, we we have $|\tau(e^{1-J})|_\infty = |\tau(e)^{1-J}|_\infty = \frac{|\tau(e)|_\infty}{|\tau(e)^J|_\infty} = 1$ for any $e \in \mathfrak{r}_K^\times$ and for any embedding $\tau : \overline{\mathbb{Q}} \hookrightarrow \mathbb{C}$. Here we remark that the complex conjugation J is well-defined independently of τ because K is a CM-field. By Lemma 3.13, we have $e^{1-J} \in \mu_K$ and we have the following exact sequence:

$$\{\mathbf{1}\} \longrightarrow \mathfrak{r}_{K+}^\times \longrightarrow \mathfrak{r}_K^\times \xrightarrow{e \mapsto e^{1-J}} \mu_K.$$

When $x \in \mu_K$, we have $x^{1-J} = \frac{x}{x^{-1}} = x^2$. Hence, the image of $\mathfrak{r}_K^\times \xrightarrow{e \mapsto e^{1-J}} \mu_K$ contains μ_K^2. Note also that $\mu_K \bigcap \mathfrak{r}_{K+}^\times = \{\pm 1\}$. Hence, we have

$$\{\mathbf{1}\} \longrightarrow \mathfrak{r}_{K+}^\times/\{\pm 1\} \longrightarrow \mathfrak{r}_K^\times/\{\pm 1\} \xrightarrow{e \mapsto e^{1-J}} \mu_K, \tag{3.5}$$

where $\mathfrak{r}_{K+}^\times/\{\pm 1\}$ is a free abelian group of finite rank and the image of $\mathfrak{r}_K^\times/\{\pm 1\} \xrightarrow{e \mapsto e^{1-J}} \mu_K$ contains also μ_K^2. Since μ_K^2 is of index 2 in μ_K, this completes the proof of Proposition 3.12. □

LEMMA 3.14. *For any finite algebraic number field K and for any non negative integer n, we have $H^1(K_{n+1}^{\rm cyc}/K_n^{\rm cyc}, \mu_{K_{n+1}^{\rm cyc}}) = 0$. That is, for any element $\zeta \in \mu_{K_{n+1}^{\rm cyc}}$ satisfying $N(\zeta) = 1$, there exists an element $\zeta' \in K_{n+1}^{\rm cyc}$ such that $\zeta = \zeta'^{g-1}$ (Here $N : (K_{n+1}^{\rm cyc})^\times \longrightarrow (K_n^{\rm cyc})^\times$ is the norm map and g is a topological generator of $\mathrm{Gal}(K_{n+1}^{\rm cyc}/K_n^{\rm cyc})$).*

PROOF. In the case $\zeta_p \notin K$, we have $\mu_{p^\infty} \bigcap \mu_{K_{n+1}^{\rm cyc}} = \{\mathbf{1}\}$. Thus we have $N(\zeta) = \zeta^p$. The condition $N(\zeta) = 1$ implies $\zeta^p = 1$, hence $\zeta = 1$ by the assumption $\mu_{p^\infty} \bigcap \mu_{K_{n+1}^{\rm cyc}} = \{\mathbf{1}\}$. It suffices to take $\zeta' = 1$.

In the case $\zeta_p \in K$, let us take the largest integer $i \in \mathbb{Z}_{\geq 1}$ such that $\zeta_{p^i} \in K$. Then we have a natural number t prime to p such that $\mu_{K_{n+1}^{\rm cyc}} \cong \mu_{p^{n+i+1}} \times \mu_t$. If $\zeta \in \mu_{K_{n+1}^{\rm cyc}}$, there exist integers $a, b \in \mathbb{Z}$ such that $\zeta = \zeta_{p^{n+i+1}}^a \zeta_t^b$. If a is divisible by p, we have $\zeta \in K_n^{\rm cyc}$. If a is not divisible by p, we have $\zeta \notin K_n^{\rm cyc}$ and the set of conjugates of ζ over $K_n^{\rm cyc}$ is $\{\zeta, \zeta\zeta_p, \zeta\zeta_p^2, \ldots, \zeta\zeta_p^{p-1}\}$. In both cases, we have $N(\zeta) = \zeta^p$. Hence, the condition $N(\zeta) = 1$ implies that ζ is a p-th root of unity. When $\zeta = 1$, it suffices to take $\zeta' = 1$ as in the previous case. When $\zeta \neq 1$, ζ is a primitive p-th root of unity. Note that $\zeta_{p^{n+i+1}}^{g-1}$ is also a primitive p-th root of unity. Let us take $c \in \{1, 2, \ldots, p-2\}$ so that $\zeta = (\zeta_{p^{n+i+1}}^{g-1})^c$. We complete the proof by setting $\zeta' = \zeta_{p^{n+i+1}}^c$. □

Suppose that K is a CM-field which is of finite degree over $\mathbb{Q}$. Then the n-th layer $K_n^{\rm cyc}$ of the cyclotomic $\mathbb{Z}_p$-extension of K is also a CM-field for every nonnegative integer n. The complex conjugate J acts on the ideal class group $\mathrm{Cl}(K_n^{\rm cyc})$ by functoriality. Since the fixed prime p is assumed to be odd, we have a decomposition $\mathrm{Cl}(K_n^{\rm cyc}) = \mathrm{Cl}(K_n^{\rm cyc})[p^\infty]^+ \oplus \mathrm{Cl}(K_n^{\rm cyc})[p^\infty]^-$ according to the eigenvalues ± 1 of the action of J.

PROPOSITION 3.15. *Let K be a CM-field which is of finite degree over $\mathbb{Q}$. Then the natural map $i_{n,n+1} : \mathrm{Cl}(K_n^{\rm cyc})[p^\infty]^- \longrightarrow \mathrm{Cl}(K_{n+1}^{\rm cyc})[p^\infty]^-$ is injective for every nonnegative integer n.*

PROOF. Let $\mathfrak{A}$ be a fractional ideal of $K_n^{\rm cyc}$ whose class $[\mathfrak{A}]$ belongs to $\mathrm{Ker}(i_{n,n+1})$. Since $\mathfrak{A}$ becomes a principal ideal on $K_{n+1}^{\rm cyc}$, there exists an element $a \in K_{n+1}^{\rm cyc}$ such that we have $\mathfrak{A} = (a)$ as fractional ideals of $K_{n+1}^{\rm cyc}$. Since $\mathfrak{A}$ was a fractional ideal on K_n^{cyc}, we have $(a^g) = (a)$ for a generator $g \in \mathrm{Gal}(K_{n+1}^{\mathrm{cyc}}/K_n^{\mathrm{cyc}})$. Let us set $u := a^{g-1} \in K_{n+1}^{\mathrm{cyc}}$. We can check that $u \in (\mathfrak{r}_{K_{n+1}^{\mathrm{cyc}}})^\times$. On the other hand, $\mathfrak{A}^{1+J}$ is a principal fractional ideal of K_n^{cyc} since $[\mathfrak{A}] \in \mathrm{Cl}(K_n^{\mathrm{cyc}})[p^\infty]^-$. Hence, there exists an element $a' \in K_n^{\mathrm{cyc}}$ such that $\mathfrak{A}^{1+J} = (a')$. Let us set

$$u' := \frac{a^1 + J}{a'} \in (\mathfrak{r}_{K_{n+1}^{\mathrm{cyc}}})^\times,$$
$$\epsilon := (\tfrac{a^2}{u'})^{g-1} = \frac{u^2}{(u')^{g-1}} \in (\mathfrak{r}_{K_{n+1}^{\mathrm{cyc}}})^\times.$$

We have

$$\epsilon^{1+J} = \left(\frac{(a^2)^{1+J}}{u'^{1+J}}\right)^{g-1} = \frac{(a'u')^{2(g-1)}}{(u'^{1+J})^{g-1}} = (u'^{g-1})^{1-J}. \tag{3.6}$$

Here the last equality is due to the fact that we have $(a')^g = a'$ since $a' \in K_n^{\mathrm{cyc}}$. We can check that

$$|\tau(u'^{g-1})^{1-J}|_\infty = \frac{|\tau(u'^{g-1})|_\infty}{|\tau(u'^{g-1})^J|_\infty} = 1$$

for any embedding $\tau : \overline{\mathbb{Q}} \hookrightarrow \mathbb{C}$. By (3.6), we have

$$|\tau(\epsilon)|_\infty^2 = |\tau(\epsilon)|_\infty \cdot |\tau(\epsilon^J)|_\infty = |\tau(\epsilon^{1+J})|_\infty = 1$$

for any embedding $\tau : \overline{\mathbb{Q}} \hookrightarrow \mathbb{C}$. By Lemma 3.13, we deduce that $\epsilon \in \mu_{K_{n+1}^{\mathrm{cyc}}}$. By the definition $\epsilon := (\frac{a^2}{u'})^{g-1}$, we have $N(\epsilon) = 1$. By Lemma 3.14, there exists an element $\epsilon' \in \mu_{K_{n+1}^{\mathrm{cyc}}}$ such that $\epsilon = \epsilon'^{g-1}$. This implies that $(\frac{a^2}{u'})^{g-1} = \epsilon'^{g-1}$, hence $(\frac{a^2}{u'\epsilon'})^g = \frac{a^2}{u'\epsilon'}$. We have the following equality between fractional ideals of K_{n+1}^{cyc}:

$$\mathfrak{A}^2 = (a^2) = (a^2/u') = (\tfrac{a^2}{u'\epsilon'}).$$

Since we have $(\alpha/\epsilon') \in K_n^{\mathrm{cyc}}$, $\mathfrak{A}^2$ is already a principal ideal on K_n^{cyc}. We have $[\mathfrak{A}] = 0$ because $p \neq 2$. This completes the proof. □

From now on throughout the book, we denote the Iwasawa algebra $\mathbb{Z}_p[[\Gamma_{\mathrm{cyc},K}]]$ by $\Lambda_{\mathrm{cyc},K}$. When K is a CM-field which is of finite degree over $\mathbb{Q}$, $X_{K_\infty^{\mathrm{cyc}}}$ is decomposed as $X_{K_\infty^{\mathrm{cyc}}} = X^+_{K_\infty^{\mathrm{cyc}}} \oplus X^-_{K_\infty^{\mathrm{cyc}}}$ according to the eigenvalues ± 1 of the action of J. The following result is known for the $\Lambda_{\mathrm{cyc},K}$-module $X_{K_\infty^{\mathrm{cyc}}}$.

THEOREM 3.16. *Let K be a CM-field which is of finite degree over $\mathbb{Q}$. Then $X^-_{K_\infty^{\mathrm{cyc}}}$ has no nontrivial pseudo-null $\Lambda_{\mathrm{cyc},K}$-module.*

REMARK 3.17. Since $X^+_{K_\infty^{\mathrm{cyc}}}$ is isomorphic to $X_{(K^+)_\infty^{\mathrm{cyc}}}$, Conjecture 3.26 in the next subsection predicts that $X^+_{K_\infty^{\mathrm{cyc}}}$ is a pseudo null $\Lambda_{\mathrm{cyc},K}$-module. The module $X^+_{K_\infty^{\mathrm{cyc}}}$ can be nonzero in general. However, we have Conjecture 3.25 in the next subsection for a special case.

PROOF OF THEOREM 3.16. In order to prove the theorem by contradiction, we assume that $X^-_{K_\infty^{\mathrm{cyc}}}$ has a nontrivial pseudo-null $\Lambda_{\mathrm{cyc},K}$-module. By assumption, there exists a nonzero element $x \in X^-_{K_\infty^{\mathrm{cyc}}}$ of order p. For each natural number n, we denote by $p_n : X^-_{K_\infty^{\mathrm{cyc}}} \longrightarrow \mathrm{Cl}(K_n^{\mathrm{cyc}})[p^\infty]^-$ the natural projection map. For a sufficiently large n, $p_n(x) \in \mathrm{Cl}(K_n^{\mathrm{cyc}})[p^\infty]^-$ is a nontrivial element of order p. On the other hand, any sufficiently small open subgroup $U \subset \Gamma_{\mathrm{cyc},K}$ acts nontrivially on x.

Replacing n by a larger natural number if necessary, we may suppose that $(\Gamma_{\mathrm{cyc}})^{p^n}$ acts nontrivially on $p_n(x)$. Let $N_{n+1,n} : \mathrm{Cl}(K_{n+1}^{\mathrm{cyc}})[p^\infty]^- \longrightarrow \mathrm{Cl}(K_n^{\mathrm{cyc}})[p^\infty]^-$ be the norm map, $i_{n,n+1} : \mathrm{Cl}(K_n^{\mathrm{cyc}})[p^\infty]^- \longrightarrow \mathrm{Cl}(K_{n+1}^{\mathrm{cyc}})[p^\infty]^-$ the map induced by the inclusion map $K_n^{\mathrm{cyc}} \hookrightarrow K_{n+1}^{\mathrm{cyc}}$. Then we have

$$i_{n,n+1} \circ p_n(x) = i_{n,n+1} \circ N_{n+1,n} \circ p_{n+1}(x) = \sum_{g \in \Gamma_{\mathrm{cyc}}^{p^n}/\Gamma_{\mathrm{cyc}}^{p^{n+1}}} p_{n+1}(x)^g.$$

Since $(\Gamma_{\mathrm{cyc}})^{p^n}$ acts nontrivially on $p_{n+1}(x)$ by the assumption that n is sufficiently large, we have $i_{n,n+1} \circ p_n(x) = p_{n+1}(px) = 0$. By Proposition 3.15, $i_{n,n+1} : \mathrm{Cl}(K_n^{\mathrm{cyc}})[p^\infty]^- \longrightarrow \mathrm{Cl}(K_{n+1}^{\mathrm{cyc}})[p^\infty]^-$ is injective. Hence, we have $p_n(x) = 0$, which is a contradiction to our assumption. This completes the proof. $\square$

3.1.4. Some results and conjecture on ideal class groups of cyclotomic fields. In this subsection, we present some known conjectures and results on Iwasawa invariants.

As explained in Section 1.3, the "analogy between algebraic number fields and function fields of one variable over a finite field" is an important philosophy in number theory. Let us try to understand the meaning of Iwasawa invariants through this philosophy. Let K be a function field of one variable over a finite field $\mathbb{F}_q$ of characteristic p and $C = C_K/\mathbb{F}_q$ a unique proper smooth algebraic curve whose function field is K. The divisor class group $\mathrm{Cl}_C(\mathbb{F}_q)$ is an analogue of the ideal class group. Let us discuss the behavior of the ℓ-power part of the divisor class group $\mathrm{Cl}_C(\mathbb{F})[\ell^\infty]$ for a fixed prime $\ell \neq p$ when the field of constants $\mathbb{F}/\mathbb{F}_q$ varies. Over the algebraically closure $\overline{\mathbb{F}}_q$, we have an isomorphism $\mathrm{Cl}_C(\overline{\mathbb{F}}_q)[\ell^\infty] \cong (\mathbb{Q}_\ell/\mathbb{Z}_\ell)^{\oplus 2g}$ where $g = g(C)$ is the genus of C. For simplicity, we assume that $\mathrm{Cl}_C(\mathbb{F}_q)[\ell] = \mathrm{Cl}_C(\overline{\mathbb{F}}_q)[\ell]$ replacing $\mathbb{F}_q$ by a sufficiently large finite extension. Since we have $\mathrm{Cl}_C(\overline{\mathbb{F}}_q)[\ell] \cong (\mathbb{Z}/\ell\mathbb{Z})^{\oplus 2g}$, this implies that $\mathrm{Cl}_C(\mathbb{F}_q)[\ell] \cong (\mathbb{Z}/\ell\mathbb{Z})^{\oplus 2g}$. Recall that $\mathrm{Gal}(\overline{\mathbb{F}}_q/\mathbb{F}_q) \cong \mathbb{Z}_\ell \times \prod_{\ell' \neq \ell} \mathbb{Z}_{\ell'}$. Let us recall the following facts whose proofs are left to the reader[6].

(1) Let $\mathbb{F}$ be any intermediate field of the unique $\mathbb{Z}_{\ell'}$-extension of $\mathbb{F}_q$ where ℓ' is a prime different from ℓ. Then $\mathrm{Cl}_C(\mathbb{F})[\ell^\infty]$ is isomorphic to $\mathrm{Cl}_C(\mathbb{F}_q)[\ell^\infty]$.
(2) There exists a constant $\nu \in \mathbb{Z}$ such that we have $\#\mathrm{Cl}_C(k_n)[\ell^\infty] = \ell^{2gn+\nu}$ for any natural number n where k_n is the n-th layer of the unique $\mathbb{Z}_\ell$-extension of $\mathbb{F}_q$.

We observe that, in the case of function fields, there appears no μ-invariant for the growth formula of the order of the divisor class group $\#\mathrm{Cl}_C(k_n)[\ell^\infty]$. In the case of algebraic number fields, we can give an example where the μ-invariant is nonzero (see Remark 3.19).

The notion of a field of coefficients does not make sense in algebraic number fields. Thus the extension of the field of coefficients which played an important role in the case of function fields does not make sense in the case of algebraic number fields. However an extension of the field of coefficients in the case of function fields is also interpreted as an extension obtained by adjoining roots of unity. Hence, the cyclotomic $\mathbb{Z}_p$-extension $K_\infty^{\mathrm{cyc}}/K$ of a finite algebraic number field K is a proper analogue of the $\mathbb{Z}_\ell$-extension of a function field of one variable over a finite field $\mathbb{F}_q$ obtained by the unique $\mathbb{Z}_\ell$-extension of $\mathbb{F}_q$.

---

[6] We can apply the argument of the proof of Iwasawa class number formula given earlier by replacing the role of X_{K_∞} by the Pontrjagin dual of $\mathrm{Cl}_C(\overline{\mathbb{F}}_q)[\ell^\infty]$ and by replacing the role of a topological generator $\gamma \in \Gamma$ by that of a Frobenius element of the absolute group of $\mathbb{F}/\mathbb{F}_q$.

The above philosophy saying that the cyclotomic $\mathbb{Z}_p$-extension is a proper analogue of the extension of the field of constants of function fields justifies the following conjecture.

CONJECTURE 3.18 (Iwasawa). *Let K be a finite algebraic number field. Then we have $\mu(K_\infty^{\mathrm{cyc}}) = 0$ where $K_\infty^{\mathrm{cyc}}/K$ is the cyclotomic $\mathbb{Z}_p$-extension.*

REMARK 3.19. Iwasawa [**52**] shows that, for any natural number m, there exists a finite algebraic number field K and a $\mathbb{Z}_p$-extension K_∞/K such that $\mu(K_\infty) \geq m$. His example is based on the observation on the anti-cyclotomic $\mathbb{Z}_p$-extension of an imaginary quadratic field explained below. Let M be an imaginary quadratic field. There is a unique $\mathbb{Z}_p$-extension M_∞^{ac}/M such that M_∞^{ac} is Galois over $\mathbb{Q}$ and $\mathrm{Gal}(M_\infty^{\mathrm{ac}}/\mathbb{Q})$ is not abelian[7]. This is called the **anti-cyclotomic $\mathbb{Z}_p$-extension** of M. The anti-cyclotomic $\mathbb{Z}_p$-extension M_∞^{ac}/M is also characterized as the unique $\mathbb{Z}_p$-extension of M such that the n-th layer M_n^{ac} of M_∞^{ac}/M is Galois over $\mathbb{Q}$ and $\mathrm{Gal}(M_n^{\mathrm{ac}}/\mathbb{Q})$ is a dihedral group of order $2p^n$ for any natural number n. As shown in [**52**, §2], any prime $l \neq p$ which is undecomposed in $M/\mathbb{Q}$ is totally decomposed in M_∞^{ac}/M. Now we take prime numbers $l_1, \ldots, l_{m+3}$ which are unramified and undecomposed in M and we set $a = l_1 \cdots l_{m+3}$. We consider $K = M(\sqrt[p]{a})$ and we define K_n to be KM_n^{ac}. For each i with $1 \leq i \leq m+3$, l_i is totally decomposed into p^n ideals in M_n^{ac} and all of them are ramified in the quadratic extension K_n/M_n^{ac}. Since the genus theory implies that $p^{mn}\#\mathrm{Cl}(M_n^{\mathrm{ac}})[p^\infty] \mid \#\mathrm{Cl}(K_n)[p^\infty]$, we conclude that $\mu(K_\infty) \geq m$. For further detail of the argument, the reader is encouraged to look into the article [**52**].

Iwasawa himself proved that Conjecture 3.18 holds true when K is a cyclic extension of degree p (see [**52**, Theorem 3]). In general, the following theorem is the strongest result which is known at the moment.

THEOREM 3.20 (Ferrero–Washington). *Let K be an abelian extension of $\mathbb{Q}$. Then, $\mu(K_\infty^{\mathrm{cyc}}) = 0$ holds true.*

Their method of the proof is not related directly to the study of ideal class groups. In fact, the "principle of the analytic class number formula" which is given in Theorem 3.74 assures that the validity of the vanishing $\mu(K_\infty^{\mathrm{cyc}}) = 0$ for all abelian fields K is equivalent to the validity of the vanishing of μ-invariants of analytic p-adic L-functions for various Dirichlet characters. Hence, the precise statement of the theorem of Ferrero–Washington concerns the μ-invariants of analytic p-adic L-functions (see Theorem 3.60).

Let $C = C_K/\mathbb{F}_q$ a unique proper smooth algebraic curve of a function field K of one variable over a finite field $\mathbb{F}_q$ of characteristic p. For every prime $\ell \neq p$, ℓ-adic Galois representations $V_{C,\ell} := (\varprojlim_n \mathrm{Cl}_C(\overline{\mathbb{F}}_q)[p^n]) \otimes_{\mathbb{Z}_\ell} \mathbb{Q}_\ell$ associated to the divisor class group Cl_C is known to be semi-simple as $\mathbb{Q}_\ell[\mathrm{Gal}(\overline{\mathbb{F}}_q/\mathbb{F}_q)]$-module. More generally, an ℓ-adic Galois representation obtained by a factor of étale cohomology associated to a proper smooth variety over a field k which is finitely generated over its prime field is expected to be semi-simple as $\mathbb{Q}_\ell[\mathrm{Gal}(\overline{k}/k)]$-module (Tate

[7]Since the Leopoldt conjecture holds true for imaginary quadratic fields, the Galois group $\mathrm{Gal}(\widetilde{M}_\infty/M)$ of the composite $\widetilde{M}_\infty$ of $\mathbb{Z}_p$-extensions of M is a free $\mathbb{Z}_p$-module of rank two (see Remark 2.8). Since the multiplicity of the eigenvalue -1 of the action of $\mathrm{Gal}(M/\mathbb{Q})$ on $\mathrm{Gal}(\widetilde{M}_\infty/M)$ is one, we cut out a $\mathbb{Z}_p$-extension M_∞^{ac}/M such that the conjugate action of $\mathrm{Gal}(M/\mathbb{Q})$ on $\mathrm{Gal}(M_\infty^{\mathrm{ac}}/M)$ is nontrivial.

conjecture). As an analogue of this philosophy for algebraic number fields, the following conjecture is widely believed.

CONJECTURE 3.21 (Semi-simplicity conjecture). *Let K be a finite algebraic number field, $K_\infty^{\mathrm{cyc}}/K$ the cyclotomic $\mathbb{Z}_p$-extension of K. Then, the action of $\Gamma_{\mathrm{cyc},K}$ on the finite dimensional $\mathbb{Q}_p$-vector space $X_{K_\infty^{\mathrm{cyc}}} \otimes_{\mathbb{Z}_p} \mathbb{Q}_p$ should be semi-simple.*

REMARK 3.22. If we consider the action of the Galois group of a noncyclotomic $\mathbb{Z}_p$-extension K_∞/K on the finite dimensional $\mathbb{Q}_p$-vector space $X_{K_\infty} \otimes_{\mathbb{Z}_p} \mathbb{Q}_p$ is not true and some counter-examples are constructed.

When $K/\mathbb{Q}$ is abelian extension of $\mathbb{Q}$, Conjecture 3.23 below is also believed. Conjecture 3.23 is stronger than Conjecture 3.21 in the sense that Conjecture 3.23 implies Conjecture 3.21.

CONJECTURE 3.23 (Multiplicity one conjecture). *Let K be a finite algebraic number field which is an abelian extension of $\mathbb{Q}$ and let $K_\infty^{\mathrm{cyc}}/K$ be the cyclotomic $\mathbb{Z}_p$-extension of K. Then, for each character η of $\mathrm{Gal}(K/\mathbb{Q})$, the characteristic polynomial of the action of a topological generator of $\Gamma_{\mathrm{cyc},K}$ on $(X_{K_\infty^{\mathrm{cyc}}} \otimes_{\mathbb{Z}_p} \mathbb{Q}_p)^\eta$ has no double roots.*

Recall that the Iwasawa λ-invariant of the Iwasawa module X_{K_∞} associated to the ideal class group over a $\mathbb{Z}_p$-extension of a finite algebraic number field K is an analogue of the genus of the smooth proper curve over a finite field of characteristic p. Thanks to this analogy, it will be a fruitful idea to look for its counterpart on the study of Iwasawa λ-invariants each time we find an interesting mathematical statement on the genus of smooth proper curves. As an example, we remember the Riemann–Hurwitz formula which describes a relation between the genus $g(C)$ and the genus $g(C')$ for a covering map $C' \twoheadrightarrow C$ of smooth proper curves[8]. When we discuss an analogue of this formula on algebraic number fields, covering maps of smooth proper curves over a finite field of characteristic p correspond to finite extensions of algebraic number fields. Since extensions of prime-to-p degree of a function field K of one variable over a finite field do not effect on the order of the p-part of divisor class groups of the model of K, we restrict ourselves to the case where K'/K is a finite extension of p-power degree. Also, by Remark 3.17, a meaningful invariant to study is the λ-invariant of the $(-)$-part of the Iwasawa module associated to the ideal class group.

The theorem below is an Iwasawa theoretic analogue of the Riemann–Hurwitz formula for algebraic curves, which is called the **Riemann–Hurwitz formula (or Kida's formula)**.

THEOREM 3.24 (Riemann–Hurwitz formula). *Let K be a CM-field which is of finite degree over $\mathbb{Q}$. We consider a CM finite extension K' such that $[K' : K]$ is a power of p. Then*

(1) *We have $\mu(K_\infty^{\mathrm{cyc}}) = 0$ if and only if $\mu({K'}_\infty^{\mathrm{cyc}}) = 0$.*

[8] If the covering degree $[C' : C]$ is prime to the field of definition of the curves C and C', the Riemann–Hurwitz formula is known as the equality:

$$2g(C') - 2 = [C' : C](2g(C) - 2) + \sum_{x \in C'} (e_x - 1)$$

where e_x is the ramification degree for each point $x \in C'$.

(2) *Suppose that* $\mu(K_\infty^{\mathrm{cyc}})=0$. *Then the following formula holds true:*

$$2\lambda(X^-_{K'^{\mathrm{cyc}}_\infty})-2\delta_{K'} = [K'^{\mathrm{cyc}}_\infty : K^{\mathrm{cyc}}_\infty]\,(2\lambda(X^-_{K^{\mathrm{cyc}}_\infty})-2\delta_K)+\sum_{\substack{v'\nmid p\\ v'_+:\ \text{split at } K'^{\mathrm{cyc}}_\infty}}(e(v'/v)-1).$$

Here, for each prime v' *of* K'^{cyc}_∞, v'_+ *(resp.* v*) denotes its restriction to* $(K'^{\mathrm{cyc}}_\infty)^+$ *resp.* K^{cyc}_∞*) and* $e(v'/v)$ *denotes the ramification index. Also,* $\delta_{K'}$ *(resp.* δ_K*) is defined to be* 1 *if* $\zeta_p\in K'$ *(resp.* $\zeta_p\in K$*) and to be* 0 *otherwise*[9].

Theorem 3.24 was first obtained by Kida [**63**]. Kida's original proof is based on a calculation of $\#\mathrm{Cl}(K'^{\mathrm{cyc}}_n)^-[p^\infty]$ and $\#\mathrm{Cl}(K^{\mathrm{cyc}}_n)^-[p^\infty]$ by the genus theory. He obtains the desired relation between $\lambda(X^-_{K'^{\mathrm{cyc}}_\infty})$ and $\lambda(X^-_{K^{\mathrm{cyc}}_\infty})$ by observing the approximative behavior of these orders when n goes to infinity (see Corollary 3.7). Later, Iwasawa [**54**] gives a different proof of Theorem 3.24 based on the method which studies the $\mathrm{Gal}(K'^{\mathrm{cyc}}_\infty/K^{\mathrm{cyc}}_\infty)$-module $X^-_{K'^{\mathrm{cyc}}_\infty}$ by means of the theory of linear representations of finite groups. Kuz'min [**68**] also obtains a similar result independently by a method which is close to that of [**54**]. When K and K' are abelian extensions of $\mathbb{Q}$, by using the Iwasawa main conjecture (Theorem 3.66), Kida's formula (Theorem 3.24) is translated to "Kida's formula for the analytic p-adic L-function", which is the same type of formula comparing the Iwasawa invariants of the analytic p-adic L-function for K and the analytic p-adic L-function for K'. Gras and Sinnott prove the Kida's formula for the analytic p-adic L-function, which gives an analytic proof of Theorem 3.24 in this specific situation.

Up to this point, we presented some conjectures and results which are motivated from the view point of the analogy between number fields and function fields. At the end of this section, we also present some conjectures which do not have analogous statements on the case of function fields case.

CONJECTURE 3.25 (Vandiver conjecture). *Let* $K=\mathbb{Q}(\zeta_p)^+$. *Then, we have* $X_{K^{\mathrm{cyc}}_\infty}=0$.

Vandiver's conjecture is verified numerically for a certain amount of primes. However it is true that there are no reasons to support this conjecture other than numerical experiments. We also have the conjecture below which concerns general totally real finite algebraic number fields K. The conjecture below is an immediate consequence of the Vandiver's conjecture when $K=\mathbb{Q}(\zeta_p)^+$.

CONJECTURE 3.26 (Greenberg's conjecture). *Let* K *be a totally real finite algebraic number field. Then we have* $\lambda(K^{\mathrm{cyc}}_\infty)=\mu(K^{\mathrm{cyc}}_\infty)=0$. *In other words,* $X_{K^{\mathrm{cyc}}_\infty}$ *is a finite abelian group.*

This conjecture originates from a question on the existence of a totally real finite algebraic number field K with $\lambda(K^{\mathrm{cyc}}_\infty)=\mu(K^{\mathrm{cyc}}_\infty)>0$ as is discussed in the end of §11 of a celebrated paper [**53**] by Iwasawa. This problem was proposed for the theme of the PhD thesis of Ralph Greenberg by his thesis advisor Iwasawa. Since this problem was discussed in the first page of Greenberg's paper [**33**], which was based on his PhD thesis, the problem is finally now known as a conjecture which

[9]Since $[K':K]$ is a power of p, we have $\delta_{K'}=\delta_K$.

insists the nonexistence of such a totally real field. When K is a real quadratic field, various active attempts on this conjecture were done to give some criteria to check the validity of the conjecture or some numerical examples. However we have not yet proved the conjecture even for real quadratic fields and we do not find any counter-examples.

3.2. Analytic aspect (p-adic L-function)

As before, we set $\Gamma_{\mathrm{cyc}} = \mathrm{Gal}(\mathbb{Q}_\infty/\mathbb{Q})$ and we denote by $\mathcal{O}$ the ring of integers of a finite extension of $\mathbb{Q}_p$. Throughout the book, we will denote by $\Lambda_{\mathrm{cyc},\mathcal{O}}$ the Iwasawa algebra $\mathcal{O}[[\Gamma_{\mathrm{cyc}}]]$ (we will denote $\Lambda_{\mathrm{cyc},\mathbb{Z}_p}$ by Λ_{cyc} for short). The aim of §3.2 is to construct a p-adic L-function in $\Lambda_{\mathrm{cyc},\mathcal{O}}$ which interpolates the special values at negative integers of a Dirichlet L-function in Theorem 3.30. For Theorem 3.30, we will give two different proofs, the one by Iwasawa's construction based on Stickelberger element (see §3.2.3) and the one by Coleman's construction based on cyclotomic units (see §3.2.4).

3.2.1. Main statement on analytic p-adic L-function.

DEFINITION 3.27. (1) Let M be a natural number. A map $\eta : \mathbb{Z} \longrightarrow \mathbb{C}$ is called a **Dirichlet character** modulo M if the following conditions hold:
 (a) For $a, b \in \mathbb{Z}$ satisfying $a \equiv b \bmod M$, we have $\eta(a) = \eta(b)$.
 (b) For any $a, b \in \mathbb{Z}$, we have $\eta(ab) = \eta(a)\eta(b)$.
 (c) We have $\eta(a) \neq 0$ if and only if $(a, M) = 1$.

(2) Let η' be a Dirichlet character modulo N and M a natural number divisible by N. Then a map η defined by

$$\eta(a) = \begin{cases} \eta'(a) & \text{when } (a, M) = 1, \\ 0 & \text{when } (a, M) \neq 1, \end{cases}$$

is a Dirichlet character modulo N. Then we say that the character η is induced from η'.

(3) Let η be a Dirichlet character modulo M. If η is not induced from any other Dirichlet character η' modulo N with $N < M$, η is called a **primitive Dirichlet character** and M is called the conductor of η. If η is a nonprimitive Dirichlet character, there is a unique primitive character η_0 which induces the character η. We call the conductor N_0 of η_0 the **conductor** of η.

(4) We define the trivial Dirichlet character $\mathbf{1}$ to be the map $\mathbf{1} : \mathbb{Z} \longrightarrow \mathbb{C}$ defined by $\mathbf{1}(a) = 1$ for any $a \in \mathbb{Z}$.

REMARK 3.28. Recall that there is a canonical isomorphism $\mathrm{Gal}(\mathbb{Q}(\mu_M)/\mathbb{Q}) \xrightarrow{\sim} (\mathbb{Z}/M\mathbb{Z})^\times$ for any natural number M (see Proposition 2.1 for the proof of the case $M = p^n$ and Remark 2.2 for the general case). Through this isomorphism, we have a natural identification as follows:

$$\{\text{Dirichlet characters modulo } M\} \overset{\sim}{\longleftrightarrow} \{\text{characters of } \mathrm{Gal}(\mathbb{Q}(\mu_M)/\mathbb{Q}) \text{ valued in } \mathbb{C}^\times\}.$$

When η and η' are Dirichlet characters modulo M, the product $\eta\eta'$ is defined by setting $(\eta\eta')(a) = \eta(a)\eta'(a)$. With this product, the set of Dirichlet characters modulo M naturally becomes an abelian group whose identity element is the

Dirichlet character $\mathbf{1}_N$ defined by

$$\mathbf{1}_N(a) = \begin{cases} 1 & \text{when } (a, N) = 1, \\ 0 & \text{when } (a, N) \neq 1. \end{cases}$$

We would also like to consider the product $\eta\eta'$ when η and η' are primitive Dirichlet characters modulo different natural numbers M and M' respectively. However, the natural product $\eta\eta'$ defined in the same way as the above product is naturally a Dirichlet character defined modulo $\mathrm{LCM}(M, M')$ but not necessarily primitive even if η and η' are primitive. In order to give a structure of an abelian group on the set of all primitive characters, we recall the following lemma.

LEMMA 3.29. *Suppose that η_1 (resp. η_2) is a Dirichlet character modulo M_1 (resp. M_2). Take a natural number M which is a multiple of both M_1 and M_2. If the Dirichlet character modulo M induced from η_1 coincides with the Dirichlet character modulo M induced from η_2, there exists a unique Dirichlet character η_0 modulo $\mathrm{GCD}(M_1, M_2)$ such that η_1 and η_2 are both induced from η_0.*

For primitive Dirichlet characters η and η' whose conductors M and M' are not necessarily equal, we first consider the product $\widetilde{\eta}\widetilde{\eta'}$ of Dirichlet characters $\widetilde{\eta}$ and $\widetilde{\eta'}$ defined modulo $\mathrm{LCM}(M, M')$ induced by η and η'. Then we define a product of η and η' to be the primitive Dirichlet character associated to $\widetilde{\eta}\widetilde{\eta'}$. Lemma 3.29 assures that this product gives a well-defined structure of a group on the set of all primitive Dirichlet characters. The verification of this fact is left to the reader as an exercise.

From now on, we only consider primitive characters and the product between two primitive Dirichlet characters are taken in the above manner. When η is a primitive Dirichlet character, we define the Dirichlet L-function $L(\eta, s)$ by the following Dirichlet series

$$L(\eta, s) = \sum_{n=1}^{\infty} \frac{\eta(n)}{n^s}$$

which is also equal to the following Euler product[10]:

$$L(\eta, s) = \prod_{\ell:\text{prime}} (1 - \eta(\ell)\ell^{-s})^{-1}$$

in its domain of absolute convergence $\mathrm{Re}(s) > 1$. When $\eta \neq \mathbf{1}$, this series is convergent at $\mathrm{Re}(s) > 0$. As is proved later in Theorem 3.36, $L(\eta, s)$ is meromorphically continued to the whole $\mathbb{C}$-plane (see [**80**, Chap. VII Theorem (2.8)] for example). When $\eta \neq \mathbf{1}$ is holomorpihc at the whole $\mathbb{C}$-plane. When $\eta = \mathbf{1}$, $L(\mathbf{1}, s)$ coincides with the Riemann's zeta function $\zeta(s)$. The function $\zeta(s)$ has a simple pole at $s = 1$ and it is holomorphic for $s \neq 1$. As we will show in Theorem 3.36, we have $L(\eta, j) \in \overline{\mathbb{Q}}$ for any Dirichlet character η and for any integer $j \leq 0$.

From now on throughout the book, for a Dirichlet character ψ, we denote by $\mathbb{Z}_p[\psi]$ the subring of $\overline{\mathbb{Q}}_p$ obtained by adjoining the values of ψ to $\mathbb{Z}_p$ and we denote by $\Lambda_{\mathrm{cyc},\psi}$ the Iwasawa algebra $\mathbb{Z}_p[\psi][[\Gamma_{\mathrm{cyc}}]]$. Note that $\Lambda_{\mathrm{cyc},\mathbf{1}}$ is nothing but Λ_{cyc}.

---

[10] In fact, we can define $L(\eta, s)$ for any non primitive Dirichlet character η modulo M in the same way as above. However, $L(\eta, s)$ can be obtained from $L(\eta_0, s)$ by removing Euler factors at primes ℓ dividing M and not dividing M_0. Thus, essentially, it will be sufficient to study the case of primitive Dirichlet characters.

We now state the following theorem obtained by Kubota–Leopoldt, Iwasawa, Coleman et al[11].

THEOREM 3.30 (Kubota–Leopoldt, Iwasawa, Coleman et al). *Let ψ be a primitive Dirichlet character of conductor Np^e where N is a natural number prime to p and e is an integer which is equal to 0 or 1. We assume that ψ is an even Dirichlet character (i.e. $\psi(-1) = 1$)[12]. Then, there exists a unique element*

$$L_p(\psi) \in \begin{cases} \Lambda_{\mathrm{cyc},\psi} & \text{when } \psi \neq \mathbf{1}, \\ \dfrac{1}{\gamma - \kappa_{\mathrm{cyc}}(\gamma)}\Lambda_{\mathrm{cyc}} & \text{when } \psi = \mathbf{1}, \end{cases}$$

which satisfies the following equality:

$$\kappa_{\mathrm{cyc}}^{1-r}\phi(L_p(\psi)) = \left(1 - \frac{(\psi\omega^{-r}\phi^{-1})(p)}{p^{1-r}}\right) L(\psi\omega^{-r}\phi^{-1}, 1-r) \tag{3.7}$$

for any integer $r \geq 1$ and for any $\overline{\mathbb{Q}}_p^\times$-valued character ϕ of Γ_{cyc} of finite order such that $r \neq 1$ or $\phi \neq \mathbf{1}$ if $\psi = \mathbf{1}$[13]. Here, γ is a topological generator of Γ_{cyc} and we note that the module $\frac{1}{\gamma - \kappa_{\mathrm{cyc}}(\gamma)}\Lambda_{\mathrm{cyc}}$ is independent of the choice of a topological generator γ of Γ_{cyc}.

As we commented at the beginning of §3.2, we give some necessary preparations at §3.2.2 and at the beginning of §3.2.3. Then we give a proof of Theorem 3.30 at the end of §3.2.3 and another proof at the end of §3.2.4.

REMARK 3.31. Let us choose a topological generator γ of Γ_{cyc} so that $\kappa_{\mathrm{cyc}}(\gamma) = 1 + p$ and we fix an isomorphism $\Lambda_{\mathrm{cyc},\psi} \xrightarrow{\sim} \mathbb{Z}_p[\psi][[T]]$, $\gamma \mapsto 1 + T$. Note that we have the following commutative diagram:

$$\begin{array}{ccc} \Lambda_{\mathrm{cyc},\psi} & \xrightarrow{\gamma \mapsto \kappa_{\mathrm{cyc}}^r(\gamma)} & \mathbb{Z}_p[\psi] \\ \downarrow & & \| \\ \mathbb{Z}_p[\psi][[T]] & \xrightarrow[T \mapsto (1+p)^r - 1]{} & \mathbb{Z}_p[\psi]. \end{array}$$

LEMMA 3.32. *Let $A(T) \in \mathbb{Z}_p[\psi][[T]]$. Then, we have*

$$A(T)|_{T=(1+p)^r-1} \equiv A(T)|_{T=(1+p)^{r'}-1} \mod p^{n+1}$$

for $r \equiv r' \mod p^n$.

[11]The result by Kubota–Leopoldt is the oldest, but the one constructed by Kubota–Leopoldt is not an element of Iwasawa algebra and it can not interpolate the specialization by finite characters ϕ as stated in Theorem 3.30 below (see Remark 3.33 (1) for more precise explanations).

[12]When ψ is an odd Dirichlet character (i.e. $\psi(-1) = -1$), the values in the right-hand side of (3.7) is zero for any ϕ and for any r. Thus, by the identity theorem (Proposition 2.21), an element $L_p(\psi)$ satisfying the interpolation property (3.7) must be zero. For a Dirichlet character ψ' whose conductor is divisible by p^2, we get an element $L_p(\psi')$ with desired interpolation by modifying $L_p(\psi)$ for ψ whose conductor is not divisible by p^2. Hence, the restriction on the conductor of ψ is not essential.

[13]Let us denote by $(\Gamma_{\mathrm{cyc}})^{p^{n-1}}$ the kernel of ϕ. Then, $\Gamma_{\mathrm{cyc}}/(\Gamma_{\mathrm{cyc}})^{p^{n-1}}$ is a quotient of $\mathrm{Gal}(\mathbb{Q}(\mu_{p^n})/\mathbb{Q})$ which is canonically identified with $(\mathbb{Z}/p^n\mathbb{Z})^\times$ by the class field theory. We thus regard ϕ as a primitive Dirichlet character of conductor p^n.

PROOF. By the binomial expansion, $((1+p)^r-1)-((1+p)^{r'}-1)$ is divisible by p^{n+1} if $r-r'$ is divisible by p^n. Hence, for any $i\geq 0$, $((1+p)^r-1)^i-((1+p)^{r'}-1)^i$ is divisible by p^{n+1} if $r-r'$ is divisible by p^n. This implies the desired conclusion. □

By the interpolation property of $L_p(\psi)$ given in Theorem 3.30 combined with Remark 3.31 and Lemma 3.32, we have

$$\begin{aligned}(3.8)\qquad &\left(1-\frac{(\psi\omega^{-r})(p)}{p^{1-r}}\right)L(\psi\omega^{-r},1-r)\\ &\equiv\left(1-\frac{(\psi\omega^{-r'})(p)}{p^{1-r'}}\right)L(\psi\omega^{-r'},1-r') \mod p^{n+1}\end{aligned}$$

for any positive integers r,r' such that $r\equiv r' \mod p^n$.

REMARK 3.33. (1) As we will prove in Theorem 3.36, special values of Dirichlet L-functions are expressed by using generalized Bernoulli numbers. By Theorem 3.36 and the equation (3.8), we have

$$\begin{aligned}(3.9)\qquad &\left(1-\frac{(\psi\omega^{-r})(p)}{p^{1-r}}\right)\frac{B_{r,\psi\omega^{-r}}}{r}\\ &\equiv\left(1-\frac{(\psi\omega^{-r'})(p)}{p^{1-r'}}\right)\frac{B_{r',\psi\omega^{-r'}}}{r'} \mod p^{n+1}\end{aligned}$$

for any positive integers r,r' such that $r\equiv r' \mod p^n$. Such p-adic congruence relations between Bernoulli numbers go back to Kummer. Conversely, such p-adic congruence relations allow us to construct a p-adic L-function. In fact, it was Kubota–Leopoldt [**65**] who constructed a p-adic L-function first through the study of p-adic congruence relations of Bernoulli numbers. However, p-adic L-functions obtained in [**65**] is not exactly the same object as that of Theorem 3.30. In fact, [**65**] proves that, for each character ϕ, there exists a p-adic analytic function $f(\psi\phi,s)$ for variable $s\in\mathbb{C}_p$ (contained in a certain domain of convergence) satisfying

$$f(\psi\phi,1-r)=\left(1-\tfrac{(\psi\phi\omega^{-r})(p)}{p^{1-r}}\right)L(\psi\phi\omega^{-r},1-r)$$

for any integer $r\geq 1$. If we assume the existence of $L_p(\psi)\in\Lambda_{\mathrm{cyc},\psi}$ stated in Theorem 3.30, we obtain $f(\psi\phi,s)$ for each character ϕ by setting $f(\psi\phi,s)=\chi^s_{\mathrm{cyc}}\phi^{-1}(L_p(\psi))$. On the other hand, it is not possible to obtain $L_p(\psi)\in\Lambda_{\mathrm{cyc},\psi}$ assuming the existence of $f(\psi\phi,s)$ for every character ϕ. In this sense, Theorem 3.30 is stronger than the original result of Kubota–Leopoldt [**65**].

(2) In order to formulate the Iwasawa main conjecture, it is convenient to formulate a p-adic L-function as an element of Iwasawa algebra. Especially, if our setting is generalized to "Iwasawa theory of p-adic Galois deformations" discussed in Part III of the book, the language of the characteristic ideal and the Iwasawa algebra is more appropriate to discuss the Iwasawa main conjecture. Also, the formulation of the p-adic L-function $L_p(\psi)$ as an element of Iwasawa algebra allows us to specialize at finite characters ϕ. For these reasons, throughout the book, we do not take the formulation of a p-adic L-functions as a p-adic analytic function on a certain p-adic

domain as in Kubota–Leopoldt [**65**], but we only take the formulation of a p-adic L-function as an element of an Iwasawa algebra.

(3) As for $L_p(\psi)$ in Theorem 3.30, we have $(\gamma - \kappa_{\mathrm{cyc}}(\gamma))L_p(\mathbf{1}) \in \Lambda_{\mathrm{cyc}}^{\times}$ when $\psi = \mathbf{1}$. We will verify this by the interpolation property. In fact, since the numerator of $L(\omega^{-1}, 0) = \frac{1}{p}\sum_{a=1}^{p} \omega^{-1}(a)a$ is a p-adic unit, we have

$$|\mathbf{1}(L_p(\mathbf{1}))|_p = |L(\omega^{-1}, 0)|_p = \left|\frac{1}{p}\right|_p.$$

On the other hand, we have

$$|\mathbf{1}(\gamma - \kappa_{\mathrm{cyc}}(\gamma))|_p = |1 - \kappa_{\mathrm{cyc}}(\gamma)|_p = |p|_p.$$

Hence, we have $|\mathbf{1}(\gamma - \kappa_{\mathrm{cyc}}(\gamma)L_p(\mathbf{1}))|_p = 1$, which implies $(\gamma - \kappa_{\mathrm{cyc}}(\gamma))L_p(\mathbf{1}) \in \Lambda_{\mathrm{cyc}}^{\times}$.

3.2.2. Bernoulli number and special values of Dirichlet L-function. Let us recall some basics on **Bernoulli numbers** and on special values of Dirichlet L-functions.

DEFINITION 3.34. For each nonnegative integer r, we define the r-th Bernoulli number B_r to be the constant term of r-th derivative of $\frac{z}{e^z-1}$. That is, they are rational numbers satisfying

$$\frac{z}{e^z - 1} = \sum_{r=0}^{\infty} B_r \frac{z^r}{r!}.$$

Similarly, for a Dirichlet character η of conductor N and for each nonnegative integer r, the r-th generalized Bernoulli number $B_{r,\eta}$ is defined to be an algebraic number satisfying

$$f_\eta(z) := \sum_{i=1}^{N} \frac{\eta(i) z e^{iz}}{e^{Nz} - 1} = \sum_{r=0}^{\infty} B_{r,\eta} \frac{z^r}{r!}. \tag{3.10}$$

REMARK 3.35. (1) When η is equal to the trivial Dirichlet character $\mathbf{1}$, we have

$$\sum_{i=1}^{N} \frac{\eta(i) z e^{iz}}{e^{Nz} - 1} = \frac{z e^z}{e^z - 1} = \frac{z}{e^z - 1} + z.$$

This equality implies that we have $B_{r,\mathbf{1}} = B_r$ for $r \geq 2$. Hence, we can say that generalized Bernoulli numbers $B_{r,\eta}$ are natural generalizations of usual Bernoulli numbers B_r.

(2) We see that $f_\eta(-z) = \eta(-1) f_\eta(z)$ from the definition of $f_\eta(z)$. By observing the right-and side of (3.10), we have $B_{r,\eta} = 0$ if $(-1)^r \neq \eta(-1)$.

Let us recall the following expression of the special values of Dirichlet L-functions, which is used essentially in the construction of the p-adic L-function $L_p(\psi)$.

THEOREM 3.36. *For any primitive Dirichlet character η, $L(\eta, s)$ is meromorphically continued to the whole $\mathbb{C}$-plane. Moreover, $L(\eta, s)$ is is holomorpihc at the whole $\mathbb{C}$-plane if $\eta \neq \mathbf{1}$. For any integer $r \geq 1$ and for any primitive Dirichlet character η satisfying $\eta(-1) = (-1)^r$, we have $L(\eta, 1 - r) = -\frac{B_{r,\eta}}{r}$.*

PROOF. Let $\log z$ be a branch of the logaritmic function on the complex plane $\mathbb{C}$ with the branch cut on $\mathbb{R}_{>0}$. We consider a path S_ϵ in $\mathbb{C}$ on which $z \in \mathbb{C}$ moves as indicated in the image below, along P_ϵ^+, C_ϵ, and P_ϵ^-

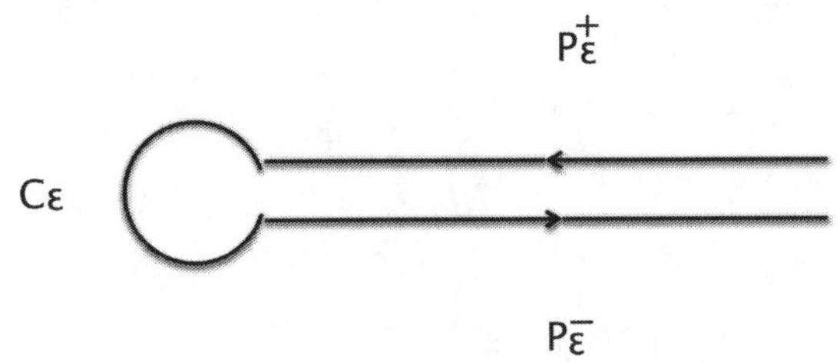

where

P_ϵ^+: the path $(+\infty, \epsilon)$ on the upper branch,

$\longrightarrow C_\epsilon$: the anticlockwise path along the circle of radius ϵ centered at $z = 0$,

$\longrightarrow P_\epsilon^-$: the path $(\epsilon, +\infty)$ on the lower branch.

Let $s \in \mathbb{C}$ and let us assume that $\mathrm{Re}(s) > 1$ for the moment. Then, we have $\lim_{\epsilon \to 0} \int_{C_\epsilon} f_\eta(z) z^{s-2} dz = 0$ since $f_\eta(z) z^{s-2}$ as no pole at $z = 0$ as a function on z. This implies that $\lim_{\epsilon \to 0} \int_{S_\epsilon} f_\eta(z) z^{s-2} dz = \lim_{\epsilon \to 0} \int_{P_\epsilon^+ + P_\epsilon^-} f_\eta(z) z^{s-2} dz$. Note that

$$\begin{aligned}\int_0^\infty z^{s-1} e^{-(Nn-i)z} dz &= \frac{1}{(Nn-i)^s} \int_0^\infty ((Nn-i)z)^{s-1} e^{-(Nn-i)z} d((Nn-i)z) \\ &= \Gamma(s) \frac{1}{(Nn-i)^s},\end{aligned}$$

and recall that we have

$$\sum_{i=1}^{N} \sum_{n=1}^{\infty} \frac{\eta(-i)}{(Nn-i)^s} = L(\eta, s)$$

for $\mathrm{Re}(s) > 1$. Hence, we have

$$\begin{aligned}\lim_{\epsilon \to 0} \int_{S_\epsilon} f_\eta(z) z^{s-2} dz &= (e^{2\pi\sqrt{-1}s} - 1) \int_0^\infty \sum_{i=1}^{N} \frac{\eta(i) z^{s-1} e^{iz}}{e^{Nz} - 1} dz \\ &= (-1)^r (e^{2\pi\sqrt{-1}s} - 1) \sum_{i=1}^{N} \eta(-i) \int_0^\infty \sum_{n=1}^{\infty} z^{s-1} e^{-(Nn-i)z} dz \\ &= (-1)^r (e^{2\pi\sqrt{-1}s} - 1) \Gamma(s) L(\eta, s),\end{aligned}$$

where N is the conductor of η and the infinite sum and the integral in the second line is checked to be exchanged. Since $L(\eta, s)$ is meromorphically continued to the whole $\mathbb{C}$-plane, we have the following equality when s moves in the whole $\mathbb{C}$-plane by the identity theorem:

$$\lim_{\epsilon \to 0} \int_{S_\epsilon} f_\eta(z) z^{s-2} dz = (-1)^r (e^{2\pi\sqrt{-1}s} - 1) \Gamma(s) L(\eta, s). \tag{3.11}$$

Let us evaluate the equality (3.11) at $s = 1 - r$ where r is a positive integer. Since $e^{2\pi\sqrt{-1}s} = 1$ at $s = 1 - r$, we have

$$\lim_{\epsilon \to 0} \int_{S_\epsilon} f_\eta(z) z^{s-2} dz = \lim_{\epsilon \to 0} \int_{C_\epsilon} f_\eta(z) z^{s-2} dz.$$

Applying Cauchy's residue theorem, we have

$$\int_{C_\epsilon} f_\eta(z) z^{1-r-2} dz = \frac{B_{r,\eta}}{r!} \times 2\pi\sqrt{-1}. \tag{3.12}$$

We also recall that $\Gamma(s)$ has a simple pole at $s = 1-r$ with residue $\frac{(-1)^{r-1}}{(r-1)!}$. Hence, we have

$$\lim_{s \to 1-r} (e^{2\pi\sqrt{-1}s} - 1)\Gamma(s) = (-1)^{r-1} \frac{2\pi\sqrt{-1}}{(r-1)!}. \tag{3.13}$$

We complete the proof by putting (3.11), (3.12), (3.13) altogether. □

For the later use, we will formulate a result of Theorem 3.36 in a different way as follows.

COROLLARY 3.37. *For any natural number $r \geq 1$ and for any Dirichlet character η satisfying $\eta(-1) = (-1)^r$, we have*

$$L(\eta, 1 - r) = -\frac{1}{r} \left(\frac{d}{dz} \right)^r f_\eta(z) \bigg|_{z=0},$$

where f_η is as defined in (3.10).

PROOF. We have $\left(\frac{d}{dz}\right)^r f_\eta(z)\Big|_{z=0} = B_{r,\eta}$ by the equation (3.10). This implies Theorem 3.36 immediately and we complete the proof. □

3.2.3. Construction of p-adic L-function à la Iwasawa. Inspired by the work of Kubota–Leopoldt [**65**], Iwasawa [**50**] gave a very clear construction of the p-adic L-function of Theorem 3.30 by using Stickelberger elements (see also a book [**51**] by Iwasawa for this construction).

Let N be a natural number such that $(N, p) = 1$. Throughout Section 3.2.3, we put $\Gamma_n = \Gamma_{\mathrm{cyc}}/(\Gamma_{\mathrm{cyc}})^{p^n}$, $q_n = Np^{n+1}$ and $c = 1 + Np$. By ψ, we will denote a primitive Dirichlet character of conductor Np^e ($e = 0$ or $e = 1$). We denote by $\kappa_{\mathrm{cyc},n} : \Gamma_n \xrightarrow{\sim} (1 + p\mathbb{Z}_p)/(1 + p^n\mathbb{Z}_p)$ the restriction of the isomorphism $\chi_{\mathrm{cyc},n}$: $\mathrm{Gal}(\mathbb{Q}(\mu_{p^n})/\mathbb{Q}) \xrightarrow{\sim} (\mathbb{Z}/p^n\mathbb{Z})^\times$ to the p-Sylow subgroup.

DEFINITION 3.38. We define an element $\Xi_n(\psi) \in \mathbb{Q}[\psi\omega^{-1}][\Gamma_n]$ to be

$$\begin{aligned}\Xi_n(\psi) &= -\frac{1}{q_n} \sum_{\substack{0<i<q_n \\ (i,Np)=1}} \langle i \rangle \psi(i) \gamma_n(i)^{-1} \\ &= -\frac{1}{q_n} \sum_{\substack{0<i<q_n \\ (i,Np)=1}} i(\psi\omega^{-1})(i) \gamma_n(i)^{-1},\end{aligned}$$

where $\langle\ \rangle$ denotes the projection map $\langle\ \rangle : \mathbb{Z}_p^\times \twoheadrightarrow 1+p\mathbb{Z}_p$ to the pro-p part and $\gamma_n(i) \in \Gamma_n$ denotes the inverse image of $\langle i\rangle \bmod p^n \in (1+p\mathbb{Z}_p)/(1+p^n\mathbb{Z}_p)$ via the isomorphism $\kappa_{\mathrm{cyc},n} : \Gamma_n \xrightarrow{\sim} (1+p\mathbb{Z}_p)/(1+p^n\mathbb{Z}_p)$. We call $\Xi_n(\psi)$ a **Stickelberger element**[14].

Note that we have $\mathbb{Q}[\psi\omega^{-1}][\Gamma_n] \subset \mathbb{Q}_p[\psi][\Gamma_n]$ since the values of ω are contained in $\mathbb{Q}_p$. Later we will regard $\Xi_n(\psi)$ as an element of $\mathbb{Q}_p[\psi][\Gamma_n]$.

LEMMA 3.39. *Let ψ, N and e be as we set at the beginning of this subsection.*
(1) *We have $(\gamma_n(c) - c)\,\Xi_n(\psi) \in \mathbb{Z}[\psi\omega^{-1}][\Gamma_n]$.*
(2) *We have $\Xi_n(\psi) \in \mathbb{Z}[\psi\omega^{-1}][\Gamma_n]$ for $\psi \neq \mathbf{1}$.*
(3) *When $\psi(-1) = 1$, $\Xi_{n+1}(\psi)$ is mapped to $\Xi_n(\psi)$ via the natural ring homomorphism $\mathbb{Q}[\psi\omega^{-1}][\Gamma_{n+1}] \longrightarrow \mathbb{Q}[\psi\omega^{-1}][\Gamma_n]$ induced by the quotient map $\Gamma_{n+1} \twoheadrightarrow \Gamma_n$.*

PROOF. (1) Since $\gamma_n(c) \in \Gamma_n \subset \mathbb{Z}[\psi\omega^{-1}][\Gamma_n]^\times$, it suffices to show that

$$(1 - c\gamma_n(c)^{-1}) \cdot \Xi_n(\psi) \in \mathbb{Z}[\psi\omega^{-1}][\Gamma_n]. \tag{3.14}$$

Since $(\psi\omega^{-1})(c) = 1$ by the choice of c, we have

$$\begin{aligned}(1 - c\gamma_n(c)^{-1}) &\cdot \Xi_n(\psi)\\ &= \frac{1}{q_n}\sum_{\substack{0<i<q_n\\(i,Np)=1}} ic(\psi\omega^{-1})(ic)\gamma_n(ic)^{-1} - \frac{1}{q_n}\sum_{\substack{0<i<q_n\\(i,Np)=1}} i\psi\omega^{-1}(i)\gamma_n(i)^{-1}.\end{aligned}$$

For each integer i satisfying $0 < i < q_n$ and $(i, Np) = 1$, we denote by $s_i^{(n)}$ and $t_i^{(n)}$ unique integers satisfying $ic = s_i^{(n)} + q_n t_i^{(n)}$, $0 < s_i^{(n)} < q_n$ and $(s_i^{(n)}, Np) = 1$. This correspondence gives us a bijection:

$$\{0 < i < q_n \mid (i, Np) = 1\} \longrightarrow \left\{0 < s_i^{(n)} < q_n \,\middle|\, (s_i^{(n)}, Np) = 1\right\}. \tag{3.15}$$

This fact combined with the above calculation implies

$$(1 - c\gamma_n(c)^{-1}) \cdot \Xi_n(\psi) = \sum_{\substack{0<i<q_n\\(i,Np)=1}} t_i^{(n)}\psi\omega^{-1}(ic)\gamma_n(ic)^{-1}. \tag{3.16}$$

Since the right-hand side of (3.16) is an element of $\mathbb{Z}[\psi\omega^{-1}][\Gamma_n]$, we complete the proof of the assertion (1).

(2) By putting $\Delta_{Np} = (\mathbb{Z}/Np\mathbb{Z})^\times$, we have $(\mathbb{Z}/q_n\mathbb{Z})^\times \simeq \Delta_{Np} \times \Gamma_n$. For each integer i satisfying $(i, Np) = 1$, we denote by $\delta(i) \in \Delta_{Np}$ the image via $\mathbb{Z} \longrightarrow \mathbb{Z}/Np\mathbb{Z}$. We remark that, for integers i, i' satisfying $0 < i, i' < q_n$, we have $\gamma_n(i) = \gamma_n(i')$ in Γ_n if and only if we have $\langle i\rangle \equiv \langle i'\rangle \bmod q_n$ in $\mathbb{Z}_p$. We can also check that we have $\langle i\rangle \equiv \langle i'\rangle \bmod q_n$ in $\mathbb{Z}_p$ in and only if we have $i\omega^{-1}(i) \equiv i'\omega^{-1}(i') \bmod q_n$

[14] Recall that we are assuming that the fixed prime number p is odd for simplicity throughout the book. The construction in the case $p = 2$ works in the same way except that we need some minor modifications. See [**51**], [**117**, §7.2] for the precise statement for $p = 2$.

in $\mathbb{Z}[\psi][\omega]$. Thus we have

$$\begin{aligned} q_n\Xi_n(\psi) &= -\sum_{\substack{0<i<q_n \\ (i,Np)=1}} \langle i\rangle\psi(i)\gamma_n(i)^{-1} \\ &\equiv -\left(\sum_{\gamma\in\Gamma_n}\sum_{\substack{0<i<q_n \\ \gamma_n(i)=\gamma}}\psi(i)\right)\left(\sum_{\delta\in\Delta_{Np}}\sum_{\substack{0<i<q_n \\ \delta(i)=\delta}} i\omega^{-1}(i)\gamma_n(i)^{-1}\right) \mod q_n. \end{aligned}$$

By the assumption $\psi \neq \mathbf{1}$, we have $\sum_{0<i<q_n,\gamma_n(i)=\gamma}\psi(i) = 0$, which implies $\Xi_n(\psi) \in \mathbb{Z}_p[\psi\omega^{-1}][\Gamma_n]$ and complete the proof of the assertion (2).

(3) Note that the inverse image of the element $\gamma_n(i) \in \Gamma_n$ via the surjection $\Gamma_{n+1} \twoheadrightarrow \Gamma_n$ is equal to the set $\{\gamma_{n+1}(i + jq_n)\}_{0\leq j\leq p-1}$. Hence, the image of $\Xi_{n+1}(\psi) \in \mathbb{Q}[\psi\omega^{-1}][\Gamma_{n+1}]$ via the surjection $\mathbb{Q}[\psi\omega^{-1}][\Gamma_{n+1}] \twoheadrightarrow \mathbb{Q}[\psi\omega^{-1}][\Gamma_n]$ is equal to

$$\begin{aligned} &-\frac{1}{q_{n+1}}\sum_{j=0}^{p-1}\sum_{\substack{0<i<q_n \\ (i,Np)=1}} (i + jq_n)\psi\omega^{-1}(i)\gamma_n(i)^{-1} \\ &= \Xi_n(\psi) - \frac{1}{q_{n+1}}\sum_{j=0}^{p-1}\sum_{\substack{0<i<q_n \\ (i,Np)=1}} jq_n\psi\omega^{-1}(i)\gamma_n(i)^{-1} \\ &= \Xi_n(\psi) - \frac{1}{p}\frac{p(p-1)}{2}\sum_{\substack{0<i<q_n \\ (i,Np)=1}} \psi\omega^{-1}(i)\gamma_n(i)^{-1}. \end{aligned}$$

By the assumption that ψ is a even character, $\psi\omega^{-1}$ is an odd character. Hence, we have $(\psi\omega^{-1})(i) = -(\psi\omega^{-1})(q_n - i)$. On the other hand, we have $\gamma_n(i) = \gamma_n(q_n - i)$. This implies that the second term of the last line of the above calculation is equal to 0. We thus complete the proof of the assertion (3). □

LEMMA 3.40. *Let r be a positive integer, η a primitive Dirichlet character such that $\eta(-1) = (-1)^r$. We assume that the conductor of η is of the form Np^s with s a nonnegative integer where N is a positive integer prime to p and s is a nonnegative integer. Then we have the following equality in $\mathbb{Q}_p[\eta]$:*

$$\lim_{n\to\infty}\frac{1}{q_n}\sum_{\substack{0<i<q_n \\ (i,Np)=1}} \eta(i)i^r = (1 - \eta(p)p^{r-1})B_{r,\eta}.$$

PROOF. For each nonnegative integer r, we define the Bernoulli polynomial $B_r(X) \in \mathbb{Q}[X]$ as follows:

$$\frac{ze^{Xz}}{e^z - 1} = \sum_{r=0}^{\infty} B_r(X)\frac{z^r}{r!}. \tag{3.17}$$

We have $B_r(1) = B_r$ for every $r \geq 2$. If the conductor N of a Dirichlet character η divides a natural number M, we claim the equality

$$B_{r,\eta} = M^{r-1}\sum_{i=1}^{M}\eta(i)B_r\left(\frac{i}{M}\right) \tag{3.18}$$

in $\mathbb{Q}[\eta]$ for every nonnegative integer r. In fact, if we substitute $X = \frac{i}{M}$, $z = Mz$ in (3.17) and we multiply it by $\eta(i)$ on the left-hand side and the right-hand side, we obtain

$$\frac{\eta(i)ze^{iz}}{e^{Mz}-1} = \eta(i)\sum_{r=0}^{\infty} B_r\left(\frac{i}{M}\right)\frac{M^{r-1}z^r}{r!}$$

for each i. If we add them for $i = 1, \ldots, M$, we obtain

$$\sum_{i=1}^{M}\frac{\eta(i)ze^{iz}}{e^{Mz}-1} = \sum_{i=1}^{M}\eta(i)\sum_{r=0}^{\infty} B_r\left(\frac{i}{M}\right)\frac{M^{r-1}z^r}{r!}$$

in $\mathbb{Q}[\eta][[z]]$. Since $B_{r,\eta}$ is the coefficient of $\frac{z^r}{r!}$ in the left-hand side of the above equality, (3.18) follows by comparing the coefficients of $\frac{z^r}{r!}$ in the both sides. On the other hand, note that we have $\frac{z}{e^z-1} = \sum_{r=0}^{\infty} B_r\frac{z^r}{r!}$ and $e^{Xz} = \sum_{r=0}^{\infty} X^r\frac{z^r}{r!}$ in the left-hand side of (3.17). By comparing the coefficient of $\frac{z^r}{r!}$ in both sides of (3.17), we have

$$B_r(X) = \sum_{j=0}^{r}\binom{r}{j}B_jX^{r-j}. \tag{3.19}$$

By combining (3.18) and (3.19), we obtain

$$B_{r,\eta} = M^{r-1}\sum_{i=1}^{M}\eta(i)\left(\frac{i^r}{M^r}B_0 + r\frac{i^{r-1}}{M^{r-1}}B_1 + \frac{r(r-1)}{2}\frac{i^{r-2}}{M^{r-2}}B_2 + \cdots\right).$$

Recall that the theorem of von Staudt and Clausen stated in §1.2 says that the prime number p divides the denominator of Bernoulli number B_j at most once for every j. Recall that $\mathrm{Cond}(\eta)$ is of the form Np^s and we have $n \geq s$ and $q_n := M = Np^{n+1}$. Hence, we have the following congruence

$$B_{r,\eta} \equiv \sum_{i=1}^{q_n}\eta(i)\left(\frac{i^r}{q_n} - ri^{r-1}\right) \mod \frac{q_n}{p} \tag{3.20}$$

in $\mathbb{Z}[\eta]$. Since $\eta(-1) = (-1)^r$, we have

$$\eta(i)i^{r-1} \equiv -\eta(q_n - i)(q_n - i)^{r-1} \mod q_n$$

and hence,

$$\sum_{i=1}^{q_n}\eta(i)i^{r-1} \equiv 0 \mod q_n.$$

This implies that the contribution of the second term of (3.20) is zero and we have deduced

$$B_{r,\eta} \equiv \frac{1}{q_n}\sum_{i=1}^{q_n}\eta(i)i^r \mod \frac{q_n}{p}.$$

By multiplying $1 - \eta(p)p^{r-1}$ on both sides, we obtain

$$(1-\eta(p)p^{r-1})B_{r,\eta} \equiv \frac{1}{q_n}\left(\sum_{i=1}^{q_n}\eta(i)i^r - \sum_{i=1}^{q_{n-1}}\eta(pi)(pi)^r\right) \mod \frac{q_n}{p}. \tag{3.21}$$

Taking the projective limit with respect to n of this equality, we complete the desired equation. in $\mathbb{Q}_p[\eta]$. □

Let us return to the proof of Theorem 3.30.

PROOF OF THEOREM 3.30. Let us define $L_p(\psi)$ to be

$$L_p(\psi) = \varprojlim_n \Xi_n(\psi). \tag{3.22}$$

By Lemma 3.39, we have

$$L_p(\psi) \in \begin{cases} \Lambda_{\mathrm{cyc},\psi} & \text{when } \psi \neq \mathbf{1}, \\ \dfrac{1}{\gamma - \kappa_{\mathrm{cyc}}(\gamma)}\Lambda_{\mathrm{cyc}} & \text{when } \psi = \mathbf{1}. \end{cases}$$

We will show that $L_p(\psi)$ satisfies the desired interpolation property (3.7). Let ϕ be a finite order character of Γ_{cyc}. Further, when we have $r = 1$ and $\psi = \mathbf{1}$, we assume that $\phi \neq \mathbf{1}$. Let us calculate the value $(\kappa_{\mathrm{cyc}}^{1-r}\phi)(L_p(\psi))$. Let $s_i^{(n)}$, $t_i^{(n)}$ be the symbols which appeared in the proof of Lemma 3.39 and let us define $\gamma(c) \in \Gamma_{\mathrm{cyc}}$ to be $\gamma(c) = \varprojlim_n \gamma_n(c)$ for $c = 1 + Np$. By (3.16) and (3.22), we have

$$\begin{aligned} &\kappa_{\mathrm{cyc}}^{1-r}\phi\left((1 - c\gamma(c)^{-1})L_p(\psi)\right) \\ &\qquad \equiv \sum_{\substack{0<i<q_n \\ (i,Np)=1}} t_i^{(n)}\psi\omega^{-r}\phi^{-1}(ic)(ic)^{r-1} \mod q_n. \end{aligned} \tag{3.23}$$

On the other hand, for any integer satisfying $0 < i < q_n$ and $(i, Np) = 1$, we have

$$(ic)^r = (s_i^{(n)} + t_i^{(n)}q_n)^r \equiv (s_i^{(n)})^r + r(ic)^{r-1}t_i^{(n)}q_n \mod q_n^2. \tag{3.24}$$

Hence, we have

$$\begin{aligned} &rq_n\kappa_{\mathrm{cyc}}^{1-r}\phi\left((1 - c\gamma(c)^{-1})L_p(\psi)\right) \\ &\equiv \sum_{\substack{0<i<q_n \\ (i,Np)=1}} r(ic)^{r-1}t_i^{(n)}q_n\psi\omega^{-r}\phi^{-1}(ic) \mod q_n^2 \\ &\equiv \sum_{\substack{0<i<q_n \\ (i,Np)=1}} \left((ic)^r - (s_i^{(n)})^r\right)\psi\omega^{-r}\phi^{-1}(ic) \mod q_n^2 \\ &\equiv \psi\omega^{-r}\phi^{-1}(c)c^r \sum_{\substack{0<i<q_n \\ (i,Np)=1}} \psi\omega^{-r}\phi^{-1}(i)i^r \\ &\qquad\qquad - \sum_{\substack{0<i<q_n \\ (i,Np)=1}} \psi\omega^{-r}\phi^{-1}(s_i^{(n)})(s_i^{(n)})^r \mod q_n^2 \\ &\equiv -(1 - \psi\omega^{-r}\phi^{-1}(c)c^r) \sum_{\substack{0<i<q_n \\ (i,Np)=1}} \psi\omega^{-r}\phi^{-1}(i)i^r \mod q_n^2. \end{aligned}$$

Here, the first congruence is nothing but (3.23) multiplied by rq_n on both sides. The second congruence follows from (3.24). For the third congruence, we note that we have $\psi\omega^{-r}\phi^{-1}(ic) = \psi\omega^{-r}\phi^{-1}(s_i^{(n)})$ since $ic \equiv s_i^{(n)} \mod q_n$. The last congruence follows from the bijectivity of (3.15).

By dividing both sides of the above congruence by $rq_n\kappa_{\mathrm{cyc}}^{1-r}\phi(1-c\gamma(c)^{-1}) = rq_n(1-\psi\omega^{-r}\phi^{-1}(c)c^r)$ [15], we obtain the following congruence

$$(\kappa_{\mathrm{cyc}}^{1-r}\phi)(L_p(\psi)) \equiv -\frac{1}{r}\frac{1}{q_n}\sum_{\substack{0<i<q_n\\(i,Np)=1}} \psi\omega^{-r}\phi^{-1}(i)i^r \mod \frac{q_n}{r}.$$

By taking the limit $n \to \infty$ of this congruence, we deduce the following equality in $\overline{\mathbb{Q}}_p$ for any integer $r \geq 1$ and for any finite order character ϕ of Γ_{cyc}:

$$\begin{aligned}(\kappa_{\mathrm{cyc}}^{1-r}\phi)(L_p(\psi)) &= -(1-\psi\omega^{-r}\phi^{-1}(p)p^{r-1})\frac{1}{r}B_{r,\psi\omega^{-r}\phi^{-1}} \\ &= (1-\psi\omega^{-r}\phi^{-1}(p)p^{r-1})L(\psi\omega^{-r}\phi^{-1},1-r).\end{aligned}$$

Here, we note that the first equality is obtained by applying Lemma 3.40 for $\eta = \psi\omega^{-r}\phi^{-1}$ and the second equality is obtained by applying Theorem 3.36 for the same character η. This completes the proof of Theorem 3.30. □

3.2.4. Construction of p-adic L-function à la Coleman. In this subsection, we will obtain another proof of Theorem 3.30 and construct p-adic L-functions $L_p(\psi)$ by applying the theory of Coleman map to the system of cyclotomic units. This another construction is important because this method leads to the proof of the Iwasawa main conjecture using the Euler system of cyclotomic units (see §3.3.3). We also remark that the method of proving the Iwasawa main conjecture by the theory of Euler system and the theory of Coleman map is quite valid when we study the Iwasawa main conjecture for more general Galois deformations. We will discuss this in the second volume of the book.

Let N be a natural number such that $(N,p) = 1$ and let us put $\Delta_N = \mathrm{Gal}(\mathbb{Q}(\zeta_N)/\mathbb{Q})$, $\Delta_p = \mathrm{Gal}(\mathbb{Q}(\zeta_p)/\mathbb{Q})$. Note that we have $\mathrm{Gal}(\mathbb{Q}(\zeta_{Np^{n+1}})/\mathbb{Q}) \cong \Delta_N \times \Delta_p \times \Gamma_{\mathrm{cyc}}/(\Gamma_{\mathrm{cyc}})^{p^n}$ as well as the following ring isomorphism:

$$\mathbb{Z}[\zeta_N]\otimes_{\mathbb{Z}}\mathbb{Z}_p \cong \prod_{\mathfrak{p}\mid(p)}\widehat{\mathbb{Z}[\zeta_N]}_{\mathfrak{p}}, \tag{3.25}$$

where $\mathfrak{p}$ runs through the primes of $\mathbb{Z}[\zeta_N]$ over p and $\widehat{\mathbb{Z}[\zeta_N]}_{\mathfrak{p}}$ denotes the completion of $\mathbb{Z}[\zeta_N]$ at $\mathfrak{p}$. The group Δ_N acts on (3.25) and both sides are free $\mathbb{Z}_p[\Delta_N]$-modules of rank one. On the other hand, we have the following ring isomorphism:

$$\mathbb{Z}[\zeta_{p^{n+1}}]\otimes_{\mathbb{Z}}\mathbb{Z}_p \cong \mathbb{Z}_p[\zeta_{p^{n+1}}]. \tag{3.26}$$

The group $\Delta_p\times\Gamma_{\mathrm{cyc}}/(\Gamma_{\mathrm{cyc}})^{p^n}$ acts on (3.26), but these modules are not free $\mathbb{Z}_p[\Delta_p\times\Gamma_{\mathrm{cyc}}/(\Gamma_{\mathrm{cyc}})^{p^n}]$-modules in general. In fact, they are free $\mathbb{Z}_p[\Delta_p\times\Gamma_{\mathrm{cyc}}/(\Gamma_{\mathrm{cyc}})^{p^n}]$-modules only when $n=0$ (tamely ramified case).

Recall that $\omega_n(Z) = (1+Z)^{p^n}-1$ and we have an isomorphism

$$\prod_{\mathfrak{p}\mid(p)}\widehat{\mathbb{Z}[\zeta_N]}_{\mathfrak{p}}[[Z]]/(\omega_n(Z)) = \prod_{\mathfrak{p}\mid(p)}\widehat{\mathbb{Z}[\zeta_N]}_{\mathfrak{p}}[C_{p^n}], \tag{3.27}$$

where C_{p^n} is a cyclic group of order p^n generated by $1+\overline{Z}$ with $\overline{Z} := Z \bmod \omega_n(Z)$. Let $[a]_n : \prod_{\mathfrak{p}\mid(p)}\widehat{\mathbb{Z}[\zeta_N]}_{\mathfrak{p}}[C_{p^n}] \longrightarrow \prod_{\mathfrak{p}\mid(p)}\widehat{\mathbb{Z}[\zeta_N]}_{\mathfrak{p}}[C_{p^n}]$ be the $\prod_{\mathfrak{p}\mid(p)}\widehat{\mathbb{Z}[\zeta_N]}_{\mathfrak{p}}$-linear endomorphism of the group ring $\prod_{\mathfrak{p}\mid(p)}\widehat{\mathbb{Z}[\zeta_N]}_{\mathfrak{p}}[C_{p^n}]$ extending the group morphism $C_{p^n} \longrightarrow C_{p^n}$, $g \mapsto g^a$. Let $G_{\mathrm{cyc}} := \Delta_p\times\Gamma_{\mathrm{cyc}}$.

[15] Note that we have $\psi(c)=1$ and $\omega(c)=1$ since $c = 1+Np$.

Below, we define some important operators[16].

DEFINITION 3.41. (1) For any $a \in \mathbb{Z}_p$, we define $[a] : \prod_{\mathfrak{p}|(p)} \widehat{\mathbb{Z}[\zeta_N]_\mathfrak{p}}[[Z]] \longrightarrow \prod_{\mathfrak{p}|(p)} \widehat{\mathbb{Z}[\zeta_N]_\mathfrak{p}}[[Z]]$ to be the endomorphism obtained as the projective limit of $[a]_n$ defined above. Here we recall that we have $\prod_{\mathfrak{p}|(p)} \widehat{\mathbb{Z}[\zeta_N]_\mathfrak{p}}[[Z]] = \varprojlim_n \prod_{\mathfrak{p}|(p)} \widehat{\mathbb{Z}[\zeta_N]_\mathfrak{p}}[[Z]]/(\omega_n(Z))$. The map $[a]$ is an injective $\prod_{\mathfrak{p}|(p)} \widehat{\mathbb{Z}[\zeta_N]_\mathfrak{p}}$-linear map which sends Z to $(1+Z)^a - 1$.

(2) Let σ be the map on the coefficient ring $\prod_{\mathfrak{p}|(p)} \widehat{\mathbb{Z}[\zeta_N]_\mathfrak{p}} \longrightarrow \prod_{\mathfrak{p}|(p)} \widehat{\mathbb{Z}[\zeta_N]_\mathfrak{p}}$ which sends $x \otimes 1 \in \mathbb{Z}[\zeta_N] \otimes_\mathbb{Z} \mathbb{Z}_p \cong \prod_{\mathfrak{p}|(p)} \widehat{\mathbb{Z}[\zeta_N]_\mathfrak{p}}$ to $\mathrm{frob}_p(x) \otimes 1$ where frob_p is the p-th power frobenius element in Δ_N. We define $\varphi : \prod_{\mathfrak{p}|(p)} \widehat{\mathbb{Z}[\zeta_N]_\mathfrak{p}}[[Z]] \longrightarrow \prod_{\mathfrak{p}|(p)} \widehat{\mathbb{Z}[\zeta_N]_\mathfrak{p}}[[Z]]$ to be $\varphi = \sigma \circ [p] = [p] \circ \sigma$ [17].

(3) We define a $\prod_{\mathfrak{p}|(p)} \widehat{\mathbb{Z}[\zeta_N]_\mathfrak{p}}$-linear action of the group G_{cyc} on $\prod_{\mathfrak{p}|(p)} \widehat{\mathbb{Z}[\zeta_N]_\mathfrak{p}}[[Z]]$ by setting $g \cdot (1+Z) = (1+Z)^{\chi_{\mathrm{cyc}}(g)}$ for $g \in G_{\mathrm{cyc}}$. By the operator σ defined above, this action is extended to a $\prod_{\mathfrak{p}|(p)} \widehat{\mathbb{Z}[\zeta_N]_\mathfrak{p}}$-semilinear action of the group $\Delta_N \times G_{\mathrm{cyc}} = \Delta_{Np} \times \Gamma_{\mathrm{cyc}}$ on $\prod_{\mathfrak{p}|(p)} \widehat{\mathbb{Z}[\zeta_N]_\mathfrak{p}}[[Z]]$.

PROPOSITION 3.42. *Let $G(Z) \in \prod_{\mathfrak{p}|(p)} \widehat{\mathbb{Z}[\zeta_N]_\mathfrak{p}}[[Z]]$. Then there exists an element $H(Z) \in \prod_{\mathfrak{p}|(p)} \widehat{\mathbb{Z}[\zeta_N]_\mathfrak{p}}[[Z]]$ such that $G(Z) = [p](H(Z))$ if and only if we have $G(Z) = G(\zeta(1+Z)-1)$ in $\prod_{\mathfrak{p}|(p)} \widehat{\mathbb{Z}[\zeta_N]_\mathfrak{p}}[\zeta_p][[Z]]$ for any $\zeta \in \mu_p$.*

PROOF. The necessity is clear. We will prove the sufficiency. Recall that the map $[p]$ is the projective limit of

$$[p]_n : \prod_{\mathfrak{p}|(p)} \widehat{\mathbb{Z}[\zeta_N]_\mathfrak{p}}[C_{p^n}] \longrightarrow \prod_{\mathfrak{p}|(p)} \widehat{\mathbb{Z}[\zeta_N]_\mathfrak{p}}[C_{p^n}] \tag{3.28}$$

induced by $C_{p^n} \longrightarrow C_{p^n}$, $x \mapsto x^p$. For a given $G(Z) \in \prod_{\mathfrak{p}|(p)} \widehat{\mathbb{Z}[\zeta_N]_\mathfrak{p}}[[Z]]$, we denote $G(Z) \bmod \omega_n(Z)$ by $\overline{G}_n(\overline{Z})$ for any natural number n. Since $1+\overline{Z}$ is a generator of C_{p^n} through the isomorphism (3.27), we have the following expansion

$$\overline{G}_n(\overline{Z}) = \sum_{i=0}^{p^n-1} a_i (1+\overline{Z})^i \in \prod_{\mathfrak{p}|(p)} \widehat{\mathbb{Z}[\zeta_N]_\mathfrak{p}}[C_{p^n}]. \tag{3.29}$$

It is easy to see that there exists an element $H_n(Z) \in \prod_{\mathfrak{p}|(p)} \widehat{\mathbb{Z}[\zeta_N]_\mathfrak{p}}[[Z]]$ such that $G_n(Z) = [p]_n(H_n(Z))$ if and only if we have $a_i = 0$ for any i with $p \nmid i$ in (3.29). We can also check that the last condition holds if and only if $\overline{G}_n(\overline{Z}) = \overline{G}_n(\zeta(1+\overline{Z})-1)$ holds for any $\zeta \in \mu_p$. Suppose that we have $G(Z) = G(\zeta(1+Z)-1)$ in $\prod_{\mathfrak{p}|(p)} \widehat{\mathbb{Z}[\zeta_N]_\mathfrak{p}}[\zeta_p][[Z]]$ for any $\zeta \in \mu_p$. Then, for each natural number n, there exists a unique element $\overline{H}_n(\overline{Z}) \in \prod_{\mathfrak{p}|(p)} \widehat{\mathbb{Z}[\zeta_N]_\mathfrak{p}}[[Z]]/(\omega_n(Z))$ such

[16] See the paper [**16**] by Coleman for the definitions of these operators in a more general setting.

[17] It is clear by definition that $[p]$ commutes with σ.

that $\overline{G}_n(\overline{Z}) = [p]_n(\overline{H}_n(\overline{Z}))$. Since $\{\overline{H}_n(\overline{Z})\}_{n\geq 1}$ is a projective system, we put $H(Z) := \varprojlim_n \overline{H}_n(\overline{Z})$. By construction, $H(Z)$ satisfies $G(Z) = [p](H(Z))$. This completes the proof. □

PROPOSITION 3.43. (1) *There exists a unique continuous multiplicative homomorphism* $\mathrm{Nr} : \prod_{\mathfrak{p}|(p)} \widehat{\mathbb{Z}[\zeta_N]_{\mathfrak{p}}}[[Z]] \longrightarrow \prod_{\mathfrak{p}|(p)} \widehat{\mathbb{Z}[\zeta_N]_{\mathfrak{p}}}[[Z]]$ *which is characterized by the following equality:*

$$([p] \circ \mathrm{Nr})(F(Z)) = \prod_{\zeta \in \mu_p} F(\zeta(1+Z)-1). \tag{3.30}$$

(2) *There exists a unique continuous additive homomorphism*

$$\mathrm{Tr} : \prod\nolimits_{\mathfrak{p}|(p)} \widehat{\mathbb{Z}[\zeta_N]_{\mathfrak{p}}}[[Z]] \longrightarrow \prod\nolimits_{\mathfrak{p}|(p)} \widehat{\mathbb{Z}[\zeta_N]_{\mathfrak{p}}}[[Z]]$$

which is characterized by the following equality:

$$([p] \circ \mathrm{Tr})(F(Z)) = \sum_{\zeta \in \mu_p} F(\zeta(1+Z)-1). \tag{3.31}$$

REMARK 3.44. Let us define a submodule $\mathcal{P} \subset \prod_{\mathfrak{p}|(p)} \widehat{\mathbb{Q}(\zeta_N)_{\mathfrak{p}}}[[Z]]$ with

$$\prod\nolimits_{\mathfrak{p}|(p)} \widehat{\mathbb{Z}[\zeta_N]_{\mathfrak{p}}}[[Z]] \subset \mathcal{P}$$

to be

$$\mathcal{P} = \left\{ G(Z) = a_0 + a_1 Z + \ldots \in \prod_{\mathfrak{p}|(p)} \widehat{\mathbb{Q}(\zeta_N)_{\mathfrak{p}}}[[Z]] \;\middle|\; na_n \in \prod_{\mathfrak{p}|(p)} \widehat{\mathbb{Z}[\zeta_N]_{\mathfrak{p}}} \ \forall n \geq 0,\ p|a_0 \right\}.$$

It is known that the operator φ and the operator Tr on $\prod_{\mathfrak{p}|(p)} \widehat{\mathbb{Z}[\zeta_N]_{\mathfrak{p}}}[[Z]]$ obtained in Proposition 3.43 (2) are extended to the operators $\varphi : \mathcal{P} \longrightarrow \mathcal{P}$ and $\mathrm{Tr} : \mathcal{P} \longrightarrow \mathcal{P}$ respectively. This fact is used later in the proof of Theorem 3.57 but we do not prove it here. For the proof of this fact, we refer the reader to [**87**, §1.1][18].

PROOF OF PROPOSITION 3.43. As in the proof of Proposition 3.42, we apply the argument of taking modulo $\omega_n(Z)$. For each n, $\prod_{\mathfrak{p}|(p)} \widehat{\mathbb{Z}[\zeta_N]_{\mathfrak{p}}}[[Z]]/(\omega_n(Z))$ is a free $\mathrm{Im}([p]_n)$-module of rank p generated by $1, 1+\overline{Z}, \ldots, (1+\overline{Z})^{p-1}$. Hence, $\prod_{\mathfrak{p}|(p)} \widehat{\mathbb{Z}[\zeta_N]_{\mathfrak{p}}}[[Z]]$ is a free $\mathrm{Im}([p])$-module of rank p generated by $1, 1+Z, \ldots, (1+Z)^{p-1}$. Thus we can define the norm map and the trace map of $\prod_{\mathfrak{p}|(p)} \widehat{\mathbb{Z}[\zeta_N]_{\mathfrak{p}}}[[Z]]$ regarded as a free $\prod_{\mathfrak{p}|(p)} \widehat{\mathbb{Z}[\zeta_N]_{\mathfrak{p}}}[[Z]]$-module of rank p via the map $[p]$. The equations (3.30) and (3.31) are nothing but the definitions of the norm map and the trace map. Since the map $[p]$ is injective they are also characterizations of the norm map and the trace map respectively. This completes the proof. □

DEFINITION 3.45. We define $\prod_{\mathfrak{p}|(p)} \widehat{\mathbb{Z}[\zeta_N]_{\mathfrak{p}}}$-semilinear endomorphisms $\mathcal{N}$, $\mathcal{T}$ of $\prod_{\mathfrak{p}|(p)} \widehat{\mathbb{Z}[\zeta_N]_{\mathfrak{p}}}[[Z]]$ by $\mathcal{N} = \sigma^{-1} \circ \mathrm{Nr} = \mathrm{Nr} \circ \sigma^{-1}$, $\mathcal{T} = \sigma^{-1} \circ \mathrm{Tr} = \mathrm{Tr} \circ \sigma^{-1}$.

---

[18]Note that the notation used in [**87**] is not exactly the same as the notation of this book. For example, the symbol Tr of this book is equal to $\psi \circ \sigma$ of [**87**] with a operator ψ defined in [**87**].

Similarly as in Proposition 3.43, the endomorphisms $\mathcal{N}$, $\mathcal{T}$ are characterized by the equations:

$$(\varphi \circ \mathcal{N})(F(Z)) = \prod_{\zeta \in \mu_p} F(\zeta(1+Z)-1),$$
$$(\varphi \circ \mathcal{T})(F(Z)) = \sum_{\zeta \in \mu_p} F(\zeta(1+Z)-1).$$

For each natural number n, we denote by $\mathrm{Nr}_{n+1,n}$ the natural norm map $\prod_{\mathfrak{p}|(p)} \widehat{\mathbb{Z}[\zeta_N]_{\mathfrak{p}}}[\zeta_{p^{n+2}}] \longrightarrow \prod_{\mathfrak{p}|(p)} \widehat{\mathbb{Z}[\zeta_N]_{\mathfrak{p}}}[\zeta_{p^{n+1}}]$. Throughout the section, we fix a system of primitive p^{n+1}-th roots of unity $(\zeta_{p^{n+1}})_{n\geq 0}$ such that $\zeta_{p^{k+2}}^p = \zeta_{p^{k+1}}$ for each $k \geq 0$. In general, a system of units $(u_n)_{n\geq 0}$ of $u_n \in \prod_{\mathfrak{p}|(p)} \widehat{\mathbb{Z}[\zeta_N]_{\mathfrak{p}}[\zeta_{p^{n+1}}]}^\times$ such that $\mathrm{Nr}_{k+1,k}(u_{k+1}) = u_k$ for each $k \geq 0$ is called a **norm compatible system**. For example, $(u_n)_{n\geq 0} = (\zeta_{p^{n+1}})_{n\geq 0}$ is a norm compatible system.

The following result due to Coleman plays an important role to construct a measure from a norm compatible system of units.

THEOREM 3.46 (Coleman). *There exists a unique $\Delta_N \times G_{\mathrm{cyc}}$-equivariant group isomorphism:*

$$\mathrm{Col} : \varprojlim_n \prod_{\mathfrak{p}|(p)} \widehat{\mathbb{Z}[\zeta_N]_{\mathfrak{p}}}[\zeta_{p^{n+1}}]^\times \xrightarrow{\sim} \Big(\prod_{\mathfrak{p}|(p)} \widehat{\mathbb{Z}[\zeta_N]_{\mathfrak{p}}}[[Z]]^\times\Big)^{\mathrm{Nr}=\sigma}, \quad \boldsymbol{u} \mapsto F_{\boldsymbol{u}}(Z),$$

which is characterized by the property $F_{\boldsymbol{u}}^{\sigma^{-k}}(\zeta_{p^{k+1}}-1) = u_k$ for every $k \geq 0$ where $\boldsymbol{u} = (u_n)_{n\geq 0} \in \varprojlim_n \prod_{\mathfrak{p}|(p)} \widehat{\mathbb{Z}[\zeta_N]_{\mathfrak{p}}}[\zeta_{p^{n+1}}]^\times$.

PROOF. For each natural number k, we consider the surjective ring homomorphism

$$\prod_{\mathfrak{p}|(p)} \widehat{\mathbb{Z}[\zeta_N]_{\mathfrak{p}}}[[Z]] \twoheadrightarrow \prod_{\mathfrak{p}|(p)} \widehat{\mathbb{Z}[\zeta_N]_{\mathfrak{p}}}[\zeta_{p^{k+1}}], \; F(Z) \mapsto F^{\sigma^{-k}}(\zeta_{p^{k+1}}-1). \tag{3.32}$$

The kernel of this map is a principal ideal generated by $\Psi_{p^{k+1}}(Z) = \frac{\omega_{k+1}(Z)}{\omega_k(Z)}$.

Now, for each natural number k, let us take a polynomial $G_k(Z) \in \widehat{\mathbb{Z}[\zeta_N]_{\mathfrak{p}}}[Z]$ such that $G^{\sigma^{-k}}(\zeta_{p^{k+1}}-1) = u_k$. In fact, for each component $\mathbb{Q}(\zeta_N)_{\mathfrak{p}} = \widehat{\mathbb{Z}[\zeta_N]_{\mathfrak{p}}} \otimes_{\mathbb{Z}} \mathbb{Q}$ of $(\prod_{\mathfrak{p}|(p)} \widehat{\mathbb{Z}[\zeta_N]_{\mathfrak{p}}}) \otimes_{\mathbb{Z}} \mathbb{Q}$, the extension $\mathbb{Q}(\zeta_N)_{\mathfrak{p}}(\zeta_{p^{k+1}})$ is totally ramified over $\mathbb{Q}(\zeta_N)_{\mathfrak{p}}$ and $\zeta_{p^{k+1}}-1$ is a uniformizer of the discrete valuation field $\mathbb{Q}(\zeta_N)_{\mathfrak{p}}(\zeta_{p^{k+1}})$. Hence, the existence of polynomials $G_k(Z)$ is clear. We put $F_k(Z) = \mathcal{N}^k(G_{2k}(Z))$. Then, we have

$$\begin{aligned} F_k^{\sigma^{-k}}(\zeta_{p^{k+1}}-1) &= \mathcal{N}^k(G_{2k}^{\sigma^{-2k}})(((\zeta_{p^{2k+1}})^{p^k}-1) \\ &= \prod_{\zeta \in \mu_{p^k}} G_{2k}^{\sigma^{-2k}}(\zeta\zeta_{p^{2k+1}}-1) \\ &= \mathrm{Nr}_{2k,k}(u_{2k}) \\ &= u_k. \end{aligned}$$

Here, the first equality is nothing but the definition of F and the second equality is the definition of $\mathcal{N}$. For the third equality, we note that the conjugates of $\zeta_{p^{2k+1}}$ under $\mathrm{Gal}(\mathbb{Q}(\zeta_N)_{\mathfrak{p}}(\zeta_{p^{2k+1}})/\mathbb{Q}(\zeta_N)_{\mathfrak{p}}(\zeta_{p^{k+1}}))$ are $\{\zeta\zeta_{p^{2k+1}} \mid \zeta \in \mu_{p^k}\}$.

First, we claim that

$$\frac{\mathcal{N}(G_{2k})}{G_{2k}} \equiv 1 \bmod p. \tag{3.33}$$

To prove this, we note that

$$[p]\left(\frac{\mathcal{N}(G_{2k}(Z))}{G_{2k}(Z)}\right) = \frac{\prod_{\zeta\in\mu_p} G_{2k}^{\sigma^{-1}}(\zeta(1+Z)-1)}{G_{2k}((1+Z)^p-1)} \equiv \frac{(G_{2k}^{\sigma^{-1}}(Z))^p}{G_{2k}(Z^p)} \bmod (\zeta_p - 1).$$

Since we have $(\zeta_p - 1) \cap \prod_{\mathfrak{p}|(p)} \widehat{\mathbb{Z}[\zeta_N]}_{\mathfrak{p}} = (p)\prod_{\mathfrak{p}|(p)} \widehat{\mathbb{Z}[\zeta_N]}_{\mathfrak{p}}$, this congruence implies

$$[p]\left(\frac{\mathcal{N}(G_{2k}(Z))}{G_{2k}(Z)}\right) \equiv \frac{(G_{2k}^{\sigma^{-1}}(Z))^p}{G_{2k}(Z^p)} \equiv 1 \bmod p.$$

Because the endomorphism on $\prod_{\mathfrak{p}|(p)} \widehat{\mathbb{Z}[\zeta_N]}_{\mathfrak{p}}[[Z]]/(p)$ induced by $[p]$ is injective, this proves the congruence (3.33).

Next, we claim that

$$\frac{\mathcal{N}^{2k-n}(G_{2k})}{\mathcal{N}^k(G_{2k})} \equiv 1 \bmod p^{k+1} \tag{3.34}$$

for each nonnegative integer satisfying $n \leq k$. To prove this, we note that we have $\frac{\mathcal{N}^{i+1}(G_{2k})}{\mathcal{N}^i(G_{2k})} \equiv 1 \bmod p$ $(i = 0, \dots, k-n)$ by applying the same argument as the proof of (3.33) replacing G_{2k} by $\mathcal{N}(G_{2k}), \dots, \mathcal{N}^{k-n-1}(G_{2k})$. Taking a product of all $\frac{\mathcal{N}^{i+1}(G_{2k})}{\mathcal{N}^i(G_{2k})}$ from $i = 0$ to $i = k-n-1$, we obtain

$$\frac{\mathcal{N}^{k-n}(G_{2k})}{G_{2k}} \equiv 1 \bmod p. \tag{3.35}$$

Let us take an integer $l \geq 1$ and a power series $H(Z) \in \prod_{\mathfrak{p}|(p)} \widehat{\mathbb{Z}[\zeta_N]}_{\mathfrak{p}}[[Z]]$ such that $H(Z) \equiv 1 \bmod p^l$. Then we show that

$$[p](\mathcal{N}(H(Z))) = \prod_{\zeta\in\mu_p} H^{\sigma^{-1}}(\zeta(1+Z)-1) \equiv 1 \bmod (\zeta_p - 1)p^l$$

by the same argument as above. Since we have $(\zeta_p - 1)p^l \cap \prod_{\mathfrak{p}|(p)} \widehat{\mathbb{Z}[\zeta_N]}_{\mathfrak{p}} = (p^{l+1})\prod_{\mathfrak{p}|(p)} \widehat{\mathbb{Z}[\zeta_N]}_{\mathfrak{p}}$, this implies that

$$[p](\mathcal{N}(H(Z))) \equiv 1 \bmod p^{l+1}.$$

We thus showed that

$$\mathcal{N}(H(Z)) \equiv 1 \bmod p^{l+1} \text{ when } H(Z) \equiv 1 \bmod p^l \tag{3.36}$$

because the endomorphism on $\prod_{\mathfrak{p}|(p)} \widehat{\mathbb{Z}[\zeta_N]}_{\mathfrak{p}}[[Z]]/(p^{l+1})$ induced by $[p]$ is injective. We complete the proof of the congruence (3.34) by applying (3.36) successively to $H(Z) = \frac{\mathcal{N}^{k-n}(G_{2k})}{G_{2k}}, \frac{\mathcal{N}^{k-n+1}(G_{2k})}{\mathcal{N}(G_{2k})}, \dots, \frac{\mathcal{N}^{2k-n-1}(G_{2k})}{\mathcal{N}^{k-1}(G_{2k})}$.

Now, we claim that

$$F_k(Z) - F_{k-1}(Z) \in (p, Z)^k \text{ for each } k. \tag{3.37}$$

To prove this, we divide $F_k(Z) - F_{k-1}(Z)$ by $\Psi_{p^k}(Z)$ as follows

$$F_k(Z) - F_{k-1}(Z) = Q_k(Z)\Psi_{p^k}(Z) + R_k(Z), \tag{3.38}$$

where $R_k(Z)$ is a polynomial whose degree is smaller than the degree of $\Psi_{p^k}(Z)$. For each natural number n such that $n \leq k$, we have

$$u_n = \mathrm{Nr}_{k,n}(u_k) = \mathcal{N}^{2k-n}(G_{2k}^{\sigma^{-2k}})(\zeta_{p^{n+1}} - 1).$$

By using (3.34), we have

$$F_k^{\sigma^{-n}}(\zeta_{p^{n+1}} - 1) \equiv u_n \bmod p^{k+1} \tag{3.39}$$

for each natural number n such that $n \leq k$. By applying (3.39) with $F_k^{\sigma^{-n}}$ and $F_{k-1}^{\sigma^{-n}}$ for $n = k-1$, we have

$$F_k(\zeta_{p^k} - 1) - F_{k-1}(\zeta_{p^k} - 1) \equiv 0 \bmod p^k. \tag{3.40}$$

By (3.38), (3.40) and $\Psi_{p^k}(\zeta_{p^k} - 1) = 0$, we have $R_k(\zeta_{p^k} - 1) \equiv 0 \bmod p^k$. Since the degree of $R_k(Z)$ is smaller than the degree of the minimal polynomial $\Psi_{p^k}(Z)$ of $\zeta_{p^k} - 1$, we deduce $p^k | R_k(Z)$. Next, we divide $Q_k(Z)$ by $\Psi_{p^{k-1}}(Z)$ as follows

$$Q_k(Z) = Q_{k-1}(Z)\Psi_{p^{k-1}}(Z) + R_{k-1}(Z), \tag{3.41}$$

where $R_{k-1}(Z)$ is a polynomial whose degree is smaller than the degree of $\Psi_{p^{k-1}}(Z)$. By applying (3.39) with $F_k^{\sigma^{-n}}$ and $F_{k-1}^{\sigma^{-n}}$ for $n = k-2$, we have

$$F_k(\zeta_{p^{k-1}} - 1) - F_{k-1}(\zeta_{p^{k-1}} - 1) \equiv 0 \bmod p^k. \tag{3.42}$$

By (3.38), (3.42) and by the fact $p^k | R_k(Z)$,

$$Q_k(\zeta_{p^{k-1}} - 1)\Psi_{p^k}(\zeta_{p^{k-1}} - 1) = F_k(\zeta_{p^{k-1}} - 1) - F_{k-1}(\zeta_{p^{k-1}} - 1) - R_k(\zeta_{p^{k-1}} - 1)$$

is divisible by p^k. by calculation, we can check that the p-adic valuation of the value $\Psi_{p^k}(\zeta_{p^{k-1}} - 1)$ is equal to 1. Hence, $Q_k(\zeta_{p^{k-1}} - 1)$ is divisible by p^{k-1}. This implies $p^{k-1} | R_{k-1}(\zeta_{p^{k-1}} - 1)$ since we have $R_{k-1}(\zeta_{p^{k-1}} - 1) = Q(\zeta_{p^{k-1}} - 1)$ by (3.41). Since the degree of $R_{k-1}(Z)$ is smaller than the degree of the minimal polynomial $\Psi_{p^{k-1}}(Z)$ of $\zeta_{p^{k-1}} - 1$, we deduce $p^{k-1} | R_{k-1}(Z)$. Let us observe the following equation obtained by substituting (3.41) in (3.38):

$$F_k(Z) - F_{k-1}(Z) = (Q_{k-1}(Z)\Psi_{p^{k-1}}(Z) + R_{k-1}(Z))\Psi_{p^k}(Z) + R_k(Z). \tag{3.43}$$

As we have seen above, the second term $R_{k-1}(Z)\Psi_{p^k}(Z)$ and the third term $R_k(Z)$ of (3.43) are in $(p, Z)^k$. As for the first term $Q_{k-1}(Z)\Psi_{p^{k-1}}(Z)\Psi_{p^k}(Z) \in (p, Z)^3$ of (3.43), we repeat inductive arguments by considering the division

$$Q_n(Z) = Q_{n-1}(Z)\Psi_{p^{n-1}}(Z) + R_{n-1}(Z). \tag{3.44}$$

We show that $p^{n-1} | R_{n-1}(Z)$ for every n and hence $F_k(Z) - F_{k-1}(Z)$ is expressed as a sum of k elements which belong to $(p, Z)^k$. This completes the proof of (3.37).

Now, $\{F_k(Z)\}_{k \in \mathbb{N}}$ is a Cauchy sequence with respect to the (p, Z)-adic topology. Since $\prod_{\mathfrak{p}|(p)} \widehat{\mathbb{Z}[\zeta_N]_{\mathfrak{p}}}[[Z]]$ is complete with respect to the (p, Z)-adic topology, the limit of $\{F_k(Z)\}_{k \in \mathbb{N}}$ exists. We denote the limit by $F_{\boldsymbol{u}}(Z)$. By construction, we have

$\mathrm{Nr}(F_{\boldsymbol{u}}(Z)) = F_{\boldsymbol{u}}^{\sigma}$ and the interpolation property the property $F_{\boldsymbol{u}}^{\sigma^{-k}}(\zeta_{p^{k+1}}-1) = u_k$ follows from (3.39). We thus constructed the desired map Col. The injectivity of the map Col follows from the interpolation property and identity theorem (Proposition 2.21)

To prove the surjectivity of the map Col, we take an arbitrary element $F(Z) \in (\prod_{\mathfrak{p}|(p)} \widehat{\mathbb{Z}[\zeta_N]_{\mathfrak{p}}}[[Z]]^{\times})^{\mathrm{Nr}=\sigma}$. Let us put $v_k := F^{\sigma^{-k}}(\zeta_{p^{k+1}}-1)$ for every $k \geq 0$. Then, we have

$$N_{k+1,k}(v_k) = (\mathcal{N}(F^{\sigma^{-(k-1)}}))(\zeta_{p^{k+1}}-1) = F^{\sigma^{-(k-1)}}(\zeta_{p^k}-1) = v_{k-1}.$$

This means that $F(Z)$ is the image of the norm compatible sequence $(v_n)_{n\geq 0}$, hence the map Col is surjective. This completes the proof of Theorem 3.46[19]. □

A power series $F_{\boldsymbol{u}}(Z) \in (\prod_{\mathfrak{p}|(p)} \widehat{\mathbb{Z}[\zeta_N]_{\mathfrak{p}}}[[Z]]^{\times})^{\mathrm{Nr}=\sigma}$ in Theorem 3.46 is called a **Coleman power series** associated to a norm compatible system $\boldsymbol{u}$ of units and is useful to give another construction of the p-adic L-function $L_p(\psi)$.

REMARK 3.47. In preceding references including the text [**117**, Thm. 13.38], the proof of Theorem 3.46 is not the same as the argument which we gave in our proof above. In our proof, we showed that the sequence $\{F_k(Z)\}_{k\in\mathbb{N}}$ is a Cauchy sequence by explicit calculation. However, in preceding references, we use the fact that there exists a convergent subsequence in the sequence $\{F_k(Z)\}_{k\in\mathbb{N}}$ thanks to the compactness of the Iwasawa algebra $\prod_{\mathfrak{p}|(p)} \widehat{\mathbb{Z}[\zeta_N]_{\mathfrak{p}}}[[Z]]$. The advantage of our proof is that our proof applies to an extension of Theorem 3.46 to more general p-adic fields. For example, if we replace $\prod_{\mathfrak{p}|(p)} \widehat{\mathbb{Z}[\zeta_N]_{\mathfrak{p}}}$ by $\widehat{\mathbb{Z}}_p^{\mathrm{ur}}$, a theory of Coleman power series over $\widehat{\mathbb{Z}}_p^{\mathrm{ur}}$ similar to Theorem 3.46 holds true. Classical arguments using the compactness does not work in this case since $\widehat{\mathbb{Z}}_p^{\mathrm{ur}}[[Z]]$ is not compact anymore. Our argument above still applies to this case in the same way.

In fact, as we will see later in §6.5.1 of the second volume of this book, the theory of Coleman power series over $\widehat{\mathbb{Z}}_p^{\mathrm{ur}}$ plays an essential role in transforming Euler systems to p-adic L-functions in Iwasawa theory for general motives.

We give examples of norm compatible sequences in $\varprojlim_n \prod_{\mathfrak{p}|(p)} \widehat{\mathbb{Z}[\zeta_N]_{\mathfrak{p}}}[\zeta_{p^{n+1}}]^{\times}$ and associated Coleman power series in the following lemma.

LEMMA 3.48. *Let $N \geq 1$ be a natural number such that $(N,p) = 1$.*

(1) *When $N = 1$, we choose a natural number $a \geq 1$ satisfying $(p,a) = 1$ and $a \not\equiv 1 \bmod p$. Then the following statements hold true:*
   (i) *We have $u_n(a) = \frac{\zeta_{p^{n+1}}^a - 1}{\zeta_{p^{n+1}} - 1} \in \mathbb{Z}_p[\zeta_{p^{n+1}}]^{\times}$.*
   (ii) *The system $\boldsymbol{u}(a) = (u_n(a))_{n\geq 0}$ is a norm compatible sequence in $\varprojlim_n \prod_{\mathfrak{p}|(p)} \mathbb{Z}_p[\zeta_{p^{n+1}}]^{\times}$.*
   (iii) *The Coleman power series $F_{\boldsymbol{u}(a)}(Z)$ associated to $\boldsymbol{u}(a) = (u_n(a))_{n\geq 0}$ is equal to $\frac{(1+Z)^a - 1}{(1+Z) - 1}$.*

[19]The theory of the field of norms due to Fontaine and Wintenberger (see [**29**], [**126**]) allows us reduce some problems for local fields of mixed characteristic $(0,p)$ to problems for local fields of equal characteristic p. There is another proof of Theorem 3.46 by the theory of the field of norms (see [**87**]).

(2) *When $N > 1$, we fix a primitive N-th root of unity $\zeta_N \in \overline{\mathbb{Q}}$ and embed it into $\prod_{\mathfrak{p}|(p)} \widehat{\mathbb{Z}[\zeta_N]}_{\mathfrak{p}}$ diagonally. Then the following statements hold true:*
   (i) *We have $u_{(N),n} = (\zeta_N)^{\sigma^{-n}} \zeta_{p^{n+1}} - 1 \in \prod_{\mathfrak{p}|(p)} \widehat{\mathbb{Z}[\zeta_N]}_{\mathfrak{p}}[\zeta_{p^{n+1}}]^\times$.*
   (ii) *The system $\boldsymbol{u}_{(N)} = (u_{(N),n})_{n \geq 0}$ is a norm compatible sequence in $\varprojlim_n \prod_{\mathfrak{p}|(p)} \widehat{\mathbb{Z}[\zeta_N]}_{\mathfrak{p}}[\zeta_{p^{n+1}}]^\times$.*
   (iii) *The Coleman power series $F_{\boldsymbol{u}_{(N)}}(Z)$ associated to $\boldsymbol{u}_{(N)} = (u_{(N),n})_{n \geq 0}$ is equal to $\zeta_N(1+Z) - 1$.*

PROOF. Let us prove the assertion (i). When $N = 1$, $\zeta_{p^{n+1}} - 1$ and $\zeta_{p^{n+1}}^a - 1$ are both uniformizers of the discrete valuation ring $\mathbb{Z}_p[\zeta_{p^{n+1}}]$. Hence, the ratio $\frac{\zeta_{p^{n+1}}^a - 1}{\zeta_{p^{n+1}} - 1}$ must be a unit of $\mathbb{Z}_p[\zeta_{p^{n+1}}]$. When $N > 1$, $\zeta_N^{p^{-n}}$ is a primitive N-th root of unity for any n since $(N, p) = 1$. For a prime $\mathfrak{p}$ of $\mathbb{Z}[\zeta_N]$ over p, the image of $\zeta_N^{p^{-n}} \zeta_{p^{n+1}} - 1$ modulo the uniformizer $\zeta_{p^{n+1}} - 1$ is a nonzero element of the residue field. Hence, $\zeta_N^{p^{-n}} \zeta_{p^{n+1}} - 1$ is a unit of the discrete valuation ring $\widehat{\mathbb{Z}[\zeta_N]}_{\mathfrak{p}}[\zeta_{p^{n+1}}]$ for each prime $\mathfrak{p}$ of $\mathbb{Z}[\zeta_N]$ over p. This completes the proof of (i) for both (1) and (2).

Let us prove the assertion (ii). When $N = 1$, the conjugate elements of $\zeta_{p^{n+1}}^a$ over $\mathbb{Z}[\zeta_{p^n}]$ are $\{\zeta_{p^{n+1}}^{a+ip^n}\}_{0 \leq i \leq p-1}$. Hence, we have

$$\mathrm{Nr}_{n,n-1}(u_n(a)) = \prod_{i=0}^{p-1} \frac{\zeta_{p^{n+1}}^{a+ip^n} - 1}{\zeta_{p^{n+1}}^{ip^n} - 1} = \frac{\zeta_{p^n}^a - 1}{\zeta_{p^n} - 1} = u_{n-1}(a).$$

When $N > 1$, the conjugate elements of $\zeta_N^{p^{-n}} \zeta_{p^{n+1}}$ over $\prod_{\mathfrak{p}|(p)} \widehat{\mathbb{Z}[\zeta_N]}_{\mathfrak{p}} \mathbb{Z}[\zeta_{p^n}]$ are $\{\zeta_N^{p^{-n}} \zeta_{p^{n+1}}^{1+ip^n}\}_{0 \leq i \leq p-1}$. Hence, we have

$$\mathrm{Nr}_{n,n-1}(u_{(N),n}) = \prod_{i=0}^{p-1} (\zeta_N^{p^{-n}} \zeta_{p^{n+1}}^{1+ip^n} - 1) = \zeta_N^{p^{-(n-1)}} \zeta_{p^n} - 1 = u_{(N),n-1}.$$

This completes the proof of (ii) for both (1) and (2).

For the assertion (iii), it is clear that we have $F_{\boldsymbol{u}(a)}^{\sigma^{-k}}(\zeta_{p^{k+1}} - 1) = u_k(a)$ (resp. $F_{\boldsymbol{u}_{(N)}}^{\sigma^{-k}}(\zeta_{p^{k+1}} - 1) = u_{(N),k}$) for every $k \geq 0$ when $N = 1$ (resp. $N > 1$). This completes the proof. □

LEMMA 3.49. *For any $F(Z) \in (\prod_{\mathfrak{p}|(p)} \widehat{\mathbb{Z}[\zeta_N]}_{\mathfrak{p}}[[Z]]^\times)^{\mathrm{Nr}=\sigma}$, we have*

$$\left(1 - \frac{\varphi}{p}\right) \log(F(Z)) \in \left(\prod_{\mathfrak{p}|(p)} \widehat{\mathbb{Z}[\zeta_N]}_{\mathfrak{p}}[[Z]]\right)^{\mathrm{Tr}=0}.$$

REMARK 3.50. Let $F(Z)$ be an element of $\prod_{\mathfrak{p}|(p)} \widehat{\mathbb{Z}[\zeta_N]}_{\mathfrak{p}}[[Z]]^\times$ and u the constant term of the formal power series $F(Z)$. By the same argument as the proof of Lemma 2.16, we see that $u \in (\prod_{\mathfrak{p}|(p)} \widehat{\mathbb{Z}[\zeta_N]}_{\mathfrak{p}})^\times$. Hence, there exists $G(Z) \in \prod_{\mathfrak{p}|(p)} \widehat{\mathbb{Z}[\zeta_N]}_{\mathfrak{p}}[[Z]]$ such that $F(Z) = u(1 + ZG(Z))$. For each prime $\mathfrak{p}|(p)$ of $\mathbb{Q}(\zeta_N)$, we denote by $\widehat{\mathbb{Q}(\zeta_N)}_{\mathfrak{p}}$ the completion of $\mathbb{Q}(\zeta_N)$ at $\mathfrak{p}$. Note that $\widehat{\mathbb{Q}(\zeta_N)}_{\mathfrak{p}} = \widehat{\mathbb{Z}[\zeta_N]}_{\mathfrak{p}} \otimes_{\mathbb{Z}} \mathbb{Q}$. Then we have

$$\log(F(Z)) = \log_p(u) + \sum_{n=1}^{\infty} \frac{(-1)^{n-1} Z^n G(Z)^n}{n} \tag{3.45}$$

where $\log_p$ is the p-adic logarithm which we will review $\log_p$ in Definition 3.61. This shows that $\log(F(Z))$ is well-defined as an element of $\prod_{\mathfrak{p}|(p)} \widehat{\mathbb{Q}[\zeta_N]}_{\mathfrak{p}}[[Z]]$.

Proof of Lemma 3.49. Since we have $F(Z)^p \equiv \varphi(F(Z)) \bmod p$, there exists $G(Z) \in \prod_{\mathfrak{p}|(p)} \widehat{\mathbb{Z}[\zeta_N]}_{\mathfrak{p}}[[Z]]$ such that $\frac{F(Z)^p}{\varphi(F(Z))} = 1 + pG(Z)$. Hence, we have the following expansion

$$\log\left(\frac{F(Z)^p}{\varphi(F(Z))}\right) = p\sum_{n=1}^{\infty} \frac{(-1)^{n-1} p^{n-1} G(Z)^n}{n}.$$

We can check that $\frac{p^{n-1}}{n} \in \prod_{\mathfrak{p}|(p)} \widehat{\mathbb{Q}[\zeta_N]}_{\mathfrak{p}}$ is contained in $\prod_{\mathfrak{p}|(p)} \widehat{\mathbb{Z}[\zeta_N]}_{\mathfrak{p}}$ for every positive integer n and the sequence $\left\{\frac{p^{n-1}}{n}\right\}_{n\geq 1}$ converges to zero. Hence, we have

$$\left(1 - \frac{\varphi}{p}\right) \log(F(Z)) = \frac{1}{p}\log\left(\frac{F(Z)^p}{\varphi(F(Z))}\right) \in \prod_{\mathfrak{p}|(p)} \widehat{\mathbb{Z}[\zeta_N]}_{\mathfrak{p}}[[Z]].$$

In order to show that $\left(1 - \frac{\varphi}{p}\right)\log(F(Z)) \in \prod_{\mathfrak{p}|(p)} \widehat{\mathbb{Z}[\zeta_N]}_{\mathfrak{p}}[[Z]]^{\mathrm{Tr}=0}$, it suffices to check that

$$\sum_{\zeta\in\mu_p} \log(F(\zeta(1+Z)-1)) = \frac{1}{p}\sum_{\zeta\in\mu_p} \log(\varphi(F)(\zeta(1+Z)-1)) \tag{3.46}$$

by Proposition 3.43. Recall that $\varphi(F)(\zeta(1+Z)-1))$ is equal to $\varphi(F)(Z)$ for any $\zeta \in \mu_p$ by the definition of φ. Hence, the right-hand side of (3.46) is equal to $\log(\varphi(F)(Z))$. On the other hand, we see that this value is equal to the left-hand side of (3.46) as follows

$$\log(\varphi(F)(Z)) = \log(\varphi \circ \mathcal{N}(F)(Z)) = \sum_{\zeta\in\mu_p} \log(F(\zeta(1+Z)-1))$$

where the first equality is due to our assumption $\mathcal{N}(F(Z)) = F(Z)$ and the second equality is due to the characterization of $\mathcal{N}$ which we gave just after Definition 3.45. This completes the proof. □

For a commutative algebra R, we denote by μ_R the group of roots of unity of R. Then we have the following lemma.

Lemma 3.51. *Let* $\boldsymbol{u} = \{u_n\}_{n\geq 0} \in \varprojlim_n \prod_{\mathfrak{p}|(p)} \widehat{\mathbb{Z}[\zeta_N]}_{\mathfrak{p}}[\zeta_{p^{n+1}}]^\times$ *and let* $F(Z) = F_{\boldsymbol{u}}(Z) \in (\prod_{\mathfrak{p}|(p)} \widehat{\mathbb{Z}[\zeta_N]}_{\mathfrak{p}}[[Z]]^\times)^{\mathrm{Nr}=\sigma}$. *Then we have* $\left(1 - \frac{\varphi}{p}\right)\log(F_{\boldsymbol{u}}(Z)) = 0$ *if and only if we have*

$$F_{\boldsymbol{u}}(Z) \in \prod_{\mathfrak{p}|(p)} \mu_{\widehat{\mathbb{Z}[\zeta_N]}_{\mathfrak{p}}} \times \prod_{\mathfrak{p}|(p)} (1+Z)^{\mathbb{Z}_p},$$

or equivalently

$$\boldsymbol{u} \in \prod_{\mathfrak{p}|(p)} \mu_{\widehat{\mathbb{Z}[\zeta_N]}_{\mathfrak{p}}} \times \prod_{\mathfrak{p}|(p)} \mathbb{Z}_p(1) \subset \varprojlim_n \prod_{\mathfrak{p}|(p)} \widehat{\mathbb{Z}[\zeta_N]}_{\mathfrak{p}}[\zeta_{p^{n+1}}]^\times,$$

where $\mathbb{Z}_p(1) = \varprojlim_n \mu_{p^n}$.

PROOF. Let us consider a power series $G(Z) \in \prod_{\mathfrak{p}|(p)} \widehat{\mathbb{Z}[\zeta_N]}_{\mathfrak{p}}[[Z]]$ such that $G(Z)|_{Z=0} \equiv 1 \bmod p$. By Remark 3.50 and by the fact that $\log_p$ in (3.45) is injective on the subgroup of $\prod_{\mathfrak{p}|(p)} \widehat{\mathbb{Z}[\zeta_N]}_{\mathfrak{p}}^\times$ consisting of units which are congruent to 1 modulo p, we observe that $\log(G(Z)) = 0$ if and only if $G(Z) = 1$.

Now, we remark that we have

$$(1 - \tfrac{\varphi}{p})\log(F_{\boldsymbol{u}}(Z)) = \tfrac{1}{p}\log\left(\tfrac{F_{\boldsymbol{u}}(Z)^p}{\varphi(F_{\boldsymbol{u}}(Z))}\right)$$

and $\left.\frac{F_{\boldsymbol{u}}(Z)^p}{\varphi(F_{\boldsymbol{u}}(Z))}\right|_{Z=0} \equiv 1 \bmod p$. By applying the above observation to $G(Z) = \frac{F_{\boldsymbol{u}}(Z)^p}{\varphi(F_{\boldsymbol{u}}(Z))}$, we see that $(1 - \frac{\varphi}{p})\log(F_{\boldsymbol{u}}(Z)) = 0$ if and only if $\frac{F_{\boldsymbol{u}}(Z)^p}{\varphi(F_{\boldsymbol{u}}(Z))} = 1$. Thus, in order to complete the proof of the lemma, it suffices to show that we have $\frac{F_{\boldsymbol{u}}(Z)^p}{\varphi(F_{\boldsymbol{u}}(Z))} = 1$ if and only if $F_{\boldsymbol{u}}(Z) \in \prod_{\mathfrak{p}|(p)} \mu_{\widehat{\mathbb{Z}[\zeta_N]}_{\mathfrak{p}}} \times \prod_{\mathfrak{p}|(p)} (1+Z)^{\mathbb{Z}_p}$.

It is clear that the latter condition implies the former. In the rest of the proof, we show that the former condition implies the latter. Assume that $F_{\boldsymbol{u}}(Z)^p = \varphi(F_{\boldsymbol{u}}(Z))$. By substituting $Z = \zeta_p - 1$ and $Z = 0$ respectively, we have

$$\begin{aligned} (u_0)^p &= F_{\boldsymbol{u}}(\zeta_p - 1))^p = (F_{\boldsymbol{u}}(0))^\sigma, \\ (F_{\boldsymbol{u}}(0))^p &= (F_{\boldsymbol{u}}(0))^\sigma. \end{aligned}$$

The condition $(F_{\boldsymbol{u}}(0))^p = (F_{\boldsymbol{u}}(0))^\sigma$ implies that $F_{\boldsymbol{u}}(0) \in \prod_{\mathfrak{p}|(p)} \mu_{\widehat{\mathbb{Z}[\zeta_N]}_{\mathfrak{p}}}$, hence we have $u_0 \in \prod_{\mathfrak{p}|(p)} \mu_{\widehat{\mathbb{Z}[\zeta_N]}_{\mathfrak{p}}} \times \mu_p$ thanks to the condition $(u_0)^p = (F_{\boldsymbol{u}}(0))^\sigma$. By the interpolation property satisfied by $F_{\boldsymbol{u}}$ given in Theorem 3.46, we have

$$\begin{aligned} (F_{\boldsymbol{u}}^{\sigma^{-(n+1)}}(\zeta_{p^{n+2}} - 1))^p &= u_{n+1}^p, \\ \varphi(F_{\boldsymbol{u}}^{\sigma^{-(n+1)}})(\zeta_{p^{n+2}} - 1) = F_{\boldsymbol{u}}^{\sigma^{-n}}(\zeta_{p^{n+1}} - 1) &= u_n, \end{aligned}$$

for any $n \in \mathbb{Z}_{\geq 0}$. Hence, our assumption implies $u_{n+1}^p = u_n$ for any $n \in \mathbb{Z}_{\geq 0}$. We have $u_n \in \prod_{\mathfrak{p}|(p)} \mu_{\widehat{\mathbb{Z}[\zeta_N]}_{\mathfrak{p}}} \times \mu_{p^{n+1}}$ for every $n \in \mathbb{Z}_{\geq 0}$ by an inductive argument and we conclude $\boldsymbol{u} \in \prod_{\mathfrak{p}|(p)} \mu_{\widehat{\mathbb{Z}[\zeta_N]}_{\mathfrak{p}}} \times \prod_{\mathfrak{p}|(p)} \mathbb{Z}_p(1) \subset \varprojlim_n \prod_{\mathfrak{p}|(p)} \widehat{\mathbb{Z}[\zeta_N]}_{\mathfrak{p}}[\zeta_{p^{n+1}}]^\times$. This completes the proof. □

Let us define a $\prod_{\mathfrak{p}|(p)} \widehat{\mathbb{Z}[\zeta_N]}_{\mathfrak{p}}$-linear endomorphism D on $\prod_{\mathfrak{p}|(p)} \widehat{\mathbb{Z}[\zeta_N]}_{\mathfrak{p}}[[Z]]$ by

$$D : \prod_{\mathfrak{p}|(p)} \widehat{\mathbb{Z}[\zeta_N]}_{\mathfrak{p}}[[Z]] \longrightarrow \prod_{\mathfrak{p}|(p)} \widehat{\mathbb{Z}[\zeta_N]}_{\mathfrak{p}}[[Z]],\ F(Z) \mapsto (1+Z)\frac{d}{dZ}F(Z). \tag{3.47}$$

PROPOSITION 3.52. *We have the following exact sequence*[20]*:*

$$(3.48)\quad 0 \longrightarrow \prod_{\mathfrak{p}|(p)} \mu_{\widehat{\mathbb{Z}[\zeta_N]}_{\mathfrak{p}}} \times \prod_{\mathfrak{p}|(p)} \mathbb{Z}_p(1) \xrightarrow{\alpha} \left(\prod_{\mathfrak{p}|(p)} \widehat{\mathbb{Z}[\zeta_N]}_{\mathfrak{p}}[[Z]]^\times\right)^{\mathrm{Nr}=\sigma}$$
$$\xrightarrow{(1-\frac{\varphi}{p})\circ\log} \left(\prod_{\mathfrak{p}|(p)} \widehat{\mathbb{Z}[\zeta_N]}_{\mathfrak{p}}[[Z]]\right)^{\mathrm{Tr}=0} \xrightarrow{\beta} \prod_{\mathfrak{p}|(p)} \widehat{\mathbb{Z}[\zeta_N]}_{\mathfrak{p}}/(\sigma-1) \longrightarrow 0.$$

Here, the map α is defined by sending

$$\prod_{\mathfrak{p}|(p)} x_{\mathfrak{p}} \times \prod_{\mathfrak{p}|(p)} (\varprojlim_n \zeta_{p^n})^{a_{\mathfrak{p}}} \in \prod_{\mathfrak{p}|(p)} \mu_{\widehat{\mathbb{Z}[\zeta_N]}_{\mathfrak{p}}} \times \prod_{\mathfrak{p}|(p)} \mathbb{Z}_p(1)$$

to

$$\prod_{\mathfrak{p}|(p)} x_{\mathfrak{p}}(1+Z)^{a_{\mathfrak{p}}} \in \left(\prod_{\mathfrak{p}|(p)} \widehat{\mathbb{Z}[\zeta_N]}_{\mathfrak{p}}[[Z]]^\times\right)^{\mathrm{Nr}=\sigma},$$

where $a_{\mathfrak{p}} \in \mathbb{Z}_p$ and the map β is defined by sending

$$F(Z) \in \left(\prod\nolimits_{\mathfrak{p}|(p)} \widehat{\mathbb{Z}[\zeta_N]}_{\mathfrak{p}}[[Z]]\right)^{\mathrm{Tr}=0}$$

to $D(F(Z))|_{Z=0}$ mod $\sigma - 1$.

PROOF. The exactness at the first term (the injectivity of α) is clear from the definition. The exactness at the second term follows from Lemma 3.51. For the exactness at the fourth term (the surjectivity of β) is easy to see because any element a mod $\sigma - 1$ with $a \in \prod_{\mathfrak{p}|(p)} \widehat{\mathbb{Z}[\zeta_N]}_{\mathfrak{p}}$ has a lift $a(1+Z) \in (\prod_{\mathfrak{p}|(p)} \widehat{\mathbb{Z}[\zeta_N]}_{\mathfrak{p}}[[Z]])^{\mathrm{Tr}=0}$.

In the rest of the proof, we will check the exactness at the third term. By definition, we have

$$(3.49)\quad \beta \circ \left((1-\frac{\varphi}{p})\circ \log\right)(F(Z))$$
$$= \left(D(\log(F(Z))) - \frac{1}{p} D(\log(F^\sigma((1+Z)^p-1)))\right)\Big|_{Z=0} \bmod \sigma - 1.$$

Let us recall the rule of the logarithmic derivative $D(\log(G(Z))) = (1+Z)\frac{G'(Z)}{G(Z)}$ for $G(Z) \in \prod_{\mathfrak{p}|(p)} \widehat{\mathbb{Z}[\zeta_N]}_{\mathfrak{p}}[[Z]]^\times$ where $G'(Z) = \frac{d}{dZ}G(Z)$. Let us denote the expansion of $F(Z)$ by $F(Z) = a_0 + a_1 Z + a_2 Z^2 + \cdots$. By applying this rule to $G(Z) = F(Z)$ and $G(Z) = F^\sigma(Z)$, we obtain

$$\left(D(\log(F(Z))) - \frac{1}{p} D(\log(F^\sigma((1+Z)^p-1)))\right)\Big|_{Z=0} = \frac{a_1}{a_0} - \left(\frac{a_1}{a_0}\right)^\sigma.$$

We conclude that $\beta \circ \left((1-\frac{\varphi}{p})\circ\log\right) = 0$ by combining this calculation with (3.49). Thus we have proved $\mathrm{Ker}(\beta) \supset \mathrm{Im}\left[(1-\frac{\varphi}{p})\circ\log\right]$.

[20]Note that the second group and the first factor of the first group are multiplicative groups and the third group and the fourth group are additive groups.

Let us prove $\mathrm{Ker}(\beta) \subset \mathrm{Im}\big[(1-\frac{\varphi}{p})\circ\log\big]$. In order to prove this, we consider the following commutative diagram:

$$\begin{array}{ccc} \left(\prod_{\mathfrak{p}|(p)} \widehat{\mathbb{Z}[\zeta_N]_{\mathfrak{p}}}[[Z]]^\times\right)^{\mathrm{Nr}=\sigma} & \xrightarrow{(1-\frac{\varphi}{p})\circ\log} & \left(\prod_{\mathfrak{p}|(p)} \widehat{\mathbb{Z}[\zeta_N]_{\mathfrak{p}}}[[Z]]\right)^{\mathrm{Tr}=0} \\ {\scriptstyle D\log}\big\downarrow & & \big\downarrow{\scriptstyle D} \\ \left(\prod_{\mathfrak{p}|(p)} \widehat{\mathbb{Z}[\zeta_N]_{\mathfrak{p}}}[[Z]]\right)^{\mathrm{Tr}=\sigma p} & \xrightarrow[1-\varphi]{} & \left(\prod_{\mathfrak{p}|(p)} \widehat{\mathbb{Z}[\zeta_N]_{\mathfrak{p}}}[[Z]]\right)^{\mathrm{Tr}=0} \end{array}.$$

As we will see later in Corollary 3.54, the right vertical map D of the above diagram is an isomorphism and the above diagram induces the following commutative diagram:

$$(3.50)\qquad \begin{array}{ccc} \left(\prod_{\mathfrak{p}|(p)} \widehat{\mathbb{Z}[\zeta_N]_{\mathfrak{p}}}[[Z]]^\times\right)^{\mathrm{Nr}=\sigma} & \xrightarrow{(1-\frac{\varphi}{p})\circ\log} & \mathrm{Ker}(\beta) \\ {\scriptstyle D\log}\big\downarrow & & \big\downarrow{\scriptstyle D} \\ \left(\prod_{\mathfrak{p}|(p)} \widehat{\mathbb{Z}[\zeta_N]_{\mathfrak{p}}}[[Z]]\right)^{\mathrm{Tr}=\sigma p} & \xrightarrow[1-\varphi]{} & \mathrm{Ker}(\beta), \end{array}$$

where the right vertical map D is an isomorphism. We can also check that the left vertical map $D\log$ of the above diagram is surjective[21]. Proving the desired inclusion $\mathrm{Ker}(\beta) \subset \mathrm{Im}\big[(1-\frac{\varphi}{p})\circ\log\big]$ amounts to showing that the upper horizontal map of the diagram (3.50) is surjective. Hence, it suffices to show that the lower horizontal map of the diagram (3.50) is surjective. Let us take any element $H(Z) \in \mathrm{Ker}(\beta)$. We want to find an element $G(Z) \in (\prod_{\mathfrak{p}|(p)}\widehat{\mathbb{Z}[\zeta_N]_{\mathfrak{p}}}[[Z]])^{\mathrm{Tr}=\sigma p}$ such that $(1-\varphi)(G(Z)) = H(Z)$. By the definition of the map β and by the assumption $H(Z) \in \mathrm{Ker}(\beta)$, the constant term b_0 of the expansion $H(Z) = b_0 + b_1 Z + b_2 Z^2 + \dots$ satisfies $b_0 \in (\sigma - 1)\prod_{\mathfrak{p}|(p)}\widehat{\mathbb{Z}[\zeta_N]_{\mathfrak{p}}}$. Since $\varphi(b_0) = \sigma(b_0)$ and the action of σ on $\prod_{\mathfrak{p}|(p)}\widehat{\mathbb{Z}[\zeta_N]_{\mathfrak{p}}}$ is periodic, we have

$$(1+\varphi+\varphi^2+\dots+\varphi^m+\dots)(b_0) = 0.$$

On the other hand, we have $\varphi^m(Z^n) = \big((Z+1)^{p^m}-1\big)^n \to 0$ $(m \to \infty)$ in $\prod_{\mathfrak{p}|(p)}\widehat{\mathbb{Z}[\zeta_N]_{\mathfrak{p}}}[[Z]]$ with respect to (p,Z)-adic topology for every integer $n \geq 1$. Hence,

$$G'(Z) := \left(1+\varphi+\varphi^2+\dots+\varphi^m+\dots\right)(DH(Z))$$

is convergent in $\prod_{\mathfrak{p}|(p)}\widehat{\mathbb{Z}[\zeta_N]_{\mathfrak{p}}}[[Z]]$ and it satisfies $(1-\varphi)(G'(Z)) = DH(Z)$. We can check that $\mathrm{Tr}\circ\varphi = p\sigma$ holds on $\prod_{\mathfrak{p}|(p)}\widehat{\mathbb{Z}[\zeta_N]_{\mathfrak{p}}}[[Z]]$ by calculation. Hence, an element $F(Z) \in \prod_{\mathfrak{p}|(p)}\widehat{\mathbb{Z}[\zeta_N]_{\mathfrak{p}}}[[Z]]$ satisfies $\mathrm{Tr}\left((1-\varphi)(F(Z))\right) = 0$ if and only if we have $F(Z) \in \prod_{\mathfrak{p}|(p)}\widehat{\mathbb{Z}[\zeta_N]_{\mathfrak{p}}}[[Z]]^{\mathrm{Tr}=p\sigma}$. Thus we have $G'(Z) \in \prod_{\mathfrak{p}|(p)}\widehat{\mathbb{Z}[\zeta_N]_{\mathfrak{p}}}[[Z]]^{\mathrm{Tr}=p\sigma}$.

Since the left vertical map $D\log$ of the above diagram is surjective, there exists an element $G(Z) \in (\prod_{\mathfrak{p}|(p)}\widehat{\mathbb{Z}[\zeta_N]_{\mathfrak{p}}}[[Z]]^\times)^{\mathrm{Nr}=\sigma}$ such that $D\log(G(Z)) = G'(Z)$. We

[21] See [**13**, Theorem 2.4.6] for the proof of the fact that $D\log$ is surjective. We note that they denote $D\log$ by Δ in [**13**].

have $(1-\varphi)(D\log(G(Z))) = D(H(Z))$ and the commutativity of the diagram (3.50) implies $D\big((1-\frac{\varphi}{p})\circ\log(G(Z))\big) = D(H(Z))$. Since the map D is an isomorphism, we have $(1-\frac{\varphi}{p})\circ\log(G(Z)) = H(Z)$. This completes the proof of $\mathrm{Ker}(\beta)\subset \mathrm{Im}\big((1-\frac{\varphi}{p})\circ\log\big)$ and we complete the proof of the proposition. $\square$

Recall that we denote by $\overline{Z}\in\prod_{\mathfrak{p}|(p)}\widehat{\mathbb{Z}[\zeta_N]_{\mathfrak{p}}}[[Z]]/(\omega_n(Z))$ the element Z mod $\omega_n(Z)$.

PROPOSITION 3.53. *Let $F(Z)\in\big(\prod_{\mathfrak{p}|(p)}\widehat{\mathbb{Z}[\zeta_N]_{\mathfrak{p}}}[[Z]]\big)$ and let $\overline{F}_n(\overline{Z})$ be its image in $\prod_{\mathfrak{p}|(p)}\widehat{\mathbb{Z}[\zeta_N]_{\mathfrak{p}}}[[Z]]/(\omega_n(Z))$ for each nonnegative integer n. We consider the expansion*

$$\overline{F}_n(\overline{Z}) = \sum_{i=0}^{p^n-1} a_{n,i}(1+\overline{Z})^i \in \prod_{\mathfrak{p}|(p)}\widehat{\mathbb{Z}[\zeta_N]_{\mathfrak{p}}}[[Z]]/(\omega_n(Z))$$

with $a_{n,i}\in\prod_{\mathfrak{p}|(p)}\widehat{\mathbb{Z}[\zeta_N]_{\mathfrak{p}}}$ for $i=0,1,\dots,p^n-1$. Then, we have

$$F(T)\in\Big(\prod_{\mathfrak{p}|(p)}\widehat{\mathbb{Z}[\zeta_N]_{\mathfrak{p}}}[[Z]]\Big)^{\mathrm{Tr}=0}$$

if and only if we have $a_{n,i}=0$ for every nonnegative integer n and for every nonnegative integer i which is divisible by p.

PROOF. The equivalence follows immediately by (3.31) and by the fact that $\sum_{\zeta\in\mu_p}\zeta = 0$. $\square$

COROLLARY 3.54. *If we have $F(Z)\in(\prod_{\mathfrak{p}|(p)}\widehat{\mathbb{Z}[\zeta_N]_{\mathfrak{p}}}[[Z]])^{\mathrm{Tr}=0}$, we have $D(F(Z))\in(\prod_{\mathfrak{p}|(p)}\widehat{\mathbb{Z}[\zeta_N]_{\mathfrak{p}}}[[Z]])^{\mathrm{Tr}=0}$ where D is the operator defined in (3.47). Further, D induces a $\prod_{\mathfrak{p}|(p)}\widehat{\mathbb{Z}[\zeta_N]_{\mathfrak{p}}}$-linear automorphism on $(\prod_{\mathfrak{p}|(p)}\widehat{\mathbb{Z}[\zeta_N]_{\mathfrak{p}}}[[Z]])^{\mathrm{Tr}=0}$.*

PROOF. Let us define $D_n\in\mathrm{End}_{\prod_{\mathfrak{p}|(p)}\widehat{\mathbb{Z}[\zeta_N]_{\mathfrak{p}}}}\big(\prod_{\mathfrak{p}|(p)}\widehat{\mathbb{Z}[\zeta_N]_{\mathfrak{p}}}[[Z]]/(\omega_n(Z))\big)$ by

$$D_n\Big(\sum_{i=0}^{p^n-1} a_{n,i}(1+\overline{Z})^i\Big) = \sum_{i=0}^{p^n-1} i a_{n,i}(1+\overline{Z})^i.$$

Note that $D=\varprojlim_n D_n$ by definition. Let us define a $\prod_{\mathfrak{p}|(p)}\widehat{\mathbb{Z}[\zeta_N]_{\mathfrak{p}}}$-submodule $\big(\prod_{\mathfrak{p}|(p)}\widehat{\mathbb{Z}[\zeta_N]_{\mathfrak{p}}}[[Z]]/(\omega_n(Z))\big)^{\mathrm{Tr}_n=0}$ of $\prod_{\mathfrak{p}|(p)}\widehat{\mathbb{Z}[\zeta_N]_{\mathfrak{p}}}[[Z]]/(\omega_n(Z))$ to be the image of $(\prod_{\mathfrak{p}|(p)}\widehat{\mathbb{Z}[\zeta_N]_{\mathfrak{p}}}[[Z]])^{\mathrm{Tr}=0}$ in $\prod_{\mathfrak{p}|(p)}\widehat{\mathbb{Z}[\zeta_N]_{\mathfrak{p}}}[[Z]]/(\omega_n(Z))$. By the characterization of $\big(\prod_{\mathfrak{p}|(p)}\widehat{\mathbb{Z}[\zeta_N]_{\mathfrak{p}}}[[Z]]/(\omega_n(Z))\big)^{\mathrm{Tr}_n=0}$ given in Proposition 3.53, the $\prod_{\mathfrak{p}|(p)}\widehat{\mathbb{Z}[\zeta_N]_{\mathfrak{p}}}$-submodule $\big(\prod_{\mathfrak{p}|(p)}\widehat{\mathbb{Z}[\zeta_N]_{\mathfrak{p}}}[[Z]]/(\omega_n(Z))\big)^{\mathrm{Tr}_n=0}$ is stable by D_n and D_n is injective. By Proposition 3.53, any element

$$\overline{F}_n(\overline{Z}) = \sum_{i=0}^{p^n-1} a_i(1+\overline{Z})^i \in \Big(\prod_{\mathfrak{p}|(p)}\widehat{\mathbb{Z}[\zeta_N]_{\mathfrak{p}}}[[Z]]/\omega_n(Z)\Big)^{\mathrm{Tr}_n=0}$$

is the image of $\sum_{0<i<p^n,(i,p)=1}\frac{a_i}{i}(1+\overline{Z})^i$ by D_n . Hence, D_n is also surjective. This completes the proof of the corollary since $D=\varprojlim_n D_n$ is bijective on $(\prod_{\mathfrak{p}|(p)}\widehat{\mathbb{Z}[\zeta_N]_{\mathfrak{p}}}[[Z]])^{\mathrm{Tr}=0}$. $\square$

COROLLARY 3.55. *Through the action of G_{cyc} on $(1+Z)$ given in Definition 3.41 (3), we have the following identification of $\mathbb{Z}_p[\Delta_N]$-modules:*

$$(\prod_{\mathfrak{p}\mid(p)} \widehat{\mathbb{Z}[\zeta_N]_{\mathfrak{p}}}[[Z]])^{\mathrm{Tr}=0} = \mathbb{Z}_p[\Delta_N][[G_{\mathrm{cyc}}]] \cdot (1+Z).$$

Especially, $(\prod_{\mathfrak{p}\mid(p)} \widehat{\mathbb{Z}[\zeta_N]_{\mathfrak{p}}}[[Z]])^{\mathrm{Tr}=0}$ is regarded as a free $\mathbb{Z}_p[\Delta_N][[G_{\mathrm{cyc}}]]$-module of rank one.

REMARK 3.56. As discussed in Lemma 2.19, we can identify the ring

$$\prod\nolimits_{\mathfrak{p}\mid(p)} \widehat{\mathbb{Z}[\zeta_N]_{\mathfrak{p}}}[[Z]]$$

with the ring of $\prod_{\mathfrak{p}\mid(p)} \widehat{\mathbb{Z}[\zeta_N]_{\mathfrak{p}}}$-valued measures on the additive group $\mathbb{Z}_p$.

Under this identification,

$$\left(\prod\nolimits_{\mathfrak{p}\mid(p)} \widehat{\mathbb{Z}[\zeta_N]_{\mathfrak{p}}}[[Z]]\right)^{\mathrm{Tr}=0}$$

corresponds to the $\prod_{\mathfrak{p}\mid(p)} \widehat{\mathbb{Z}[\zeta_N]_{\mathfrak{p}}}$-submodule consisting of the extensions by zero of $\prod_{\mathfrak{p}\mid(p)} \widehat{\mathbb{Z}[\zeta_N]_{\mathfrak{p}}}$-valued measures on the open subset $\mathbb{Z}_p^\times \subset \mathbb{Z}_p$.

We denote by the symbol $(\prod_{\mathfrak{p}\mid(p)} \widehat{\mathbb{Z}[\zeta_N]_{\mathfrak{p}}}/(\sigma-1))(1)$ a $\mathbb{Z}_p[\Delta_N]$-module isomorphic to $\prod_{\mathfrak{p}\mid(p)} \widehat{\mathbb{Z}[\zeta_N]_{\mathfrak{p}}}/(\sigma-1)$ on which every element $g \in G_{\mathrm{cyc}}$ acts by the cyclotomic character $\chi_{\mathrm{cyc}}(g)$.

THEOREM 3.57. *We have the following exact sequence of $\mathbb{Z}_p[\Delta_N][[G_{\mathrm{cyc}}]]$-modules:*

$$(3.51)\quad 0 \longrightarrow \prod_{\mathfrak{p}\mid(p)} \mu_{\widehat{\mathbb{Z}[\zeta_N]_{\mathfrak{p}}}} \times \prod_{\mathfrak{p}\mid(p)} \mathbb{Z}_p(1) \longrightarrow \varprojlim_n \prod_{\mathfrak{p}\mid(p)} \widehat{\mathbb{Z}[\zeta_N]_{\mathfrak{p}}}[\zeta_{p^{n+1}}]^\times$$
$$\xrightarrow{(1-\frac{\varphi}{p})\circ\log\circ\mathrm{Col}} \mathbb{Z}_p[\Delta_N][[G_{\mathrm{cyc}}]] \longrightarrow (\prod_{\mathfrak{p}\mid(p)} \widehat{\mathbb{Z}[\zeta_N]_{\mathfrak{p}}}/(\sigma-1))(1) \longrightarrow 0.$$

PROOF. The exact sequence (3.51) is obtained from the exact sequence (3.48) by replacing the second term (resp. the third term) of (3.48) by the second term (resp. the third term) of (3.51) using Theorem 3.46 (resp. Corollary 3.55). □

For a nonnegative integer n, we denote $G_{\mathrm{cyc}}/(G_{\mathrm{cyc}})^{(p-1)p^n}$ by $G_{\mathrm{cyc},n}$. When η is a Dirichlet character of conductor M, $\tau(\eta) = \sum_{i=1}^{M} \eta(i) e^{\frac{2\pi\sqrt{-1}i}{M}}$ is called a **Gauss sum** associated to η.

LEMMA 3.58. *Let $\Xi \in \mathbb{Z}_p[\Delta_{Np}][[\Gamma_{\mathrm{cyc}}]] = \mathbb{Z}_p[\Delta_N][[G_{\mathrm{cyc}}]]$ be an element in the image of $(1-\frac{\varphi}{p})\circ\log\circ\mathrm{Col}$ given in (3.51). Then, for any nonnegative integer $r \geq 0$ and for any Dirichlet character η of p-power conductor p^{n+1}, we have*

$$\tau(\eta^{-1}) \cdot (\chi_{\mathrm{cyc}}^r \eta)(\Xi) =$$
$$\begin{cases} \left(\sum_{g\in G_{\mathrm{cyc},n}} \eta(g^{-1}) D^r \left(G(\zeta_{p^{n+1}}^g(1+Z)-1) - \frac{1}{p} G^\sigma(\zeta_{p^n}^g(1+Z)-1)\right)\right)\Big|_{Z=0} & \eta \neq \mathbf{1}, \\ \left(D^r\left(G(Z) - \frac{1}{p}G^\sigma((1+Z)^p-1)\right)\right)\Big|_{Z=0} & \eta = \mathbf{1}, \end{cases}$$

where $G(Z) \in \mathcal{P}^{\mathrm{Tr}=\sigma}$ is an element satisfying

$$\left(1 - \frac{\varphi}{p}\right) G(Z) = \Xi \cdot (1+Z)$$

which is associated to Ξ by the sequence (3.48) *and its proof*

PROOF. The automorphism

$$D^r : \mathbb{Z}_p[\Delta_N][[G_{\mathrm{cyc}}]] \cdot (1+Z) \xrightarrow{\sim} \mathbb{Z}_p[\Delta_N][[G_{\mathrm{cyc}}]] \cdot (1+Z)$$

coincides with the map characterized by

$$(1+Z)^{\chi_{\mathrm{cyc}}(h)} \mapsto \chi_{\mathrm{cyc}}(h)(1+Z)^{\chi_{\mathrm{cyc}}(h)}$$

for every $h \in G_{\mathrm{cyc}}$. For any finite character $\eta : G_{\mathrm{cyc}} \longrightarrow \overline{\mathbb{Q}}_p^{\times}$, the map

$$\mathbb{Z}_p[\Delta_N][[G_{\mathrm{cyc}}]] \cdot (1+Z) \longrightarrow \overline{\mathbb{Q}}_p, \quad h \cdot (1+Z) \mapsto \tau(\eta^{-1})\eta(h)$$

coincides with the map characterized by

$$(1+Z)^{\chi_{\mathrm{cyc}}(h)} \mapsto \left(\sum_{g \in G_{\mathrm{cyc},n}} \eta^{-1}(g)(\zeta_{p^{n+1}}^{g}(1+Z))^{\chi_{\mathrm{cyc}}(h)} \right)\Bigg|_{Z=0}$$

for every $h \in G_{\mathrm{cyc}}$ where n is the nonnegative integer so that the conductor of η regarded as a Dirichlet character is equal to p^{n+1}. Here note that we have $h \cdot (1+Z) = (1+Z)^{\chi_{\mathrm{cyc}}(h)}$ for $h \in G_{\mathrm{cyc}}$. Hence, the following formula holds:

$$\tau(\eta^{-1}) \cdot (\chi_{\mathrm{cyc}}^{r}\eta)(\Xi_G) = \begin{cases} \left(\sum_{g \in G_{\mathrm{cyc},n}} \eta(g^{-1}) D^r \left(H(\zeta_{p^{n+1}}^{g}(1+Z) - 1) \right) \right)\Big|_{Z=0} & \eta \neq \mathbf{1}, \\ \left(D^r(H(Z)) \right)\big|_{Z=0} & \eta = \mathbf{1}, \end{cases}$$

for a power series $H(Z) = \Xi_G \cdot (1+Z)$. By substituting $H(Z) = \left(1 - \frac{\varphi}{p}\right)G(Z)$ to the above formula, we complete the proof. □

At the end of §3.2.3, we gave a proof of Theorem 3.30 by Iwasawa construction. Below, we will give a different proof of Theorem 3.30 based on the theory of Coleman.

ANOTHER PROOF OF THEOREM 3.30. Let us start from the case $N = 1$. Note that ψ is a power of ω since the conductor of ψ is a divisor of p in this case. Let us consider a supplementary natural number a such that $(p, a) = 1$ and $a \not\equiv 1 \mod p$. Let us take a norm compatible system $\boldsymbol{u}(a)$ and the Coleman power series $F_{\boldsymbol{u}(a)}(Z) = \frac{(1+Z)^a - 1}{(1+Z) - 1}$ associated to $\boldsymbol{u}(a)$ as given in Lemma 3.48 (1). We define an element $\Xi(a) \in \mathbb{Z}_p[[G_{\mathrm{cyc}}]]$ to be

$$\Xi(a) \cdot (1+Z) = \left(1 - \frac{\varphi}{p}\right)(G_{\boldsymbol{u}(a)}(Z))$$

where $G_{\boldsymbol{u}(a)}(Z) = \log(F_{\boldsymbol{u}(a)}(Z))$. Let $\iota : \mathbb{Z}_p[\Delta_p][[\Gamma_{\mathrm{cyc}}]] \longrightarrow \mathbb{Z}_p[\Delta_p][[\Gamma_{\mathrm{cyc}}]]$ be the map induced by an involution $\Gamma_{\mathrm{cyc}} \longrightarrow \Gamma_{\mathrm{cyc}}$, $g \mapsto g^{-1}$ and let $\gamma(a) \in \Gamma_{\mathrm{cyc}}$ be a unique element satisfying $\kappa_{\mathrm{cyc}}(\gamma(a)) = \langle a \rangle$. Now, we define $L_p(\psi) \in \Lambda_{\mathrm{cyc}} = \mathbb{Z}_p[[\Gamma_{\mathrm{cyc}}]]$ to be[22]

$$L_p(\psi) = \frac{1}{(1 - \langle a \rangle \psi(a)\gamma(a)^{-1})} \psi\omega^{-1}\left(\iota(D(\Xi(a)))\right). \tag{3.52}$$

[22]The operator D twists the action of Δ_p by ω though the action of $\Delta_p = G_{\mathrm{cyc}}/\Gamma_{\mathrm{cyc}}$ must be invariable. The modification by the character ω^{-1} appears in order to cancel the twist of the action of Δ_p by D.

When $\psi \neq \mathbf{1}$ (i.e., $N = 1$ and $e = 1$), we have $\psi(a) \neq 1$ according to the way we chose a. Hence, $(1 - \langle a \rangle \psi(a) \gamma(a)^{-1})$ is a unit of $\mathbb{Z}_p[\Delta_p][[\Gamma_{\mathrm{cyc}}]]$ and we have $L_p(\psi) \in \Lambda_{\mathrm{cyc}}$ in this case. When $\psi = \mathbf{1}$ (i.e., $N = 1$ and $e = 0$), we have $L_p(\mathbf{1}) \in \frac{1}{\gamma - \kappa_{\mathrm{cyc}}(\gamma)} \Lambda_{\mathrm{cyc}}$.

Let $r \geq 1$ be a natural number and ϕ a $\overline{\mathbb{Q}}_p^{\times}$-valued character of Γ_{cyc} of finite order. According to the construction (3.52) of $L_p(\psi)$, we have

$$(\kappa_{\mathrm{cyc}}^{1-r}\phi)(L_p(\psi)) = \frac{1}{(1 - \psi(a)\phi(\gamma(a))^{-1}\langle a \rangle^r)} \left(\psi\omega^{-r}\chi_{\mathrm{cyc}}^{r-1}\phi^{-1}\right)(D(\Xi(a)))$$

Let us set $\eta := \psi\omega^{-r}\phi^{-1}$. We will calculate the right-hand side of the above equation by applying Lemma 3.58. We have

$$D(G_{\boldsymbol{u}(a)}(Z)) = D(\log(F_{\boldsymbol{u}(a)}(Z))) = \frac{a(1+Z)^a}{(1+Z)^a - 1} - \frac{(1+Z)}{(1+Z)-1}.$$

Setting $E(Z) = \frac{a(1+Z)^a}{(1+Z)^a - 1} - \frac{(1+Z)}{(1+Z)-1}$, Lemma 3.58 gives

$$\text{(3.53)} \quad \tau(\eta^{-1}) \cdot (\chi_{\mathrm{cyc}}^{r-1}\eta)(D(\Xi(a))) =$$
$$\begin{cases} \left.\left(\sum_{g \in G_{\mathrm{cyc},n}} \eta(g^{-1}) D^{r-1} \left(E(\zeta_{p^{n+1}}^g (1+Z) - 1) - \frac{1}{p} E^{\sigma}(\zeta_{p^n}^g (1+Z) - 1) \right) \right)\right|_{Z=0} & \eta \neq \mathbf{1}, \\ \left.\left(D^{r-1} \left(E(Z) - \frac{1}{p} E^{\sigma}((1+Z)^p - 1) \right) \right)\right|_{Z=0} & \eta = \mathbf{1}. \end{cases}$$

Note that we have $(1+Z)\frac{d}{dZ} = \frac{d}{dz}$ when $Z = e^z - 1$. Hence, when $\eta \neq \mathbf{1}$, the above calculation combined with Corollary 3.37 deduce the following value[23]:

$$\begin{aligned} &\tau(\eta^{-1}) \cdot (\chi_{\mathrm{cyc}}^{r-1}\eta)(D(\Xi(a))) \\ &= \left(\frac{d}{dz}\right)^{r-1} \sum_{g \in G_{\mathrm{cyc},n}} \eta(g^{-1}) \left.\left(\frac{a \sum_{i=1}^{p^{n+1}} (\zeta_{p^{n+1}}^g e^z)^{ai}}{(e^z)^{ap^{n+1}} - 1} - \frac{\sum_{i=1}^{p^{n+1}} (\zeta_{p^{n+1}}^g e^z)^{i}}{(e^z)^{p^{n+1}} - 1} \right)\right|_{z=0} \\ &= \tau(\eta^{-1}) \cdot \left(\frac{d}{dz}\right)^{r-1} \sum_{g \in G_{\mathrm{cyc},n}} \left.\left(\frac{a \sum_{i=1}^{p^{n+1}} \eta(ai)(e^z)^{ai}}{(e^z)^{ap^{n+1}} - 1} - \frac{\sum_{i=1}^{p^{n+1}} \eta(i)(e^z)^{i}}{(e^z)^{p^{n+1}} - 1} \right)\right|_{z=0} \\ &= \tau(\eta^{-1}) \cdot \left(1 - a^r \eta^{-1}(a)\right) L(\psi\omega^{-r}\phi^{-1}, 1-r) \\ &= \tau(\eta^{-1}) \cdot \left(1 - \psi(a)\phi(\langle a \rangle)^{-1}\langle a \rangle^r\right) L(\psi\omega^{-r}\phi^{-1}, 1-r). \end{aligned}$$

When $\eta = \mathbf{1}$ (i.e. $\phi = \mathbf{1}$ and $\psi = \omega^r$), we calculate similarly as follows :

$$\begin{aligned} &\chi_{\mathrm{cyc}}^{r-1}(D(\Xi(a))) \\ &= \left(\frac{d}{dz}\right)^{r-1} \left.\left(\frac{ae^{az}}{e^{az} - 1} - \frac{e^z}{e^z - 1} - \frac{1}{p} \left(\frac{ae^{apz}}{e^{apz} - 1} - \frac{e^{pz}}{e^{pz} - 1} \right) \right)\right|_{z=0} \\ &= (1 - a^r) \left(1 - \frac{1}{p^{1-r}}\right) \zeta(1-r). \end{aligned}$$

This completes the proof of the fact that the element $L_p(\psi)$ constructed in (3.52) satisfies the desired interpolation property when $N = 1$.

---

[23] Note that the contribution of the second term of (3.53) is zero below since the conductor of η is p^{n+1} and we have $\sum_{g \in G_{\mathrm{cyc},n}} \eta(g^{-1}) D^{r-1} \left(E^{\sigma}(\zeta_{p^n}^g (1+Z) - 1) \right) = 0$.

For the case $N > 1$, we regard the element $\Xi_N \in \mathbb{Z}_p[\Delta_N][[G_{\mathrm{cyc}}]]$ to be a unique element satisfying $\Xi_N \cdot (1+Z) = \left(1 - \frac{\varphi}{p}\right)\log(\zeta_N(1+Z) - 1)$ through the Δ_N-equivariant isomorphism

$$d : \prod_{\mathfrak{p}|(p)} \widehat{\mathbb{Z}[\zeta_N]}_{\mathfrak{p}} \cong \mathbb{Z}_p[\Delta_N]$$

and we set

$$L_p(\psi) = \psi\omega^{-1}\left(\iota(D(\Xi_N))\right) \in \mathbb{Z}_p[\Delta_N][[G_{\mathrm{cyc}}]]. \tag{3.54}$$

We can check that this element satisfies the desired interpolation property by a calculation as in the case $N = 1$ though we omit it and leave it to the reader. This completes the proof also in the case $N > 1$. □

REMARK 3.59. We summarize the construction of the p-adic L-function à la Coleman as follows.

(A) The construction goes: a norm compatible system of cyclotomic units $\boldsymbol{u} \overset{(1)}{\Rightarrow}$ a Coleman power series $F_{\boldsymbol{u}}(Z) \overset{(2)}{\Rightarrow}$ a candidate element of $\mathbb{Z}_p[\Delta_N][[\Gamma_{\mathrm{cyc}}]]$.

(B) Since the operator $(1+Z)\frac{d}{dZ}$ on $F_{\boldsymbol{u}}(Z)$ coincides with the operator $\frac{d}{dz}$ on $F_{\boldsymbol{u}}(Z)|_{Z=e^z-1}$ and since $F_{\boldsymbol{u}}(Z)|_{Z=e^z-1}$ is closely related to $f_\eta(z)$, we can show by using Corollary 3.37 that the candidate element given at (A) interpolates the special values $L(\eta, 1-r)$ when η and r vary.

We constructed a candidate element from a norm compatible element at (A) and we showed that this element satisfies the desired interpolation property at (B). However, the reader might note that we obtain a rational function $F_\eta(Z)$ whose logarithmic derivative is related to the values of a Dirichlet L-function by putting $F_\eta(Z) = f_\eta(z)|_{z=\log Z}$ for $f_\eta(z)$ in (3.10) without starting from a norm compatible element. In fact, Leopoldt and Mazur gave another construction of the p-adic L-function $L_p(\psi)$ by applying the step (2) of (A) to a suitable rational function $F(Z)$ omitting the step (1) of (A) (see **[65]** for example). The construction by Coleman as well as those by Leopoldt and Mazur is called "a construction from a generating rational function". The fact that there exists a construction from a generating rational function for the p-adic L-function $L_p(\psi)$ is sometimes useful (see comments after Theorem 3.60).

The step (1) of (A) in which we prove that the rational function $F_\eta(Z)$ originates from a cyclotomic units might look unnecessary. However, for the proof of the Iwasawa main conjecture using an Euler system of cyclotomic units which will be discussed later, the fact that the p-adic L-function is related to cyclotomic units is very important (see 3.3.3).

In the end of this section, we present two important results on Kubota–Leopoldt p-adic L-function without proof.

THEOREM 3.60 (Ferrero–Washington). *Let ψ be a primitive Dirichlet character of conductor Np^e where N is a natural number prime to p and e is an integer which is equal to 0 or 1. We assume that ψ is an even Dirichlet character (i.e., $\psi(-1) = 1$). Then, the p-adic L-function $L_p(\psi) \in \mathbb{Z}_p[\psi][[\Gamma_{\mathrm{cyc}}]]$ obtained in Theorem 3.30 is not divisible by a uniformizer of $\mathbb{Z}_p[\psi]$. In other words, the μ-invariant of $L_p(\psi)$ is zero.*

This theorem was first proved by Ferrero–Washington [**28**]. The original proof of Ferrero–Washington is an analytic method based on a certain uniformness of the distribution of values of the p-adic L-function specialized at various characters. Later, Sinnott [**113**] gave a simple and algebraic another proof of this theorem as an application of "a construction from a generating rational function". Recall that the p-adic L-function $L_p(\psi)$ is a transformation of $G_{\boldsymbol{u}}(Z) = \log(F_{\boldsymbol{u}}(Z))$ where $F_{\boldsymbol{u}}(Z)$ is rational. In the another proof of Theorem 3.60 by Sinnott, we prove that, $L_p(\psi)$ is not divisible by a uniformizer ϖ of $\mathbb{Z}_p[\psi]$ if $D(G_{\boldsymbol{u}}(Z)) = (1+Z)\frac{F'_{\boldsymbol{u}}(Z)}{F_{\boldsymbol{u}}(Z)}$ is not divisible by ϖ (see [**113**, Thm. 1]). Also, it is not difficult to verify that $(1+Z)\frac{F'_{\boldsymbol{u}}(Z)}{F_{\boldsymbol{u}}(Z)}$ is not divisible by ϖ since $F_{\boldsymbol{u}}(Z)$ is a rational function of simple form. For the detail of the argument, we refer the reader to the article [**113**].

In order to present another important theorem, we prepare the following notation.

Definition 3.61. Let $\overline{\mathbb{Z}}_p$ be the ring of integers of the algebraic closure $\overline{\mathbb{Q}}_p$ of $\mathbb{Q}_p$ and let $\overline{P}$ be the maximal ideal of $\overline{\mathbb{Z}}_p$. On $1+\overline{P} \subset \overline{\mathbb{Z}}_p^{\times}$, we set

$$\log_p(1+x) = \sum_{n=1}^{\infty} \frac{(-1)^{n-1}x^n}{n} \quad \text{(for any } x \in \overline{P}\text{)}.$$

We denote by $\mu'_{\overline{\mathbb{Z}}_p}$ the group of all roots of unity in $\overline{\mathbb{Z}}_p$ whose orders are prime to p. We have a canonical isomorphism $\overline{\mathbb{Z}}_p^{\times}/(1+\overline{P}) \xrightarrow{\sim} \overline{\mathbb{F}}_p^{\times} \xrightarrow{\sim} \mu'_{\overline{\mathbb{Z}}_p}$ where the first isomorphism is the reduction modulo $\overline{P}$ and the second isomorphism is induced by the Teichmüller lifting. Since the group $1+\overline{P}$ is pro-p and the group $\overline{\mathbb{Z}}_p^{\times}/(1+\overline{P})$ as no nontrivial pro-p quotient, the exact sequence $\mathbf{1} \longrightarrow 1+\overline{P} \longrightarrow \overline{\mathbb{Z}}_p^{\times} \longrightarrow \overline{\mathbb{Z}}_p^{\times}/(1+\overline{P}) \longrightarrow \mathbf{1}$ splits. By setting $\log_p = 0$ on $\overline{\mathbb{Z}}_p^{\times}/(1+\overline{P}) \cong \mu'_{\overline{\mathbb{Z}}_p}$, the function $\log_p$ is extended uniquely to $\overline{\mathbb{Z}}_p^{\times}$. Recall that we have an exact sequence $\mathbf{1} \longrightarrow \overline{\mathbb{Z}}_p^{\times} \longrightarrow \overline{\mathbb{Q}}_p^{\times} \xrightarrow{\mathrm{val}_p} \mathbb{Q} \longrightarrow \mathbf{1}$ where val_p is the map of p-adic valuation. Now, let us fix a system of roots of p as follows:

$$\left\{ p^q \in \overline{\mathbb{Q}}_p^{\times} \,\middle|\, (p^q)^{q'} = p^{qq'} \text{ for any } q, q' \in \mathbb{Q} \right\}_{q \in \mathbb{Q}}. \tag{3.55}$$

This induces a splitting of the map val_p of the above exact sequence. By setting $\log_p(p^q) = 0$ on the set (3.55), the function $\log_p$ is extended un uniquely to $\overline{\mathbb{Q}}_p^{\times}$. If we replace the fixed system $\{p^q\}_{q\in\mathbb{Q}}$ of roots of p given at (3.55) by another system $\{(p^q)'\}_{q\in\mathbb{Q}}$, the ratio $p^q/(p^q)'$ is a system of p-power roots of unity. Hence, the function $\log_p$ on $\overline{\mathbb{Q}}_p^{\times}$ does not depend on the choice of the system (3.55). We call the function $\log_p$ thus defined a **p-adic logarithmic function**.

The the specialization range of the original interpolation property of the Kubota–Leopoldt p-adic L-function $L_p(\psi)$ covers all arithmetic characters $\kappa_{\mathrm{cyc}}^j\phi$ with $j \leq 0$ (see Theorem 3.30). As for the specialization at $j = 1$, we have the following classical theorem[24].

---

[24]Later, we discuss the specialization at $j \geq 1$ for the p-adic Beilinson conjecture (see Theorem 3.115).

THEOREM 3.62 (Leopoldt formula). *For any finite order character ϕ of Γ_{cyc}, we have*

$$\kappa_{\mathrm{cyc}}\phi(L_p(\psi)) = -\left(1 - \frac{\phi\psi(p)}{p}\right)\frac{\tau(\phi\psi)}{M}\sum_{i=1}^{M}\overline{\phi\psi}(i)\log_p(1-\zeta_M^i),$$

where M is the conductor of $\phi\psi$.

We can prove the above theorem by using a construction from a generating rational function of the p-adic L-function and by applying Lemma 3.58 with $r = 0$. However, we remark that the original proof by Leopoldt [**65**] is an argument of certain p-adic approximation which is not at all related to a generating rational function. We refer the reader to [**51**, §5], [**117**, Chap. 5, §4] for the original proof by Leopoldt.

We remark that Theorem 3.62 is a p-adic analogue of the classical class number formula for Dirichlet L-functions as follows:

THEOREM 3.63. *Let η be a Dirichlet character of conductor $M > 1$ such that $\eta(-1) = 1$. Then, we have*

$$L(\eta, 1) = -\frac{\tau(\eta)}{M}\sum_{i=1}^{M}\overline{\eta}(i)\log|1-\zeta_M^i|.$$

3.3. Bridge between the algebraic side and the analytic side (Iwasawa main conjecture)

Up to now, we have prepared the main object of the algebraic side (Selmer group) and the main object of the analytic side (Analytic p-adic L-function) of the cyclotomic Iwasawa theory of ideal class groups. In this chapter, we finally discuss the Iwasawa main conjecture which relates the algebraic side and the analytic side.

3.3.1. Main statements on the cyclotomic Iwasawa main conjecture. In Theorem C of §1.3, we gave a brief statement of the cyclotomic Iwasawa main conjecture for ideal class groups. In this section, we give the most precise forms of this main conjecture.

As for the background to follow the arguments of this section, we recommend that the readers familiarize themselves with some basics on Kummer extensions and some basics on the Dedekind zeta function such as the functional equation or the class number formula describing the residue at $s = 1$. We refer the reader to the text [**80**] for these subjects.

DEFINITION 3.64. Let $\mathcal{M}$ be a $\mathbb{Z}_p[\Delta_{Np}]$-module and η a $\overline{\mathbb{Q}}_p$-valued character of Δ_{Np}. Then we call $\mathcal{M}_\eta := \mathcal{M} \otimes_{\mathbb{Z}_p[\Delta_{Np}]} \mathbb{Z}_p[\eta]$ the η-quotient of $\mathcal{M}$. We denote by $\mathcal{M}^\vee$ the Pontrjagin dual of $\mathcal{M}$ and we call $\mathcal{M}^\eta := ((\mathcal{M}^\vee)_\eta)^\vee$ the η-part of $\mathcal{M}$.

Both $\mathcal{M}_\eta$, $\mathcal{M}^\eta$ have a natural structure of $\mathbb{Z}_p[\eta]$-module where $\mathbb{Z}_p[\eta]$ is the ring obtained by adjoining the values of η. When the order of η is prime to p, a natural map from $\mathcal{M}^\eta$ to $\mathcal{M}_\eta$ is an isomorphism. Under the same hypothesis, the functor from the category of finitely generated $\mathbb{Z}_p[\Delta_{Np}]$-modules to the category of finitely generated $\mathbb{Z}_p[\eta]$-modules of η-quotient is exact and coincides with the functor of η-part.

DEFINITION 3.65. Let η be a Dirichlet character. We denote by K_η the abelian extension of $\mathbb{Q}$ which corresponds to the kernel of the finite order character $\mathrm{Gal}(\overline{\mathbb{Q}}/\mathbb{Q}) \longrightarrow \overline{\mathbb{Q}}_p^\times$ corresponding to η by class field theory. We denote the cyclotomic $\mathbb{Z}_p$-extension of K_η by $K^{\mathrm{cyc}}_{\eta,\infty}$ and the maximal everywhere unramified abelian p-power extension of $K^{\mathrm{cyc}}_{\eta,\infty}$ (resp. the maximal abelian p-power extension of $K^{\mathrm{cyc}}_{\eta,\infty}$ which is unramified outside p) by $L^{\mathrm{cyc}}_{\eta,\infty}$ (resp. $M^{\mathrm{cyc}}_{\eta,\infty}$). We set $X_{K^{\mathrm{cyc}}_{\eta,\infty}} = \mathrm{Gal}(L^{\mathrm{cyc}}_{\eta,\infty}/K^{\mathrm{cyc}}_{\eta,\infty})$, $\mathfrak{X}_{K^{\mathrm{cyc}}_{\eta,\infty}} = \mathrm{Gal}(M^{\mathrm{cyc}}_{\eta,\infty}/K^{\mathrm{cyc}}_{\eta,\infty})$.

By the argument using Lemma 2.23 from the beginning of §3.1.1, $X_{K^{\mathrm{cyc}}_{\eta,\infty}}$, $\mathfrak{X}_{K^{\mathrm{cyc}}_{\eta,\infty}}$ are naturally regarded as compact Λ_{cyc}-modules. By Theorem 3.4, $X_{K^{\mathrm{cyc}}_{\eta,\infty}}$ is a torsion Λ_{cyc}-module.

Let us formulate the cyclotomic Iwasawa main conjecture for class groups using the terminologies which we prepared so far. As we explain later in Remark 3.68, the Iwasawa main conjecture in this situation is already proved and it is already a theorem. However since the name of the Iwasawa main conjecture is quite settled, we call it the Iwasawa main conjecture even after it is proved. The cyclotomic Iwasawa main conjecture for class groups has the $(-)$-type and the $(+)$-type, which are equivalent to each other. One of them is as follows.

THEOREM 3.66. *($(-)$-type Iwasawa main conjecture) Let N be a natural number prime to p and let ψ be a primitive Dirichlet character of conductor Np^e where e is an integer which is equal to 0 or 1. We assume that ψ is an even Dirichlet character (i.e. $\psi(-1) = 1$). We denote by $K_{\omega\psi^{-1}}$ the finite abelian extension of $\mathbb{Q}$ corresponding to the kernel of $\omega\psi^{-1}$. Then the following equality holds*[25]*:*

$$\mathrm{char}_{\Lambda_{\mathrm{cyc},\psi}}(X_{K^{\mathrm{cyc}}_{\omega\psi^{-1},\infty}})_{\omega\psi^{-1}} = \begin{cases} (L_p(\psi)) & \text{when } \psi \neq \mathbf{1}, \\ (\gamma - \kappa_{\mathrm{cyc}}(\gamma))(L_p(\mathbf{1})) & \text{when } \psi = \mathbf{1}. \end{cases}$$

If ψ is an arbitrary Dirichlet character, the Λ_{cyc}-module $\mathfrak{X}_{K^{\mathrm{cyc}}_{\psi,\infty}}$ is finitely generated but not necessarily torsion[26]. However, as we will show it later in Proposition 3.69, $(\mathfrak{X}_{K^{\mathrm{cyc}}_{\psi,\infty}})_\psi$ is a torsion $\Lambda_{\mathrm{cyc},\psi}$-module if ψ is even.

In general, for a given Λ_{cyc}-module $\mathcal{M}$, we denote by $\mathcal{M}^\bullet$ the same module whose Γ_{cyc}-action is twisted through the involution $\Gamma_{\mathrm{cyc}} \to \Gamma_{\mathrm{cyc}}$, $g \mapsto g^{-1}$ from the original one. Similarly, we denote by $\mathcal{M} \otimes \eta$ the base extension $\mathcal{M} \otimes_{\mathbb{Z}_p} \mathbb{Z}_p[\eta]$ whose Γ_{cyc}-action is twisted by the character $g \mapsto \eta(g) \cdot g$ for $g \in \Gamma_{\mathrm{cyc}}$ from the original action.

The $(-)$-type Iwasawa main conjecture stated above is equivalent to the $(+)$-type Iwasawa main conjecture as follows:

THEOREM 3.67. *($(+)$-type Iwasawa main conjecture) Let us take a primitive Dirichlet character ψ satisfying the same assumption as that of Theorem 3.66. Then the following equality holds:*

$$\mathrm{char}_{\Lambda_{\mathrm{cyc},\psi}}((\mathfrak{X}_{K^{\mathrm{cyc}}_{\psi,\infty}})_\psi)^\bullet \otimes \kappa_{\mathrm{cyc}} = \begin{cases} (L_p(\psi)) & \text{when } \psi \neq \mathbf{1}, \\ (\gamma - \kappa_{\mathrm{cyc}}(\gamma))(L_p(\mathbf{1})) & \text{when } \psi = \mathbf{1}. \end{cases}$$

[25] Recall that we assume $p > 2$ throughout the book. As for the formulation of the cyclotomic Iwasawa main conjecture for class groups for $p = 2$, we need to multiply $\frac{1}{2}$ to the p-adic L-function of the right-hand side of the equality (see [**27**] for related discussion).

[26] If K is a CM number field, it is known that $\mathfrak{X}_{K^{\mathrm{cyc}}_\infty}$ is pseudo-isomorphic to a direct sum of a free Λ_{cyc}-module of rank $r_2(K)$ and a finitely generated torsion Λ_{cyc}-module where $r_2(K) = \frac{[K:\mathbb{Q}]}{2}$ is the number of imaginary places (see [**117**, §13.5] for example).

REMARK 3.68. Here is a brief history on the proof of the cyclotomic Iwasawa main conjecture for class groups.

(1) Mazur–Wiles [**75**] proved the conjecture when $p \neq 2$ and p does not divide the order of ψ $(= [K_\psi : \mathbb{Q}])$. The method of their proof is based on a generalization of the method of Ribet where he proved the inverse of Herbrand's theorem by constructing a nontrivial unramified p-extension using p-torsion points of the jacobian variety of the modular curve $X_1(Np)$[27]. Afterwards, Wiles [**124**] generalized the result of [**75**] from the abelian extension K_ψ of $\mathbb{Q}$ corresponding to a Dirichlet character ψ to the abelian extension K_ψ of a totally real field F corresponding to a Hecke character ψ of finite order. Even when restricted to the case $F = \mathbb{Q}$, the method of [**124**] is more sophisticated than the original work [**75**] thanks to the use of Hida theory. Even for the case $F = \mathbb{Q}$, the result of [**124**] improve the result of [**75**]. For example, the result of [**124**] on the Iwasawa main conjecture covers the case where $p = 2$ but 2 does not divide the order of ψ and the case where $p \neq 2$ and p divides the order of ψ. Note that an ideal group can be regarded as a Selmer group of a Galois representation of rank one as we will discuss later in Chapter 5 of the second volume of this book. The papers [**75**] and [**124**] use modular Galois representations of rank two in order to prove the Iwasawa main conjecture of Selmer groups associated to Galois representations of rank one. Such a method using modular Galois representations of higher rank is usually called "method of the Eisenstein ideal" and this method bounds the sizes of Selmer groups from below. For the cyclotomic Iwasawa main conjecture for class groups, we note that the modular method proves the $(-)$-type Iwasawa main conjecture directly.
(2) There is a method called "method of Euler systems", which grew out of a refinement of a technique which Thaine, Kolyvagin developed to study the structures of ideal class groups and Tate–Shafarevich groups of elliptic curves. This method bounds the sizes of Selmer groups from above. For $K = \mathbb{Q}(\mu_p)$, in the Appendix of [**69**], Rubin gave another proof of the cyclotomic Iwasawa main conjecture for class groups using the Euler system of cyclotomic units over $K = \mathbb{Q}(\mu_p)$. Rubin generalized this proof in Chapter 3 of his book [**73**] and covered the proof of the cyclotomic Iwasawa main conjecture for class groups over K_ψ when $p \neq 2$ and p does not divide the order of ψ. Later, Greither [**34**] gave a proof by Euler system method of cyclotomic units for the cyclotomic Iwasawa main conjecture for class groups over K_ψ in the most general situation where we allow the case $p = 2$ and the case where p divides the order of ψ. For the cyclotomic Iwasawa main conjecture for class groups, we note that the method of Euler systems proves the $(+)$-type Iwasawa main conjecture directly[28].

[27] See §1.2 for the statement of the theorem of Herbrand–Ribet.

[28] There is an Euler system called an "Euler system of Gauss sum" and this Euler system is related to the $(-)$-type Iwasawa main conjecture, but we do not touch it in this book. We refer the reader to [**64**] for simple cases without Galois action and to [**1**] for cases with Galois action.

We will explain the proof of the cyclotomic Iwasawa main conjecture for class groups later in §3.3.2 (by the modular method) and §3.3.3 (by the method of Euler systems). Before we go into the proof, we prepare two important principles as follows:

(A)(Kummer duality) A principle which implies the equivalence between (+)-type Iwasawa main conjecture and (−)-type Iwasawa main conjecture.

(B)(Analytic class number formula) A principle which ensures the validity of the equalities of the cyclotomic Iwasawa main conjecture for class groups over K_ψ when ψ runs through even characters of $\mathrm{Gal}(\mathbb{Q}(\mu_{Np})/\mathbb{Q})$ assuming the validity of one of inclusions $\subset$ and $\supset$ of the cyclotomic Iwasawa main conjecture for class groups over K_ψ when ψ runs through even characters of $\mathrm{Gal}(\mathbb{Q}(\mu_{Np})/\mathbb{Q})$

We remark that the principle (B) is valid for both (+)-type Iwasawa main conjecture and (−)-type Iwasawa main conjecture. We also remark that the choice of $\subset$ or $\supset$ must be the same for all ψ.

Let ψ be an even Dirichlet character of conductor Np^e where N is a natural number prime to p and $e = 0$ or $e = 1$. We set $K = K_\psi(\mu_p)$. We note that $K_{\omega\psi^{-1}}$ and K_ψ are both subfields of K and that K is a finite abelian extension of $\mathbb{Q}$ satisfying $K \bigcap \mathbb{Q}_\infty = \mathbb{Q}$.

First we discuss principle (A) whose precise statement will be established in Theorem 3.73. The following result is a key to for principle (A).

PROPOSITION 3.69 (Kummer Theory). *Let us take a primitive Dirichlet character ψ satisfying the same assumption as that of Theorem 3.66. Then there exists a Γ_{cyc}-equivariant nondegenerate pairing:*

$$(3.56) \qquad (\mathfrak{X}_{K_\infty^{\mathrm{cyc}}})_\psi \times \mathrm{Cl}(K_\infty^{\mathrm{cyc}})[p^\infty]^{\omega\psi^{-1}} \xrightarrow{\sim} \mu_{p^\infty}.$$

In other words, there is an isomorphism of $\Lambda_{\mathrm{cyc},\psi}$-modules

$$(\mathfrak{X}_{K_\infty^{\mathrm{cyc}}})_\psi \cong \mathrm{Hom}_{\mathbb{Z}_p}(\mathrm{Cl}(K_\infty^{\mathrm{cyc}})[p^\infty]^{\omega\psi^{-1}}, \mathbb{Q}_p/\mathbb{Z}_p) \otimes \kappa_{\mathrm{cyc}}.$$

PROOF. Let I_∞ be the group of fractional ideals of K_∞^{cyc} [29]. For each natural number m, we set

$$\mathfrak{M}_m = \left\{ x \in (K_\infty^{\mathrm{cyc}})^\times/((K_\infty^{\mathrm{cyc}})^\times)^{p^m} \;\middle|\; (x) \in (I_\infty)^{p^m} \right\}.$$

Then we have the following short exact sequence

$$(3.57) \qquad \{\mathbf{1}\} \longrightarrow (\mathfrak{r}_{K_\infty^{\mathrm{cyc}}})^\times/((\mathfrak{r}_{K_\infty^{\mathrm{cyc}}})^\times)^{p^m} \longrightarrow \mathfrak{M}_m \longrightarrow \mathrm{Cl}(K_\infty^{\mathrm{cyc}})[p^m] \longrightarrow \{\mathbf{1}\}.$$

where the last map assigns to each element $x \in \mathfrak{M}_m$ the class $[\mathfrak{A}] \in \mathrm{Cl}(K_\infty^{\mathrm{cyc}})[p^m]$ of an ideal $\mathfrak{A} \in I_\infty$ such that $\mathfrak{A}^{p^m} = (\widetilde{x})$ with $\widetilde{x} \in K_\infty^{\mathrm{cyc}}$ a representative of x. Let us set $\mathfrak{M}_\infty = \varinjlim_m \mathfrak{M}_m$. By taking the inductive limit of (3.57) with respect to m, we obtain the following exact sequence:

$$(3.58) \qquad \{\mathbf{1}\} \longrightarrow (\mathfrak{r}_{K_\infty^{\mathrm{cyc}}})^\times \otimes_{\mathbb{Z}} \mathbb{Q}_p/\mathbb{Z}_p \longrightarrow \mathfrak{M}_\infty \longrightarrow \mathrm{Cl}(K_\infty^{\mathrm{cyc}})[p^\infty] \longrightarrow \{\mathbf{1}\}.$$

[29] We remark that K_∞^{cyc} is an algebraic number field of infinite degree. We define a fractional ideal of K_∞^{cyc} as a nonzero finitely generated $\mathfrak{r}_{K_\infty^{\mathrm{cyc}}}$-submodule of K_∞^{cyc} and we can check that the set of fractional ideals of K_∞^{cyc} becomes naturally an abelian group by multiplication. In fact, the group of fractional ideals of K_∞^{cyc} is isomorphic to the inductive limit of the group of fractional ideals of subfields of finite degree of K_∞^{cyc}.

Let us denote the maximal totally real subfield of K_∞^{cyc} by $K_\infty^{\mathrm{cyc},+}$. By Proposition 3.12, the index of the subgroup $(\mathfrak{r}_{K_\infty^{\mathrm{cyc},+}})^\times \mu_{K_\infty^{\mathrm{cyc}}}$ of $(\mathfrak{r}_{K_\infty^{\mathrm{cyc}}})^\times$ is at most 2. Hence, the natural map $(\mathfrak{r}_{K_\infty^{\mathrm{cyc},+}})^\times \mu_{K_\infty^{\mathrm{cyc}}} \otimes_{\mathbb{Z}} \mathbb{Q}_p/\mathbb{Z}_p \longrightarrow (\mathfrak{r}_{K_\infty^{\mathrm{cyc}}})^\times \otimes_{\mathbb{Z}} \mathbb{Q}_p/\mathbb{Z}_p$ is an isomorphism. Since $\mu_{K_\infty^{\mathrm{cyc}}} \otimes_{\mathbb{Z}} \mathbb{Q}_p/\mathbb{Z}_p = 0$, we have a natural isomorphism $(\mathfrak{r}_{K_\infty^{\mathrm{cyc},+}})^\times \otimes_{\mathbb{Z}} \mathbb{Q}_p/\mathbb{Z}_p \cong (\mathfrak{r}_{K_\infty^{\mathrm{cyc}}})^\times \otimes_{\mathbb{Z}} \mathbb{Q}_p/\mathbb{Z}_p$. Recall that $\omega\psi^{-1}$ is an odd character by our assumption. Hence, we obtain

$$(3.59) \qquad \left((\mathfrak{r}_{K_\infty^{\mathrm{cyc}}})^\times \otimes_{\mathbb{Z}} \mathbb{Q}_p/\mathbb{Z}_p\right)^{\omega\psi^{-1}} = \left((\mathfrak{r}_{K_\infty^{\mathrm{cyc},+}})^\times \otimes_{\mathbb{Z}} \mathbb{Q}_p/\mathbb{Z}_p\right)^{\omega\psi^{-1}} = 0.$$

By taking the part of the sequence (3.58) where the group $\Delta_N \times \Delta_p$ acts by $\omega\psi^{-1}$ and by using (3.59), we obtain

$$(3.60) \qquad (\mathfrak{M}_\infty)^{\omega\psi^{-1}} \cong \mathrm{Cl}(K_\infty^{\mathrm{cyc}})[p^\infty]^{\omega\psi^{-1}}.$$

On the other hand, we have a nondegenerate pairing

$$(3.61) \qquad \mathrm{Gal}(K_\infty^{\mathrm{cyc}}(\sqrt[p^m]{\mathfrak{M}_m})/K_\infty^{\mathrm{cyc}}) \times \mathfrak{M}_m \longrightarrow \mu_{p^\infty}, \quad (g,x) \mapsto \frac{(\sqrt[p^m]{x})^g}{\sqrt[p^m]{x}},$$

where $K_\infty^{\mathrm{cyc}}(\sqrt[p^m]{\mathfrak{M}_m})$ denotes the extension of K_∞^{cyc} obtained by adjoining $\{\sqrt[p^m]{x} \mid x \in \mathfrak{M}_m\}$. Let $M/K_\infty^{\mathrm{cyc}}$ be a cyclic Galois extension of degree p^m. Since the cyclotomic $\mathbb{Z}_p$-extension K_∞^{cyc} of $K = K_\psi(\mu_p)$ contains all p-power roots of unity, there exists an element $x \in (K_\infty^{\mathrm{cyc}})^\times/((K_\infty^{\mathrm{cyc}})^\times)^{p^m}$ such that $M = K_\infty^{\mathrm{cyc}}(\sqrt[p^m]{x})$ by Kummer theory. Let n be a natural number large enough so that we can take a representative $\widetilde{x} \in K_n^{\mathrm{cyc}}$ of x and that all primes over p are totally ramified in $K_\infty^{\mathrm{cyc}}/K_n^{\mathrm{cyc}}$. Let $\mathfrak{p}_1, \ldots, \mathfrak{p}_s$ be the prime ideals of $\mathfrak{r}_{K_n^{\mathrm{cyc}}}$ over p. We decompose the principal fractional ideal $(\widetilde{x})$ of K_n^{cyc} as $(\widetilde{x}) = \mathfrak{A}\mathfrak{p}_1^{m_1} \cdots \mathfrak{p}_s^{m_s}$ where $\mathfrak{A}$ is a fractional ideal of K_n^{cyc} whose prime decomposition does not contain any factor of the form $\mathfrak{p}_i^{l_i}$ with $l_i \in \mathbb{Z}\setminus\{0\}$. Let I_{m+n} be the group of fractional ideals of K_n^{cyc}. Note that the element of I_{m+n} given by $\mathfrak{p}_1^{m_1} \cdots \mathfrak{p}_s^{m_s}$ is contained in $(I_{m+n})^{p^m}$ since $K_{m+n}^{\mathrm{cyc}}/K_n^{\mathrm{cyc}}$ is totally ramified over p. By the local class field theory, $M/K_\infty^{\mathrm{cyc}}$ is unramified outside p if and only if there exists a fractional ideal $\mathfrak{B}$ of K_n^{cyc} such that $\mathfrak{A} = \mathfrak{B}^{p^m}$. By the above discussion, a cyclic Galois extension $M = K_\infty^{\mathrm{cyc}}(\sqrt[p^m]{x})$ of K_∞^{cyc} is unramified outside p if and only if $(x) \in (I_\infty)^{p^m}$. We thus obtained

$$(3.62) \qquad M_\infty^{\mathrm{cyc}} = \varinjlim_m K_\infty^{\mathrm{cyc}}(\sqrt[p^m]{\mathfrak{M}_m}).$$

By combining the limit of (3.61) with respect to m and (3.62), we obtain the a nondegenerate and $\Delta_N \times \Delta_p$-equivariant pairing as follows:

$$\mathfrak{X}_{K_\infty^{\mathrm{cyc}}} \times \mathfrak{M}_\infty \longrightarrow \mu_{p^\infty}.$$

Since $\Delta_N \times \Delta_p$ acts on μ_{p^∞} via the Teichmüller character ω, the above pairing induces following pairing on the ψ-quotient $(\mathfrak{X}_{K_\infty^{\mathrm{cyc}}})_\psi$ of $\mathfrak{X}_{K_\infty^{\mathrm{cyc}}}$:

$$(3.63) \qquad (\mathfrak{X}_{K_\infty^{\mathrm{cyc}}})_\psi \times (\mathfrak{M}_\infty)^{\omega\psi^{-1}} \longrightarrow \mu_{p^\infty}$$

We complete the proof by combining (3.60) and (3.63). □

In order to complete principle (A), we need to recall the theory of adjoint modules which was developed in the paper **[53]** of Iwasawa.

DEFINITION 3.70. Let $\mathcal{O}$ be the ring of integers of a finite extension of $\mathbb{Q}_p$, Γ a profinite topological group which is isomorphic to $1+p\mathbb{Z}_p$. For a finitely generated torsion $\mathcal{O}[[\Gamma]]$-module M, we take a set $\{h_\alpha\}_{\alpha\in A}$ of nonzero elements of the maximal ideal of $\mathcal{O}[[\Gamma]]$ labelled by a directed set A satisfying the following conditions:

(i) We have $(h_{\alpha'}) \subset (h_\alpha)$ if $\alpha' > \alpha$.
(ii) We have $\bigcap_{\alpha\in A}(h_\alpha) = \{0\}$.
(iii) For any $\alpha \in A$, the principal ideal (h_α) and $\mathrm{char}_{\mathcal{O}[[\Gamma]]}M$ have no common prime factor.

The modules $M/h_\alpha M$ are of finite cardinality by the condition (iii). By the condition (i), the modules $M/h_\alpha M$ form an inductive system of finite $\mathcal{O}[[\Gamma]]$-modules by setting the transition homomorphism $f_{\alpha\alpha'}$ to be

$$M/I_\alpha M \hookrightarrow M/h_{\alpha'}M,\ x \bmod (h_\alpha) \ \mapsto \frac{h_{\alpha'}}{h_\alpha}x \bmod (h_{\alpha'})$$

for each pair (α,α') with $\alpha' > \alpha$. Then, we define a $\mathcal{O}[[\Gamma]]$-module $\mathrm{Ad}(M)$ to be

$$\mathrm{Ad}(M) := \left(\varinjlim_{\alpha\in A} M/I_\alpha M\right)^\vee$$

and call it the **(Iwasawa) adjoint module** associated to M. We remark that, by Theorem 3.71, the module $\mathrm{Ad}(M)$ is independent of the choice of the system $\{h_\alpha\}_{\alpha\in A}$.

Theorem 3.71 is a principal theorem of the theory of adjoint modules. We will not prove this theorem in this book but the reader who wishes to find the proof can refer to [**81**, Chap. 5, §5] or [**117**, §15.5] for example.

THEOREM 3.71. *Let $\mathcal{O}$ be the ring of integers of a finite extension of $\mathbb{Q}_p$, Γ a profinite topological group which is isomorphic to $1+p\mathbb{Z}_p$. Then, for any finitely generated torsion $\mathcal{O}[[\Gamma]]$-modules M and M', the following statements old:*

1. *The $\mathcal{O}[[\Gamma]]$-module $\mathrm{Ad}(M)$ is independent of the choice of the choice of the system $\{h_\alpha\}_{\alpha\in A}$ and is unique up to isomorphism.*
2. *The $\mathcal{O}[[\Gamma]]$-module $\mathrm{Ad}(M)$ is canonically isomorphic to $\mathrm{Ext}^1_{\mathcal{O}[[\Gamma]]}(M,\mathcal{O}[[\Gamma]])$.*
3. *The module $\mathrm{Ad}(M)$ is a finitely generated torsion $\mathcal{O}[[\Gamma]]$-module which contains no nonzero pseudo-null $\mathcal{O}[[\Gamma]]$-submodule.*
4. *If $M \sim M'$, we have $\mathrm{Ad}(M) \sim \mathrm{Ad}(M')$.*
5. *When M is an elementary Iwasawa module in the sense of Definition 2.41, we have $\mathrm{Ad}(M) \cong M^\bullet$.*

By applying the above theorem, we obtain the following result:

PROPOSITION 3.72. *Let us take a primitive Dirichlet character ψ satisfying the same assumption as that of Theorem 3.66. Then we have the following pseudo-isomorphism of $\Lambda_{\mathrm{cyc},\psi}$-modules:*

$$\mathrm{Hom}_{\mathbb{Z}_p}(\mathrm{Cl}(K^{\mathrm{cyc}}_\infty)[p^\infty]^{\omega\psi^{-1}}, \mathbb{Q}_p/\mathbb{Z}_p)^\bullet \sim (X_{K^{\mathrm{cyc}}_\infty})_{\omega\psi^{-1}}.$$

PROOF. Since $K^{\mathrm{cyc}}_\infty/K$ is a cyclotomic $\mathbb{Z}_p$-extension, there exists an integer n_0 such that K^{cyc}_∞ is totally ramified over $K^{\mathrm{cyc}}_{n_0} = (K^{\mathrm{cyc}}_\infty)^{\Gamma^{p^{n_0}}_{\mathrm{cyc}}}$ at all primes above p. Let $\mathfrak{p}_1,\dots,\mathfrak{p}_s$ be the primes over p of $K^{\mathrm{cyc}}_{n_0}$. For each i with $1\le i\le s$, the inertia subgroup $I_{\mathfrak{p}_i}$ is isomorphic to $\mathrm{Gal}(K^{\mathrm{cyc}}_\infty/K^{\mathrm{cyc}}_{n_0}) = \Gamma^{p^{n_0}}_{\mathrm{cyc}}$ via the natural map $I_{\mathfrak{p}_i} \hookrightarrow \mathrm{Gal}(L^{\mathrm{cyc}}_\infty/K^{\mathrm{cyc}}_{n_0}) \twoheadrightarrow \Gamma^{p^{n_0}}_{\mathrm{cyc}}$. Let us fix a topological generator γ of $\Gamma^{p^{n_0}}_{\mathrm{cyc}}$.

For each i with $1 \leq i \leq s$, we denote by $\gamma_i \in I_{\mathfrak{p}_i}$ the pull-back of $\gamma^{p^{n_0}}$ by the above isomorphism and we define a closed subgroup $Y_{K_\infty^{\mathrm{cyc}}}$ of the topological group $X_{K_\infty^{\mathrm{cyc}}} = \mathrm{Gal}(L_\infty^{\mathrm{cyc}}/K_\infty^{\mathrm{cyc}})$ to be

$$Y_{K_\infty^{\mathrm{cyc}}} = \overline{\langle \omega_{n_0} X_{K_\infty^{\mathrm{cyc}}}, \gamma_2^{-1}\gamma_1, \dots, \gamma_s^{-1}\gamma_1 \rangle},$$

where we set $\omega_{n_0} = \gamma^{p^{n_0}} - 1$ and $\overline{\langle\ \rangle}$ denotes the closure of the subgroup $\langle\ \rangle$. By the argument of §3.1.2, $Y_{K_\infty^{\mathrm{cyc}}}/(\frac{\omega_n}{\omega_{n_0}})Y_{K_\infty^{\mathrm{cyc}}}$ is canonically identified with a subgroup of $\mathrm{Cl}(K_n^{\mathrm{cyc}})[p^\infty]$ for any $n > n_0$ and the order of the cokernel of canonical injection $Y_{K_\infty^{\mathrm{cyc}}}/(\frac{\omega_n}{\omega_{n_0}})Y_{K_\infty^{\mathrm{cyc}}} \hookrightarrow \mathrm{Cl}(K_n^{\mathrm{cyc}})[p^\infty]$ is uniformly bounded independently of $n > n_0$. Hence, by choosing $\{I_\alpha\}_{\alpha \in A}$ of Definition 3.70 to be $\{\frac{\omega_n}{\omega_{n_0}}\}_{n > n_0}$, we have

$$\mathrm{Ad}((Y_{K_\infty^{\mathrm{cyc}}})_{\omega\psi^{-1}}) \sim \mathrm{Hom}_{\mathbb{Z}_p}(\mathrm{Cl}(K_\infty^{\mathrm{cyc}})[p^\infty]^{\omega\psi^{-1}}, \mathbb{Q}_p/\mathbb{Z}_p) \tag{3.64}$$

by definition. By Theorem 3.71 (4), (5) and the structure theorem of Iwasawa modules (Theorem 2.39), we have $\mathrm{Ad}(M)^\bullet \sim M$ for any finitely generated $\Lambda_{\mathrm{cyc},\psi}$-module M. Hence, we have

$$\mathrm{Ad}((Y_{K_\infty^{\mathrm{cyc}}})_{\omega\psi^{-1}})^\bullet \sim (Y_{K_\infty^{\mathrm{cyc}}})_{\omega\psi^{-1}}. \tag{3.65}$$

Finally we have

$$(Y_{K_\infty^{\mathrm{cyc}}})_{\omega\psi^{-1}} \sim (X_{K_\infty^{\mathrm{cyc}}})_{\omega\psi^{-1}} \tag{3.66}$$

since $Y_{K_\infty^{\mathrm{cyc}}}$ is a subgroup of $X_{K_\infty^{\mathrm{cyc}}}$ of finite index. By combining (3.64), (3.65) and (3.66), we complete the proof. □

By putting the above results together, we obtain the following theorem.

THEOREM 3.73 (Principle of Kummer theory). *Let us take a primitive Dirichlet character ψ satisfying the same assumption as that of Theorem 3.66. Then we have the following equality:*

$$\mathrm{char}_{\Lambda_{\mathrm{cyc},\psi}}(X_{K_\infty^{\mathrm{cyc}}})_{\omega\psi^{-1}} = \mathrm{char}_{\Lambda_{\mathrm{cyc},\psi}}((\mathfrak{X}_{K_\infty^{\mathrm{cyc}}})_\psi)^\bullet \otimes \kappa_{\mathrm{cyc}}.$$

We thus established principle (A)[30]. Principle (B) is established by the following theorem.

THEOREM 3.74 (Principle of analytic class number formula). *Let $K = \mathbb{Q}(\mu_{Np})$. Then we have the following equalities:*

$$\sum_\psi \lambda((X_{K_\infty^{\mathrm{cyc}}})_{\omega\psi^{-1}}) = \sum_\psi \lambda(L_p(\psi)),$$

$$\sum_\psi \mu((X_{K_\infty^{\mathrm{cyc}}})_{\omega\psi^{-1}}) = \sum_\psi \mu(L_p(\psi)),$$

where ψ runs through all characters of $\mathrm{Gal}(K/\mathbb{Q})$ *such that $\psi(-1) = 1$ and $\psi \neq \mathbf{1}$.*

PROOF. In this proof, we denote by $\mathfrak{C}$ the set of all characters of $\mathrm{Gal}(K/\mathbb{Q})$ such that $\psi(-1) = 1$ and $\psi \neq \mathbf{1}$. Let $\mathcal{O}$ be the ring of integers of a finite extension of $\mathbb{Q}_p$ which contains values of any $\psi \in \mathfrak{C}$. We set

$$X^{\mathrm{alg}} := (X_{K_\infty^{\mathrm{cyc}}} \otimes_{\mathbb{Z}_p} \mathcal{O})^-, \quad X^{\mathrm{anal}} := \Lambda_{\mathrm{cyc},\mathcal{O}}/(\prod_{\psi \in \mathfrak{C}} L_p(\psi)),$$

[30]Recall that, in order that Kummer theory works, we established principle (A) with K obtained by adjoining ζ_p to K_ψ and $K_{\omega\psi^{-1}}$. Note that we have pseudo-isomorphisms of $\Lambda_{\mathrm{cyc},\psi}$-modules $(X_{K_\infty^{\mathrm{cyc}}})_{\omega\psi^{-1}} \sim (X_{K^{\mathrm{cyc}}_{\omega\psi^{-1},\infty}})_{\omega\psi^{-1}}$ and $(\mathfrak{X}_{K_\infty^{\mathrm{cyc}}})_\psi \sim (\mathfrak{X}_{K^{\mathrm{cyc}}_{\psi,\infty}})_\psi$.

where $(\ )^-$ denotes the part on which the complex conjugate J acts with multiplication of -1. By decomposing X^{alg} with characters, we have the following decomposition of the Iwasawa invariants on the algebraic side[31]:

$$\lambda(X^{\mathrm{alg}}) = \sum_{\psi \in \mathfrak{C}} \lambda((X_{K^{\mathrm{cyc}}_\infty})_{\omega\psi^{-1}}),\ \mu(X^{\mathrm{alg}}) = \sum_{\psi \in \mathfrak{C}} \mu((X_{K^{\mathrm{cyc}}_\infty})_{\omega\psi^{-1}}).$$

By the definition of X^{anal}, we have the following decomposition of the Iwasawa invariants on the analytic side:

$$\lambda(X^{\mathrm{anal}}) = \sum_{\psi \in \mathfrak{C}} \lambda(L_p(\psi)),\ \mu(X^{\mathrm{anal}}) = \sum_{\psi \in \mathfrak{C}} \mu(L_p(\psi)).$$

Hence, it suffices to prove that

$$\lambda(X^{\mathrm{alg}}) = \lambda(X^{\mathrm{anal}}),\ \mu(X^{\mathrm{alg}}) = \mu(X^{\mathrm{anal}}). \tag{3.67}$$

By the argument developed in §3.1.2, there exist a natural number n_0 and a real number c satisfying $0 < c \leq 1$ such that we have the following inequality for any $n \geq n_0$ [32]:

$$c \leq \frac{\#(\mathrm{Cl}(K_n^{\mathrm{cyc}})[p^\infty])/\#(\mathrm{Cl}(K_n^{\mathrm{cyc},+})[p^\infty])}{\#(X^{\mathrm{alg}}/(\frac{\omega_n}{\omega_{n_0}})X^{\mathrm{alg}})} \leq \frac{1}{c}. \tag{3.68}$$

Let us recall some classical facts concerning Dedekind's zeta functions $\zeta(K_n^{\mathrm{cyc}}, s)$ for K_n^{cyc} and $\zeta(K_n^{\mathrm{cyc},+}, s)$ for its totally real subfield $K_n^{\mathrm{cyc},+}$.

(1) By Dirichlet Class Number Formula, the order of zero of $\zeta(K_n^{\mathrm{cyc}}, s)$ at $s = 0$ is equal to $r_1(K_n^{\mathrm{cyc}}) + r_2(K_n^{\mathrm{cyc}})$ and we have

$$\lim_{s \to 0} \frac{\zeta(K_n^{\mathrm{cyc}}, s)}{s^{r_2(K_n^{\mathrm{cyc}})}} = -\frac{\#\mathrm{Cl}(K_n^{\mathrm{cyc}}) \cdot R_{K_n^{\mathrm{cyc}}}}{\#\mu_{K_n^{\mathrm{cyc}}}} \tag{3.69}$$

where $R_{K_n^{\mathrm{cyc}}}$ is the regulator of K_n^{cyc} for $n \geq 3$[33].

(2) Similarly, we have

$$\lim_{s \to 0} \frac{\zeta(K_n^{\mathrm{cyc},+}, s)}{s^{r_1(K_n^{\mathrm{cyc},+})}} = -\frac{\#\mathrm{Cl}(K_n^{\mathrm{cyc},+}) \cdot R_{K_n^{\mathrm{cyc},+}}}{\#\mu_{K_n^{\mathrm{cyc},+}}}. \tag{3.70}$$

Here we note that we have $r_2(K_n^{\mathrm{cyc}}) = r_1(K_n^{\mathrm{cyc},+})$ for $n \geq 3$.

(3) Since K_n^{cyc} is an abelian extension of $\mathbb{Q}$, we have

$$\zeta(K_n^{\mathrm{cyc}}, s) = \prod_{\eta} L(\eta, s) \tag{3.71}$$

where η runs through Dirichlet characters defined modulo Np^{n+1}.

(4) Similarly, we have

$$\zeta(K_n^{\mathrm{cyc},+}, s) = \prod_{\eta} L(\eta, s) \tag{3.72}$$

where η runs through even Dirichlet characters defined modulo Np^{n+1}.

---

[31] By using Remark 3.33 (3) and Stickelberger's theorem (Theorem 3.94), we have $\lambda((X_{K^{\mathrm{cyc}}_\infty})_\omega) = \mu((X_{K^{\mathrm{cyc}}_\infty})_\omega) = 0$ when $\psi = \mathbf{1}$. This justifies the validity of the following equality though the case $\psi = \mathbf{1}$ is excluded in the definition of $\mathfrak{C}$.

[32] By taking a sufficiently large n_0, we can avoid the contribution of trivial-zero, for example.

[33] Note that we have $r_1(K_n^{\mathrm{cyc}}) = 0$ for $n \geq 3$.

By taking the ratio of (3.69) and (3.70), we have:

$$\lim_{s\to 0}\frac{\zeta(K_n^{\mathrm{cyc}},s)}{\zeta(K_n^{\mathrm{cyc},+},s)}=\frac{2^{\frac{d_n}{2}+1}}{\#\mu_{K_n^{\mathrm{cyc}}}}\cdot\frac{\#\mathrm{Cl}(K_n^{\mathrm{cyc}})}{\#\mathrm{Cl}(K_n^{\mathrm{cyc},+})},\tag{3.73}$$

where $d_n=[K_n^{\mathrm{cyc}}:\mathbb{Q}]$. Here we used the fact that $\frac{R_{K_n^{\mathrm{cyc}}}}{R_{K_n^{\mathrm{cyc},+}}}=2^{\frac{d_n}{2}}$ and $\#\mu_{K_n^{\mathrm{cyc},+}}=2$.

On the other and, by the interpolation of the p-adic L-function $L_p(\psi)$ (Theorem 3.30) and by Remark 3.33 (3), we have

$$|\phi(L_p(\psi))|_p=\begin{cases}|L(\psi\omega^{-1}\phi^{-1},0)|_p & \text{when } \psi\neq\mathbf{1},\\ \dfrac{1}{|\phi(\gamma)-1|_p} & \text{when } \psi=\mathbf{1},\end{cases}$$

for any Dirichlet character ϕ of p-power order whose conductor divides p^{n+1} where γ is a topological generator of Γ and $\phi=\mathbf{1}$ is excluded in the latter case. Now we have the following equality for any n:

$$\left|\frac{\prod\phi(L_p(\mathbf{1}))}{\#\mu_{K_n^{\mathrm{cyc}}}}\right|_p=\left|\frac{\prod(\phi(\gamma)-1)}{p^{n+1}}\right|_p=p,$$

where ϕ runs through nontrivial characters of $\Gamma_{\mathrm{cyc}}/\Gamma_{\mathrm{cyc}}^{p^n}$. Since p is an odd prime number, (3.71), (3.72) and (3.73) imply that there exists a real number satisfying $0<c'\le 1$ such that we have the following inequality for any positive integer n:

$$c'\le\frac{\#(\mathrm{Cl}(K_n^{\mathrm{cyc}})[p^\infty])/\#(\mathrm{Cl}(K_n^{\mathrm{cyc},+})[p^\infty])}{\#(X^{\mathrm{anal}}/(\frac{\omega_n}{\omega_0})X^{\mathrm{anal}})}\le\frac{1}{c'}.\tag{3.74}$$

The asymptotic behavior of $\#(X^{\mathrm{alg}}/(\frac{\omega_n}{\omega_{n_0}})X^{\mathrm{alg}})$ and $\#(X^{\mathrm{anal}}/(\frac{\omega_n}{\omega_0})X^{\mathrm{anal}})$ when n varies obtained in (3.68), (3.74) combined with the arguments of §2.3.4 (especially Remark 2.51) implies the desired equalities of Iwasawa invariants (3.67). This completes the proof. □

By the principle of analytic class number formula established earlier, it suffices to prove only one of inclusions $\subset$ and $\supset$ of the cyclotomic Iwasawa main conjecture for class groups over K_ψ when ψ runs through even characters of $\mathrm{Gal}(\mathbb{Q}(\mu_{Np})/\mathbb{Q})$. However, X^{alg} and X^{anal} are quite different in nature and it is not easy to relate these two objects directly. So, usually we try to relate them with help of a certain intermediate object as follows:

characteristic ideal of Selmer group (algebraic object)	⋯⋯	intormodiate object	⋯⋯	a principal ideal generated by p-adic L-function . (analytic object)

There are mainly two different methods, the method of the Eisenstein ideal and the method of Euler systems. We will explain each of them in §3.3.2 and in §3.3.3 respectively.

3.3.2. Proof of the Iwasawa main conjecture by the Eisenstein ideal. In this section, we will explain the proof along the method of the Eisenstein ideal of $(-)$-type Iwasawa main conjecture (Theorem 3.66) obtained by Mazur–Wiles [**75**],

Wiles [**124**]. Their proof has its origin in the proof of the following result by Ribet [**93**][34]:

$$p|L(\omega^{i-1},-1) \Longrightarrow p|\#\mathrm{Cl}(\mathbb{Q}(\mu_p))^{\omega^i} \tag{3.75}$$

for any odd integer i satisfying $1 < i < p-1$.

Roughly speaking, their work generalizes the work of Ribet, from $\mathbb{Q}(\mu_p)$ to a system of $\mathbb{Q}(\mu_{Np^{n+1}})$ where n varies, from the divisibility by p to the divisibility by p^m. In other words, the proof of the above-mentioned work of Ribet [**93**] is a prototype of the proof of by Mazur–Wiles and Wiles. Hence, we will explain the argument of [**93**] first. Based on this, we then discuss the generalization by by Mazur–Wiles and Wiles.

The basic knowledge required to understand the topics of this subsection is more advanced than the basic knowledge required in other sections. For example, the proofs of Ribet, Mazur–Wiles, and Wiles assume modular forms, Hecke operators for which the reader can refer to [**77**] and [**39**, Chap.5, Chap.6]. It is also important to have a basic understanding of basics on the theory of p-adic Galois representations as well as the theory Galois cohomology theory to the level of [**107**, Chap.1]. Other than these subjects, the basic knowledge of the arithmetic algebraic geometry such as algebraic curves, abelian varieties and finite flat group schemes over local fields is also assumed (see [**131**] for example on such basic topics). For these reasons, this section has two $*$[35]. In fact, we only give a sketch for the proof of Wiles. If the reader is not familiar with some technical terms or some notion, he can try to grasp only the outline without following the detail of the argument. If reading in such a way is also difficult, we can skip this section.

From now on throughout the book, we fix two sets of complex and p-adic embeddings of $\overline{\mathbb{Q}}$ as follows:

(3.76)

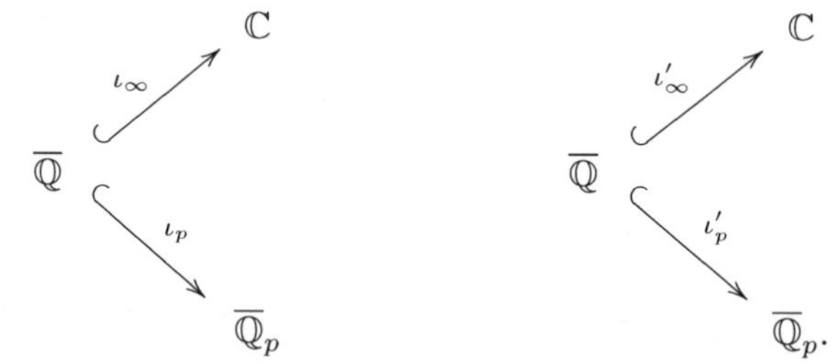

We fix a finite extension $\mathcal{K}$ of $\mathbb{Q}_p$ which is sufficiently large so that the coefficients of q-expansion of a given modular form and the values of some Dirichlet characters are contained in $\mathcal{K}$. The first set of fixed embeddings ι_∞, ι_p serves for identifying the decomposition group at a finite prime v of a number field K with the absolute Galois group of the local field K_v and the second set of fixed embeddings ι'_∞, ι'_p serves for regarding the coefficients of q-expansion of a given normalized modular form as elements in $\overline{\mathbb{Q}}_p$[36].

[34] For an integer k such that $i \equiv 1-k \bmod p-1$, we have $\zeta(1-k) \equiv L(\omega^{i-1},-1) \bmod p$. Hence, the implication $\Longrightarrow$ of the theorem of Herbrand–Ribet stated in Chapter 1 is equivalent to the implication $\Longrightarrow$ of (3.75).

[35] For the meaning of $*$, please refer to "How to read this book" in the preface of this book.

[36] After Chapter 5, the first set of fixed embeddings ι_∞, ι_p serves for fixing "the field of definition" of a given motif or a given algebraic variety and the second set of fixed embeddings ι'_∞, ι'_p serves for fixing "the field of coefficients" of a given motif.

We denote by $M_k(M,\eta;\mathbb{Z})$ (resp. $S_k(M,\eta;\mathbb{Z})$) the space of modular forms (resp. cuspforms) of weight k, level $\Gamma_1(M)$ with Neben character η whose coefficients of q-expansion are contained in $\mathbb{Z}$. We set:

$$M_k(M,\eta;\mathcal{O}_\mathcal{K}) = M_k(M,\eta;\mathbb{Z})\otimes_\mathbb{Z}\mathcal{O}_\mathcal{K},\ S_k(M,\eta;\mathcal{O}_\mathcal{K}) = S_k(M,\eta;\mathbb{Z})\otimes_\mathbb{Z}\mathcal{O}_\mathcal{K}.$$

Let ϖ be a uniformizer of $\mathcal{O}_\mathcal{K}$. Below, we will explain the proof of (3.75) given by Ribet [**93**].

(Step 1) If $\varpi | L(\omega^{i-1},-1)$, there exists a normalized cuspidal eigenform $f = \sum_{n=1}^\infty a_n(f)q^n \in S_2(p,\omega^{i-1};\mathcal{O}_\mathcal{K})$ such that we have the congruences $a_\ell(f)\equiv 1+\ell\omega^{i-1}(\ell) \bmod \varpi$ for all primes $\ell\neq p$.

Outline of the proof. Recall that, for a Dirichlet character η of conductor M and a natural number $k\geq 1$, there exists an eigenform $G_{k,\eta}\in M_k(M,\eta;\mathcal{O}_\mathcal{K})$ called an Eisenstein series given by

$$G_{k,\eta} = \frac{M^k(k-1)!}{2\tau(\eta^{-1})(2\pi\sqrt{-1})^k}\sum_{\substack{(m,n)\in\mathbb{Z}^2\\(m,n)\neq(0,0)}}\frac{\eta^{-1}(n)}{(mMz+n)^k}, \tag{3.77}$$

where we assume that $\eta\neq\mathbf{1}$ when $k\leq 2$ and $\tau(\eta^{-1})$ denotes the Gauss sum $\sum_{1\leq j\leq M}\eta^{-1}(j)e^{\frac{2\pi j\sqrt{-1}}{M}}$ [37]. The q-expansion of $G_{k,\eta}$ is given as follows:

$$G_{k,\eta} = \frac{L(\eta,1-k)}{2}+\sum_{n=1}^\infty\sigma_{k-1,\eta}(n)q^n, \tag{3.78}$$

where $\sigma_{k-1,\eta}(n)=\sum_{1\leq d\leq n, d|n}\eta(d)d^{k-1}$, the σ-function of Dedekind for the character η[38]. Note that, when $k=2$, $M=p$, $\eta=\omega^{i-1}$, the constant term of $G_{2,\omega^{i-1}}$ is $\frac{L(\omega^{i-1},-1)}{2}$. We can verify by computation that, among all pairs (a,b) of integers satisfying $1\leq a,b\leq p-1$ and $a+b\equiv i-1 \bmod p-1$, there exists a pair (a,b) such that the constant term $a_0(g(a,b))$ of $g(a,b):=G_{1,\omega^a}G_{1,\omega^b}\in M_2(p,\omega^{i-1};\mathcal{O}_\mathcal{K})$ is a p-adic unit[39].

With such a pair (a,b), we set

$$f' := G_{2,\omega^{i-1}} - \frac{L(\omega^{i-1},-1)}{2}\frac{g(a,b)}{a_0(g(a,b))}\in M_2(p,\omega^{i-1};\mathcal{O}_\mathcal{K}). \tag{3.79}$$

By the definition of f', we have $a_0(f')=0$. By the assumption $\varpi|L(\omega^{i-1},-1)$, we have $f'\equiv G_{2,\omega^{i-1}} \bmod \varpi$ where we define the congruence between modular forms

---

[37] When $k\leq 2$, the series (3.77) is not absolutely convergent. Hence, we consider the series $G_{k,\eta}(s)$ which is defined by introducing another complex variable s and by replacing $(mMz+n)^k$ in (3.77) by $(mMz+n)^k|mMz+n|^{2s}$. The series $G_{k,\eta}(s)$ is absolutely convergent for $2\mathrm{Re}(s)+k-2>0$. When $\eta\neq\mathbf{1}$, $G_{k,\eta}(s)$ is analytically continued to the whole $\mathbb{C}$-plane and we define $G_{k,\eta}$ by $G_{k,\eta} = G_{k,\eta}(s)|_{s=0}$. For the detail of the calculation and the proof, we refer the reader to [**77**, §7.2].

[38] For the proof of the formula of the q-expansion of the Eisenstein series with character, we refer the reader to [**39**, §5.1].

[39] By (3.78), the constant term of $g(a,b)$ is equal to $\frac{L(\omega^a,0)L(\omega^b,0)}{4}$. Since the p-adic valuation of the value $\prod_{j=1}^{p-1}L(\omega^j,0)$ is bounded by the p-adic valuation of $\#\mathrm{Cl}(\mathbb{Q}(\zeta_p))$ from above thanks to the class number formula, the asymptotic behavior of $\mathrm{ord}_p\#\mathrm{Cl}(\mathbb{Q}(\zeta_p))$ obtained by Carlitz implies that we have $\mathrm{ord}_p\#\mathrm{Cl}(\mathbb{Q}(\zeta_p))<\frac{p-1}{4}$ if $p>19$. For primes $p\leq 19$, we always have $\mathrm{ord}_p\#\mathrm{Cl}(\mathbb{Q}(\zeta_p))=0$. Hence, the existence of a pair (a,b) with which the constant term of $g(a,b)$ is a p-adic unit is assured for all primes p by pigeonhole principle. For more of the detail and related references, we refer the reader to [**93**, Theorem (3.3)].

with coefficients in $\mathcal{O}_{\mathcal{K}}$ by the congruence of all coefficients of their q-expansions here. In general, a modular form whose constant term vanishes as is the case for f' is called a **semi-cuspform**.

Note that f' might not necessarily be an eigenform. In order to obtain an eigenform, we use the following lemma:

LEMMA 3.75. *Let $\mathcal{M}$ be a finitely generated free module over a discrete valuation ring $\mathcal{O}_{\mathcal{K}}$ and $\mathcal{P}$ a set of $\mathcal{O}_{\mathcal{K}}$-linear endomorphisms of $\mathcal{M}$ which are mutually commutative. Suppose that $\overline{v} \in \mathcal{M}/\varpi\mathcal{M}$ is a simultaneous eigenvector of the set of endomorphisms on $\mathcal{M}/\varpi\mathcal{M}$ induced by elements of $\mathcal{P}$. Then, by replacing $\mathcal{M}$ and $\mathcal{P}$ by their base extension to a finite extension of $\mathcal{K}$ if necessary, there exists a simultaneous eigenvector $\widetilde{v} \in \mathcal{M}$ of the set of endomorphisms $\mathcal{P}$ over $\mathcal{O}_{\mathcal{K}}$ such that $\widetilde{v} \equiv \overline{v}$ mod ϖ.*

Roughly speaking, the idea is to consider the subalgebra $\mathcal{H}$ of the ring of endomorphism $\mathrm{End}_{\mathcal{O}_{\mathcal{K}}}(\mathcal{M})$ generated by $\mathcal{P}$ and to discuss the lifting of the prime ideal associated to the component of $\mathcal{H}$ modulo the maximal ideal of $\mathcal{O}_{\mathcal{K}}$ corresponding to $\overline{v}$. Since the detail of the algebraic argument of the proof of this lemma does not help us to understand the essential point of the proof of Ribet, we omit the proof of lemma. For the detail of the proof, we refer the reader to [**24**, Lemma 6.11].

We will apply Lemma 3.75 to the space of semi-cuspforms

$$\mathcal{M} = \{h \in M_2(p, \omega^{i-1}; \mathcal{O}_{\mathcal{K}}) \mid a_0(h) = 0\}$$

and the set $\mathcal{P}$ of Hecke operators in $\mathrm{End}_{\mathcal{O}_{\mathcal{K}}}(\mathcal{M})$[40]. If we put $v = f'$, v mod ϖ is an eigenvector by the definition given in (3.79)[41]. Let $\widetilde{v}$ be the eigenvector obtained from v applying Lemma 3.75 and we define $f'' = \sum_{n=0}^{\infty} a_n(f'')q^n \in M_2(p, \omega^{i-1}; \mathcal{O}_{\mathcal{K}})$ to be the form corresponding to $\widetilde{v}$. By definition, f'' is a semi-cuspidal eigenform which satisfies the congruence $f'' \equiv G_{2,\omega^{i-1}}$ mod ϖ. In general, the constant terms of a semi-cuspform does not vanish at other cusps. Hence, a semi-cuspform is not necessarily a cuspform.

Recall that we have a decomposition

$$M_2(p, \omega^{i-1}; \mathcal{K}) = S_2(p, \omega^{i-1}; \mathcal{K}) \oplus E_2(p, \omega^{i-1}; \mathcal{K}),$$

where $E_2(p, \omega^{i-1}; \mathcal{K})$ is the subspace of $M_2(p, \omega^{i-1}; \mathcal{K})$ which consists of Eisenstein series. In this situation, $E_2(p, \omega^{i-1}; \mathcal{K})$ is of dimension 2 and the coefficients of q-expansions of two eigenforms are explicitly calculated (see [**77**, §4.7][42]). One of two eigenforms is the Eisenstein series $G_{2,\omega^{i-1}}$, which has nonzero constant term and the other eigenform is a semi-cuspform. Since the Fourier coefficient at ℓ of $G_{2,\omega^{i-1}}$ is $1 + \omega^{i-1}(\ell)\ell$ and the Fourier coefficient at ℓ of the unique semi-cuspform in $E_2(p, \omega^{i-1}; \mathcal{K})$ is $\ell + \omega^{i-1}(\ell)$ by [**77**, §4.7], they are not congruent modulo ϖ to each other unless ω^{i-1} is trivial. We thus conclude that the semi-cuspidal eigenform $f'' \in S_2(p, \omega^{i-1}; \mathcal{O}_{\mathcal{K}})$ which we constructed belongs to $S_2(p, \omega^{i-1}; \mathcal{K})$. Since

[40] In general, the subspace of semi-cuspforms in the space of modular forms of certain weight and level is not necessarily stable under the action of Hecke operators. Here, since the subspace of non semi-cuspforms inside the space of Eisenstein series in $M_2(p, \omega^{i-1}; \mathcal{O}_{\mathcal{K}})$ is of dimension one, the subspace of semi-cuspforms in $M_2(p, \omega^{i-1}; \mathcal{O}_{\mathcal{K}})$ is stable under the action of Hecke operators. Thus, we can apply Lemma 3.75 in this situation.

[41] The element v mod ϖ is nonzero because the first coefficient of the q-expansion of f' is a unit of $\mathcal{O}$.

[42] We refer the reader to [**77**, Chap. 7] for a general reference of q-expansion of Eisenstein series of general weight and level.

$a_1(G_{2,\omega^{i-1}}) = 1$, we have $a_1(f'') \in 1 + \varpi\mathcal{O}_{\mathcal{K}} \subset \mathcal{O}_{\mathcal{K}}^{\times}$. Hence, the normalization $f = \frac{1}{a_1(f'')} f''$ of f'' also belongs to $S_2(p, \omega^{i-1}; \mathcal{O}_{\mathcal{K}})$[43]. □

(Step 2a) Let us denote by $V_f \cong \mathcal{K}^{\oplus 2}$ the representation space of the p-adic Galois representation ρ_f[44] of the absolute Galois group $G_{\mathbb{Q}}$ of $\mathbb{Q}$ associated to the normalized cuspidal eigenform $f = \sum_{n=0}^{\infty} a_n(f) q^n \in S_2(p, \omega^{i-1}; \mathcal{O}_{\mathcal{K}})$ obtained in Step 1. Then there exists a $G_{\mathbb{Q}}$-stable $\mathcal{O}_{\mathcal{K}}$-lattice[45] $T \subset (V_f)^*$ such that the $G_{\mathbb{Q}}$-module $T/\varpi T$ have the following extension[46]:

$$0 \longrightarrow (\mathcal{O}_{\mathcal{K}}/(\varpi))(\mathbf{1}) \longrightarrow T/\varpi T \longrightarrow (\mathcal{O}_{\mathcal{K}}/(\varpi))(\omega^{-i}) \longrightarrow 0, \tag{3.80}$$

where $(\mathcal{O}_{\mathcal{K}}/(\varpi))(\eta)$ denotes the Galois representation of rank one over $\mathcal{O}_{\mathcal{K}}/(\varpi)$ on which $G_{\mathbb{Q}}$ acts by $\eta : G_{\mathbb{Q}} \longrightarrow (\mathcal{O}_{\mathcal{K}}/(\varpi))^{\times}$.

PROOF. Let us choose a $G_{\mathbb{Q}}$-stable $\mathcal{O}_{\mathcal{K}}$-lattice $T' \subset (V_f)^*$. As we discussed in Step 1, we have $a_1(f'') \equiv 1 \bmod \varpi$. Hence, we have $a_\ell(f) \equiv a_\ell(f'') \bmod \varpi$ for any prime ℓ. Since we have

$$\mathrm{tr}(\mathrm{Frob}_l^{-1} \text{ on } T'/\varpi T') \equiv a_l(f) \equiv a_l(G_{2,\omega^{i-1}}) = 1 + l\omega^{i-1}(l) \bmod \varpi$$

for the arithmetic Frobenius element Frob_l^{-1} at each prime $\ell \neq p$, the semi-simplification $(T'/\varpi T')^{\mathrm{ss}}$ of $T'/\varpi T'$ has the following isomorphism:

$$(T'/\varpi T')^{\mathrm{ss}} \cong (\mathcal{O}_{\mathcal{K}}/(\varpi))(\mathbf{1}) \oplus (\mathcal{O}_{\mathcal{K}}/(\varpi))(\omega^{-i}) \tag{3.81}$$

by the Chebotarev density theorem. Especially, $T'/\varpi T'$ is reducible. If $T'/\varpi T'$ has a subspace which is isomorphic to $(\mathcal{O}_{\mathcal{K}}/(\varpi))(\mathbf{1})$, $T'/\varpi T'$ has an extension of the type (3.80). In this case, it suffices to take $T = T'$. If it is not the case, the $G_{\mathbb{Q}}$-module $T'/\varpi T'$ has the following extension:

$$0 \longrightarrow (\mathcal{O}_{\mathcal{K}}/(\varpi))(\omega^{-i}) \longrightarrow T'/\varpi T' \longrightarrow (\mathcal{O}_{\mathcal{K}}/(\varpi))(\mathbf{1}) \longrightarrow 0. \tag{3.82}$$

Then, we set $T = \mathrm{Ker}[T' \twoheadrightarrow (\mathcal{O}_{\mathcal{K}}/(\varpi))(\mathbf{1})]$, which is also a $G_{\mathbb{Q}}$-stable $\mathcal{O}_{\mathcal{K}}$-lattice of $(V_f)^*$. the $G_{\mathbb{Q}}$-module $T/\varpi T$ has an extension of the type (3.80) This completes the proof of Step 2a. □

(Step 2b) There exists a $G_{\mathbb{Q}}$-stable $\mathcal{O}_{\mathcal{K}}$-lattice $T \subset (V_f)^*$ such that the $G_{\mathbb{Q}}$-module $T/\varpi T$ have an extension of the type (3.80) which is not split as a short exact sequence of $\mathcal{O}_{\mathcal{K}}[G_{\mathbb{Q}}]$-modules.[47]

---

[43] The argument to this point is based on [**93**, Prop. 3.5].

[44] We refer the reader to Theorem A.1 of this book to know what ρ_f and V_f are.

[45] We refer the reader to [**107**, Chap. 1] or §5.1 of this book for the definition and the generality of Galois stable lattices of a p-adic Galois representation.

[46] The symbol $(V_f)^*$ denotes the $\mathcal{K}$-linear dual of V_f. In this book, our Galois representation V_f associated to f is cohomological. On the other hand, the argument of Ribet is based on the homological construction of the Galois representation associated to f. We refer the reader to Remark A.3 to understand what we mean by a cohomological Galois representation and a homological Galois representation.

[47] In general, given an irreducible p-adic Galois representation whose residual representation is reducible, it is important to consider a statement which assures the existence of a $G_{\mathbb{Q}}$-stable $\mathcal{O}_{\mathcal{K}}$-lattice whose mod ϖ reduction is non semi-simple. A statement of this type which generalizes [**93**, Proposition (2.1)] is called a "**Ribet's lemma**".

PROOF. Let $T' \subset (V_f)^*$ be a $G_{\mathbb{Q}}$-stable $\mathcal{O}_{\mathcal{K}}$-lattice satisfying (3.80) of Step 2a. By choosing an $\mathcal{O}_{\mathcal{K}}$-basis of T', we consider the representation

$$G_{\mathbb{Q}} \longrightarrow GL_2(\mathcal{O}_{\mathcal{K}}), \quad g \mapsto \begin{pmatrix} a'(g) & b'(g) \\ c'(g) & d'(g) \end{pmatrix} \tag{3.83}$$

by matrices relative to the fixed basis. Then each of matrix components a', b', c', d' is a map from $G_{\mathbb{Q}}$ to $\mathcal{O}_{\mathcal{K}}$. By choosing the $\mathcal{O}_{\mathcal{K}}$-basis of T' giving (3.83) so that a' modulo ϖ is the matrix of the subrepresentation $(\mathcal{O}_{\mathcal{K}}/(\varpi))(\mathbf{1}) \subset T'/\varpi T'$, we have

$$a' \equiv \mathbf{1} \bmod \varpi, \; c' \equiv 0 \bmod \varpi, \; d' \equiv \omega^{-i} \bmod \varpi. \tag{3.84}$$

Under this choice of the basis, the representation T' is non split if and only if we have $b' \not\equiv 0 \bmod \varpi$. Hence, if the representation T' satisfies (3.84) and we have $b' \not\equiv 0 \bmod \varpi$, it suffices to take $T = T'$. In the rest of the proof, we assume that the representation T' satisfies (3.84) and $b' \equiv 0 \bmod \varpi$. Starting from this T', we want to find another $G_{\mathbb{Q}}$-stable $\mathcal{O}_{\mathcal{K}}$-lattice $T \subset (V_f)^*$ which has a matrix representation

$$G_{\mathbb{Q}} \longrightarrow GL_2(\mathcal{O}_{\mathcal{K}}), \quad g \mapsto \begin{pmatrix} a(g) & b(g) \\ c(g) & d(g) \end{pmatrix}$$

satisfying

$$a \equiv \mathbf{1} \bmod \varpi, \; b \not\equiv 0 \bmod \varpi, \; c \equiv 0 \bmod \varpi, \; d \equiv \omega^{-i} \bmod \varpi.$$

Since the p-adic Galois representation V_f of $G_{\mathbb{Q}}$ is irreducible by Ribet [**94**, Prop. 4.4] [48], there exists an integer $n \geq 1$ such that $b' \equiv 0 \bmod \varpi^n$ and $b' \not\equiv 0 \bmod \varpi^{n+1}$. Let us consider a matrix $P = \begin{pmatrix} 1 & 0 \\ 0 & \varpi^n \end{pmatrix}$ and let us define a $G_{\mathbb{Q}}$-stable $\mathcal{O}_{\mathcal{K}}$-lattice $T \subset (V_f)^*$ by $T := PT' \subset (V_f)^*$, which has the following representation:

$$G_{\mathbb{Q}} \longrightarrow GL_2(\mathcal{O}_{\mathcal{K}}), \quad g \mapsto \begin{pmatrix} a(g) & b(g) \\ c(g) & d(g) \end{pmatrix} = P \begin{pmatrix} a'(g) & b'(g) \\ c'(g) & d'(g) \end{pmatrix} P^{-1}.$$

We can check that $a(g) = a'(g)$, $b(g) = \frac{b'(g)}{\varpi^n}$, $c(g) = \varpi^n c'(g)$, $d(g) = d'(g)$ by calculation. The matrix components a, b, c, d satisfies (3.84) and $b \not\equiv 0 \bmod \varpi$. We thus constructed the desired $G_{\mathbb{Q}}$-stable $\mathcal{O}_{\mathcal{K}}$-lattice $T \subset (V_f)^*$. □

(Step 3) Let $T \subset (V_f)^*$ be a $G_{\mathbb{Q}}$-stable $\mathcal{O}_{\mathcal{K}}$-lattice obtained in Step 2b. Then the sequence (3.80) splits as an exact sequence of $G_{\mathbb{Q}_p(\zeta_p+\zeta_p^{-1})}$-modules.

PROOF. Since f is of weight two, $(V_f)^*$ is a subrepresentation of $T_pX_1(p) \otimes_{\mathbb{Z}_p} \mathbb{Q}_p$ where $T_pX_1(p)$ denotes the p-adic Tate module of $X_1(p)$ [49]. Let $B = J_1(p)/J_0(p)$ be the quotient abelian variety where $J_1(p)$ and $J_0(p)$ are the Jacobian varieties of the modular curves $X_1(p)$ and $X_0(p)$ respectively. Since the Neben character ω^{i-1} of f is nontrivial, $(V_f)^*$ is a subrepresentation of $T_pB \otimes_{\mathbb{Z}_p} \mathbb{Q}_p$.

THEOREM 3.76. *The abelian variety B has good reduction over $\mathbb{Q}_p(\zeta_p + \zeta_p^{-1})$.*

---

[48] See Theorem A.1 for V_f.

[49] See the comments after Remark A.3 for the fact that the Galois representation V_f associated to cuspforms of weight two is constructed from the p-adic Tate module of a Jacobian variety.

We can prove this theorem by applying the local Langlands correspondence proved by Carayol [**10**]. Here we use this theorem by omitting the proof, but we refer the reader to [**22**, Chap. V] for the proof of this theorem[50]. Let $\mathbb{Q}_f$ be the field obtained by adjoining all coefficients of the q-expansion of f to the field of rationals and $\mathbb{Z}_f$ the ring of integers. Then $\mathcal{O}_\mathcal{K}$ is a local component of a semi-local ring $\mathbb{Z}_f \otimes_\mathbb{Z} \mathbb{Z}_p$. Let us take an abelian variety B' isogenious to B with $\mathbb{Z}_f \hookrightarrow \mathrm{End}(B')$ such that $T = T_pB' \otimes_{\mathbb{Z}_f \otimes_\mathbb{Z} \mathbb{Z}_p} \mathcal{O}_\mathcal{K}$. Since abelian varieties in the same isogeny class have the same reduction type, B' also has good reduction over $\mathbb{Q}_p(\zeta_p + \zeta_p^{-1})$. Let $\mathcal{B}'$ be the Néron model of B' over $\mathbb{Z}_p[\zeta_p + \zeta_p^{-1}]$. Since $\mathcal{B}'$ is an abelian scheme over $\mathbb{Z}_p[\zeta_p+\zeta_p^{-1}]$ and the multiplication by p map $\mathcal{B}' \xrightarrow{p} \mathcal{B}'$ is finite flat, the kernel $\mathcal{B}'[p]$ of the multiplication by p is a finite flat group scheme over $\mathbb{Z}_p[\zeta_p + \zeta_p^{-1}]$. We note that the ring $\mathbb{Z}_f \otimes_\mathbb{Z} \mathbb{Z}_p$ acts on $\mathcal{B}'[p]$. In general, given a finite flat group scheme $\mathcal{G}$ defined over the ring of integers $\mathcal{O}_k$ of a p-adic field k, the group of $\overline{\mathbb{Q}}_p$-valued points $\mathcal{G}(\overline{\mathbb{Q}}_p)$ is a G_k-module. Let $\mathcal{G} \subset \mathcal{B}'[p]$ be the finite flat subgroup scheme over $\mathbb{Z}_p[\zeta_p + \zeta_p^{-1}]$ obtained by cutting out the part where the ring $\mathbb{Z}_f \otimes_\mathbb{Z} \mathbb{Z}_p$ acts through the local component $\mathcal{O}_\mathcal{K}$. For $k = \mathbb{Q}_p(\zeta_p + \zeta_p^{-1})$, we have an isomorphism of G_k-modules

$$T/\varpi T \cong \mathcal{G}(\overline{\mathbb{Q}}_p). \tag{3.85}$$

The following theorem is due to Raynaud (see [**92**, Thm. 3.3.3, Cor. 3.3.6.] for the proof).

THEOREM 3.77. *Let k be a p-adic field whose absolute ramification index $e(k)$ satisfies $e(k) < p - 1$. Then the following statements hold for any finite flat group schemes $\mathcal{G}$, $\mathcal{G}'$ which are defined over $\mathcal{O}_k$ and annihilated by p.*

(1) *If we have an isomorphism of G_k-modules $\mathcal{G}(\overline{\mathbb{Q}}_p) \cong \mathcal{G}'(\overline{\mathbb{Q}}_p)$, we have an isomorphism $\mathcal{G} \cong \mathcal{G}'$.*
(2) *We have an injection $\mathrm{Hom}_{\mathbb{F}_p[G_k]}(\mathcal{G}(\overline{\mathbb{Q}}_p), \mathcal{G}'(\overline{\mathbb{Q}}_p)) \hookrightarrow \mathrm{Hom}(\mathcal{G}, \mathcal{G}')$.*

By the basic theory on finite flat group schemes (see [**115**] for example), we have a standard short exact sequence: of finite flat group schemes

$$0 \longrightarrow \mathcal{G}^{\mathrm{conn}} \longrightarrow \mathcal{G} \longrightarrow \mathcal{G}^{\text{ét}} \longrightarrow 0, \tag{3.86}$$

where $\mathcal{G}^{\mathrm{conn}}$ is a connected finite flat group scheme and $\mathcal{G}^{\text{ét}}$ is an étale finite flat group scheme.

Let us get back to the situation of (3.85). By Step 2a, we have a $G_\mathbb{Q}$-stable short exact sequence (3.80). Since this sequence is stable by the subgroup $G_k \subset G_\mathbb{Q}$ with $k = \mathbb{Q}_p(\zeta_p + \zeta_p^{-1})$, the G_k-module $T/\varpi T$ of rank two over $\mathcal{O}_\mathcal{K}/(\varpi)$ contains a G_k-module $(\mathcal{O}_\mathcal{K}/(\varpi))(\mathbf{1})$ of rank one over $\mathcal{O}_\mathcal{K}/(\varpi)$ with trivial action. Since we have $e(k) = \frac{p-1}{2} < p - 1$, we have an injective morphism of finite flat group scheme from a nonzero constant finite flat group scheme over $\mathcal{O}_k$ into $\mathcal{G}$ by Theorem 3.77. On the other hand, the representation $\mathcal{G}(\overline{\mathbb{Q}}_p) \cong T/\varpi T$ of G_k is ramified since ω^i is a nontrivial ramified character of G_k by the assumption that i is odd. This implies that $\mathcal{G}^{\mathrm{conn}} \neq 0$. Since $\mathcal{G}$ contains a nonzero constant finite flat group scheme, $\mathcal{G}$ is not a connected finite flat group scheme over $\mathcal{O}_k$. Hence, $\mathcal{G}^{\mathrm{conn}}$ is not equal to the whole of $\mathcal{G}$ and $\mathcal{G}^{\mathrm{conn}}$ is a finite flat group scheme over $\mathcal{O}_k$ such that $\mathcal{G}^{\mathrm{conn}}(\overline{\mathbb{Q}}_p)$ is

[50] See also [**61**, Chap. 14] for an extension of this theorem to the case of the general level.

of rank one over $\mathcal{O}_{\mathcal{K}}/(\varpi)$. We thus proved that the free $\mathcal{O}_{\mathcal{K}}/(\varpi)$-module $T/\varpi T$ of rank two have two different G_k-stable subspaces of rank one. This implies that the short exact sequence (3.80) of G_k-modules splits. □

(Step 4) The cohomology class obtained by the extension (3.80) satisfied by the reduction $T/\varpi T$ of an $\mathcal{O}_{\mathcal{K}}$-lattice $T \subset (V_f)^*$ modified as in Step 2b gives a nonzero element in $\mathrm{Cl}(\mathbb{Q}(\zeta_p))[p]^{\omega^i}$.

PROOF. By applying the twist $\otimes\omega^i$ with the character ω^i to (3.82), we have the following exact sequence:

$$(3.87) \qquad 0 \longrightarrow (\mathcal{O}_{\mathcal{K}}/(\varpi))(\omega^i) \longrightarrow (T/\varpi T)\otimes\omega^i \longrightarrow (\mathcal{O}_{\mathcal{K}}/(\varpi))(\mathbf{1}) \longrightarrow 0.$$

Let x be a generator of the rightmost term $(\mathcal{O}_{\mathcal{K}}/(\varpi))(\mathbf{1})$ of (3.87) and take a lift $\widetilde{x} \in (T/\varpi T)\otimes\omega^i$. For any $g \in G_{\mathbb{Q}}$, we have

$$g(\widetilde{x}) - \widetilde{x} \in \mathrm{Ker}[(T/\varpi T)\otimes\omega^i \twoheadrightarrow (\mathcal{O}_{\mathcal{K}}/(\varpi))(\mathbf{1})] = (\mathcal{O}_{\mathcal{K}}/(\varpi))(\omega^i)$$

and we can check that the element $g(\widetilde{x}) - \widetilde{x}$ is independent of the choice of the lift $\widetilde{x}$. This defines a map

$$(3.88) \qquad G_{\mathbb{Q}} \longrightarrow (\mathcal{O}_{\mathcal{K}}/(\varpi))(\omega^i),\ g \mapsto g(\widetilde{x}) - \widetilde{x},$$

and we can check that this is a group homomorpihsm. By Step 2b, the group homomorphism (3.88) is nontrivial. Since the order of the group $(\mathcal{O}_{\mathcal{K}}/(\varpi))(\omega^i)$ is a p-power and the index $[G_{\mathbb{Q}} : G_{\mathbb{Q}(\zeta_p)}]$ is prime to p, a group homomorphism

$$(3.89) \qquad G_{\mathbb{Q}(\zeta_p)} \longrightarrow (\mathcal{O}_{\mathcal{K}}/(\varpi))(\omega^i),$$

which is obtained by restricting (3.88) is also nontrivial. Since the abelian variety B' has good reduction outside p, the general theory of étale cohomology assures that the homomorphism (3.89) restricted to the inertia subgroup I_λ is trivial for any prime λ of $\mathbb{Q}(\zeta_p)$ which is not lying over p. By Step 3, the homomorphism (3.89) restricted to the decomposition group $D_{\mathfrak{p}}$ at the unique prime $\mathfrak{p}$ of $\mathbb{Q}(\zeta_p)$ over p is trivial. Thus, there exists a nontrivial $\mathrm{Gal}(\mathbb{Q}(\zeta_p)/\mathbb{Q})$-equivariant homomorphism $\mathrm{Cl}(\mathbb{Q}(\zeta_p))[p] \longrightarrow (\mathcal{O}_{\mathcal{K}}/(\varpi))(\omega^i)$ by the unramified global class field theory. This completes the proof of (3.75). □

Let us proceed to the proof of the Iwasawa main conjecture for ideal class groups (Theorem 3.66). Theorem 3.66 was first proved by Mazur–Wiles [**75**] when the order of the Dirichlet caracter ψ associated to the base field $K_{\omega\psi^{-1}}$ is prime to p and Wiles [**124**] removed the assumption on the order of ψ later. The proofs given by Mazur–Wiles [**75**] and Wiles [**124**] are both generalizations of the method of Ribet explained above, but the proof of [**124**] refines and simplifies the proof of [**75**] even when the order of ψ is prime to p. The first remarkable idea of [**124**] is the use of Hida theory on families of p-adic modular over local domains $\mathbb{I}$ which are finite flat over Λ_{cyc}. We localize Λ_{cyc} at every height-one prime $\mathfrak{I}$ dividing the p-adic L-function $L_p(\psi)$ and bound from below the length of the torsion $(\Lambda_{\mathrm{cyc}})_{\mathfrak{I}}$ module obtained by localizing $(X_{K^{\mathrm{cyc}}_{\omega\psi^{-1}},\infty})_{\omega\psi^{-1}}$ at $\mathfrak{I}$ with help of Hida theory[51].

[51] It is thanks to Hida theory and the localization that [**124**] removed the assumption of the order of ψ being prime to p.

Before explaining the proof of [**124**], we will recall and fix some notation on Hida theory. Let $\mathbb{I}$ be a local domain which is finite flat over Λ_{cyc}. By fixing a natural number N satisfying $(N,p) = 1$, Hida theory provides us a certain $\mathbb{I}$-module $\mathcal{M}_{Np^\infty}(\mathbb{I})$ called the space of Hida's $\mathbb{I}$-adic modular forms of level Np^∞. We list up some of important facts on the space $\mathcal{M}_{Np^\infty}(\mathbb{I})$ below[52]:

(1) The $\mathbb{I}$-module $\mathcal{M}_{Np^\infty}(\mathbb{I})$ is a finitely generated $\mathbb{I}$-submodule of $\mathbb{I}[[q]]$ where q is a formal variable. Hence, every element of $\mathcal{M}_{Np^\infty}(\mathbb{I})$ is given as a formal q-expansion $\mathbb{F} = \Sigma_{n=0}^{\infty} A_n(\mathbb{F})q^n$ where $A_n(\mathbb{F})$ is an element of $\mathbb{I}$ for every nonnegative integer n.

(2) Let $\mathbb{F}$ be an element of $\mathcal{M}_{Np^\infty}(\mathbb{I})$. For any homomorphism $\kappa : \mathbb{I} \longrightarrow \overline{\mathbb{Q}}_p$ satisfying $\kappa|_{\Lambda_{\mathrm{cyc}}} = \kappa_{\mathrm{cyc}}^{k-2}\phi$ with a natural number $k \geq 2$ and a character of finite order $\phi : \Gamma_{\mathrm{cyc}} \longrightarrow \overline{\mathbb{Q}}^\times$, there exists a modular form $f_\kappa \in M_k(\Gamma_0(Np^*), \psi_{\mathbb{F}}\omega^{2-k}\phi; \kappa(\mathbb{I}))$ such that the specialization

$$\kappa(\mathbb{F}) := \Sigma_{n=0}^{\infty}\kappa(A_n(\mathbb{F}))q^n$$

is equal to the q-expansion of $f_\kappa = f_\kappa(z)$ when we identify the formal variable q of $\kappa(\mathbb{F})$ with $\exp(2\pi\sqrt{-1}z)$[53]. Here, $\psi_{\mathbb{F}}$ is a Dirichlet character associated to $\mathbb{F}$ whose conductor is a divisor of Np. We call the character $\psi_{\mathbb{F}}$ the Neben character of the $\mathbb{I}$-adic modular form $\mathbb{F}$.

(3) An $\mathbb{I}$-adic modular form $\mathbb{F}$ is called an $\mathbb{I}$-adic cuspform if the specialization f_κ belongs to the space of cuspforms $S_k(\Gamma_0(Np^*), \psi_{\mathbb{F}}\omega^{2-k}\phi; \kappa(\mathbb{I})) \subset M_k(\Gamma_0(Np^*), \psi_{\mathbb{F}}\omega^{2-k}\phi; \kappa(\mathbb{I}))$ for any homomorphism $\kappa : \mathbb{I} \longrightarrow \overline{\mathbb{Q}}_p$ satisfying the same condition as in (2). We denote by $\mathcal{S}_{Np^\infty}(\mathbb{I}) \subset \mathcal{M}_{Np^\infty}(\mathbb{I})$ the $\mathbb{I}$-submodule which consists of $\mathbb{I}$-adic cuspforms.

(4) As in the theory of classical modular forms, we have Hecke operators $T_2, T_3, \ldots$ acting on the modules $\mathcal{S}_{Np^\infty}(\mathbb{I})$ and $\mathcal{M}_{Np^\infty}(\mathbb{I})$. Since these Hecke operators commute to each other, it makes sense to consider a simultaneous eigenvector with respect to all Hecke operators. We call such a simultaneous eigenvector a Hecke eigenform.

(5) Let $\mathbb{F}$ be an $\mathbb{I}$-adic cuspidal eigenform of level Np^∞. We denote by $\mathbb{K}$ the field of fractions of $\mathbb{I}$. Then there exists an irreducible Galois representation $\rho_{\mathbb{F}} : G_{\mathbb{Q}} \longrightarrow GL_2(\mathbb{K})$ which satisfies the following conditions:
   (tr) At any prime ℓ not dividing Np, the representation $\rho_{\mathbb{F}}$ is unramified and we have the equality $A_\ell(\mathbb{F}) = \mathrm{tr}(\rho_{\mathbb{F}}(\mathrm{Frob}_\ell))$.
   (det) The representation $\det\rho_{\mathbb{F}} : G_{\mathbb{Q}} \longrightarrow GL_1(\mathbb{K})$ is isomorphic to the character $\widetilde{\kappa}^{-1}\kappa_{\mathrm{cyc}}^{-1}\omega^{-1}\psi_{\mathbb{F}}^{-1}$ where $\widetilde{\kappa}$ is the universal character $G_{\mathbb{Q}} \twoheadrightarrow \Gamma_{\mathrm{cyc}} \hookrightarrow \mathbb{Z}_p[[\Gamma_{\mathrm{cyc}}]]^\times \hookrightarrow \mathbb{I}^\times$.

   We note that $\rho_{\mathbb{F}}$ is uniquely characterized modulo isomorphism by the above properties. The representation $\rho_{\mathbb{F}}$ is continuous in the sense that the representation space $V_{\mathbb{F}} \cong \mathbb{K}^{\oplus 2}$ of $\rho_{\mathbb{F}}$ has a finitely generated $\mathbb{I}$-submodule $\mathbb{T} \subset V_{\mathbb{F}}$ such that $\mathbb{T} \otimes \mathbb{K} \cong V_{\mathbb{F}}$ which is stable by the action of $G_{\mathbb{Q}}$ and the action of $G_{\mathbb{Q}}$ on $\mathbb{T}$ is continuous with respect to the topology of $\mathbb{T}$ induced by the maximal ideal of $\mathbb{I}$.

---

[52]In this section, we only give a brief summary on Hida theory to outline the proof of Theorem 3.66. However, we will give a more detailed explanation of Hida theory later in §7.2 of the second volume of this book.

[53]We have $* = \mathrm{ord}_p(\mathrm{Cond}(\phi))$ when $\phi \neq \mathbf{1}$ and we have $* = 1$ when $\phi = \mathbf{1}$.

Let us explain the $\mathbb{I}$-adic Eisenstein series, which is the most important example of $\mathbb{I}$-adic modular form.

An $\mathbb{I}$-adic Eisenstein series is a p-adic family of classical Eisenstein series. Let ψ be a Dirichlet character of conductor Np^e ($e=0$ or $e=1$) such that $\psi(-1)=1$ and $\psi \neq \mathbf{1}$. We set $\mathbb{I} = \Lambda_{\mathrm{cyc},\psi}$. Let us fix a topological generator γ of Γ_{cyc}. For any natural number d, we define $s(d) \in \mathbb{Z}_p$ to be a unique element characterized by $\kappa_{\mathrm{cyc}}(\gamma)^{s(d)} = \langle d \rangle$. We define $\mathrm{Tw}_{\kappa_{\mathrm{cyc}}} : \Lambda_{\mathrm{cyc},\psi} \longrightarrow \Lambda_{\mathrm{cyc},\psi}$ to be the $\mathbb{Z}_p[\psi]$-linear map induced by $\gamma \mapsto \kappa_{\mathrm{cyc}}(\gamma)\gamma$. Then, we define $\mathbb{G}_\psi = \sum_{n=0}^{\infty} A_n(\mathbb{G}_\psi) q^n \in \mathcal{M}_{Np^\infty}(\mathbb{I})$ by

$$A_n(\mathbb{G}_\psi) = \begin{cases} \mathrm{Tw}_{\kappa_{\mathrm{cyc}}}\left(\frac{\iota(L_p(\psi))}{2}\right) & \text{when } n=0, \\ \sum_{\substack{1\le d\le n \\ d\mid n, (d,p)=1}} \psi(d) d \gamma^{s(d)} & \text{when } n>0, \end{cases} \tag{3.90}$$

where we recall that ι is the involution map on $\Lambda_{\mathrm{cyc},\psi}$ induced by $\Gamma_{\mathrm{cyc}} \longrightarrow \Gamma_{\mathrm{cyc}}$, $g \mapsto g^{-1}$. For any homomorphism $\kappa : \mathbb{I} \longrightarrow \overline{\mathbb{Q}}_p$ satisfying $\kappa|_{\Lambda_{\mathrm{cyc}}} = \kappa_{\mathrm{cyc}}^{k-2}\phi$ with a natural number $k \geq 2$ and a character of finite order $\phi : \Gamma_{\mathrm{cyc}} \longrightarrow \overline{\mathbb{Q}}^\times$, we have $\kappa(\mathbb{G}_\psi) = G^{(p)}_{k,\psi\omega^{2-k}\phi}$ where $G^{(p)}_{k,\psi\omega^{2-k}\phi}$ is a classical cuspidal eigenform given by the following q-expansion[54]:

$$G^{(p)}_{k,\psi\omega^{2-k}\phi}(q) = G_{k,\psi\omega^{2-k}\phi}(q) - p^{k-1}\psi\omega^{2-k}\phi(p) G_{k,\psi\omega^{2-k}\phi}(q^p).$$

Thus $\mathbb{G}_\psi$ is an element of $\mathcal{M}_{Np^\infty}(\mathbb{I})$[55].

Using these tools from Hida theory, we will explain the proof of Theorem 3.66 by Wiles in which we carry out a systematic generalization of the construction of unramified extensions by Ribet. By the principle of the analytic class number formula (Theorem 3.74), it suffices to show the inequality

$$\mathrm{ord}_{\iota(\mathfrak{I})} \mathrm{Tw}_{\kappa_{\mathrm{cyc}}^{-1}}(L_p(\psi)) \leq \mathrm{ord}_{\iota(\mathfrak{I})}\left(\mathrm{char}_{\Lambda_{\mathrm{cyc},\psi}}(X_{K^{\mathrm{cyc}}_{\omega\psi^{-1},\infty}})_{\omega\psi^{-1}} \otimes \kappa_{\mathrm{cyc}}^{-1}\right) \tag{3.91}$$

for each height-one prime ideal $\mathfrak{I} \subset \Lambda_{\mathrm{cyc},\psi}$ which contains the element $A_0(\mathbb{G}_\psi) = \mathrm{Tw}_{\kappa_{\mathrm{cyc}}}(\frac{\iota(L_p(\psi))}{2}) \in \Lambda_{\mathrm{cyc},\psi}$. In addition, by the vanishing theorem of the μ-invariant (Theorem 3.60), it suffices to prove (3.91) assuming that the prime $\mathfrak{I}$ is not over p (In other words, $\Lambda_{\mathrm{cyc},\psi}/\mathfrak{I}$ is of characteristic 0). From now, in the rest of the proof, we fix such a height-one prime ideal $\mathfrak{I} \subset \Lambda_{\mathrm{cyc},\psi}$ and we set

$$l = \mathrm{ord}_{\iota(\mathfrak{I})} \mathrm{Tw}_{\kappa_{\mathrm{cyc}}^{-1}}(L_p(\psi)) = \mathrm{ord}_{\mathfrak{I}} A_0(\mathbb{G}_\psi). \tag{3.92}$$

Our goal is to prove (3.91). We will do this in 4 steps[56].

(Step I) There exists a local domain $\mathbb{I}$ which is finite flat over Λ_{cyc}, an $\mathbb{I}$-adic cuspidal eigenform $\mathbb{F} \in \mathcal{S}_{Np^\infty}(\mathbb{I})$ and a height-one prime ideal $\widetilde{\mathfrak{I}} \subset \mathbb{I}$ over the height-one prime ideal $\mathfrak{I} \subset \Lambda_{\mathrm{cyc},\psi}$ such that $\mathbb{F} \equiv \mathbb{G}_\psi \bmod \widetilde{\mathfrak{I}}^l$.

---

[54] The construction by which we obtain $G^{(p)}_{k,\psi\omega^{2-k}\phi}(q)$ from $G_{k,\psi\omega^{2-k}\phi}(q)$ as below is usually called a p-stabilization of a modular form. It is known that, at any prime different from p, the Hecke eigenvalue of the p-stabilization $G^{(p)}_{k,\psi\omega^{2-k}\phi}(q)$ is the same as that of $G_{k,\psi\omega^{2-k}\phi}(q)$.

[55] We refer the reader to the text [**39**, §7.1] for a more detailed explanation on the construction of $\mathbb{G}_\psi$ under the assumption that $N=1$.

[56] The steps I, II, III, and IV below are respectively the analogues of steps 1, 2, 3, and 4 of the proof of Ribet explained earlier. It will help to understand the arguments deeper if we always keep this analogy in mind.

PROOF. If we try to prove Step I by just imitating the proof of Step 1 in the proof of Ribet, we have to start from showing the existence of an $\Lambda_{\mathrm{cyc},\psi}$-adic modular form $\mathbb{G} \in \mathcal{M}_{Np^\infty}(\Lambda_{\mathrm{cyc},\psi})$ with the same Neben character ψ as $\mathbb{G}_\psi$ whose constant term $A_0(\mathbb{G})$ is a unit of $\Lambda_{\mathrm{cyc},\psi}$. Since N is arbitrary, it is impossible to check this with help of numerical calculations of the asymptotic behavior of class numbers as we did in the proof of Ribet for $N = 1$. We sketch a brief idea on how to avoid this difficulty.

The second important idea of Wiles is that it suffices to show only the existence of an $\Lambda_{\mathrm{cyc},\psi}$-adic modular form $\mathbb{G} \in \mathcal{M}_{Np^\infty}(\Lambda_{\mathrm{cyc},\psi})$ with the same Neben character ψ as $\mathbb{G}_\psi$ whose constant term $A_0(\mathbb{G})$ has no common zero with constant term $A_0(\mathbb{G}_\psi)$ of $\mathbb{G}_\psi$. We will find this by a twisting technique which is explained below.

For a character ρ of Γ_{cyc} of finite order, we set $\mathbb{G}^{(\rho)}_\psi = G_{1,\omega^{-1}\rho^{-1}} \cdot \mathrm{Tw}_{\kappa_{\mathrm{cyc}}^{-1}\rho}(\mathbb{G}_\psi)$. By construction, $\mathbb{G}^{(\rho)}_\psi$ is an element of $\mathcal{M}_{Np^\infty}(\Lambda_{\mathrm{cyc},\psi})$ whose Neben character is equal to that of $\mathbb{G}_\psi$[57]. Note that $A_0(\mathbb{G}^{(\rho)}_\psi)$ has only finitely many zeros by p-adic Weierstrass preparation theorem (Theorem2.15). Hence, if we choose ρ so that the order of ρ is sufficiently large, $A_0(\mathbb{G}^{(\rho)}_\psi)$ and $A_0(\mathbb{G}_\psi)$ have no common zeros. By the definition of the ideal $\mathcal{I}$ given above, $A_0(\mathbb{G}^{(\rho)}_\psi)$ is a unit of the localization $(\Lambda_{\mathrm{cyc},\psi})_{\mathfrak{I}}$ under this choice of ρ. Let us set

$$\mathbb{F}' := \mathbb{G}_\psi - \frac{A_0(\mathbb{G}_\psi)}{A_0(\mathbb{G}^{(\rho)}_\psi)}\mathbb{G}^{(\rho)}_\psi \in (\Lambda_{\mathrm{cyc},\psi})_{\mathfrak{I}}[[q]].$$

Then we have $\mathbb{F}' \equiv \mathbb{G}_\psi$ modulo $\mathfrak{I}^l$. Since the constant term of $\mathbb{F}'$ is zero, $\mathbb{F}'$ is an $\Lambda_{\mathrm{cyc},\psi}$-adic semi-cuspform.

If we follow the analogy with the strategy of the Ribet's proof which is the prototype of this proof, we might try to find a sufficiently large finite flat extension $\mathbb{I}$ of $\Lambda_{\mathrm{cyc},\psi}$ and an $\mathbb{I}$-adic semi-cuspidal eigenform which is congruent to $\mathbb{F}'$ modulo $\mathfrak{I}^l$. However, since N is arbitrary, the rank of the space of $\mathbb{I}$-adic Eisenstein series is large and the space of semi-cuspforms is not stable by the action of Hecke operators in general. So it is not possible to follow exactly the same strategy as the Ribet's proof. The third important idea of Wiles is to study carefully the action of Hecke operators on the space of $\mathbb{I}$-adic Eisenstein series. We denote by $\mathcal{E}_{Np^\infty}(\mathbb{I})$ the space of $\mathbb{I}$-adic Eisenstein series with Neben character ψ. Then we have the decomposition as follows stable by the action of Hecke operators:

$$\mathcal{M}_{Np^\infty}(\mathbb{I}) = \mathcal{S}_{Np^\infty}(\mathbb{I}) \oplus \mathcal{E}_{Np^\infty}(\mathbb{I}).$$

The space $\mathcal{E}_{Np^\infty}(\mathbb{I})$ is spanned by forms $\mathbb{E}(\psi',\psi'')(q^m)$ labelled with Dirichlet characters ψ' and ψ'' satisfying $\psi = \psi'\psi''$ and m dividing $\frac{Np}{\mathrm{Cond}(\psi)}$ where $\mathbb{E}(\psi',\psi'')$ is the $\mathbb{I}$-adic Eisenstein series interpolating classical Eisenstein series $E_k(\psi',\psi'')$ when k varies. It is known that $\mathbb{E}(\psi',\psi'')$ is a semi-cuspform if ψ' is not primitive and $\mathbb{E}(\psi',\psi'')$ is annihilated by

$$\lim_{n\to\infty} T_p^{n!} T_N \prod_{l\mid N} (T_l^{\varphi(Np)} - (\gamma^{s(l)} l)^{\varphi(Np)})$$

[57]For any integer $k \geq 2$ and for any finite order character $\phi : \Gamma_{\mathrm{cyc}} \to \overline{\mathbb{Q}}^\times$, the specialization of $\mathrm{Tw}_{\kappa_{\mathrm{cyc}}^{-1}\rho}(\mathbb{G}_\psi)$ by $\kappa_{\mathrm{cyc}}^{k-2}\phi$ is a modular form in $M_{k-1}(\Gamma_0(Np^*), \psi\omega^{3-k}\rho\phi; \mathbb{Z}_p[\psi\rho\phi])$. Multiplied by $G_{1,\omega^{-1}\rho^{-1}} \in M_{k-1}(\Gamma_0(Np^*), \omega^{-1}\rho^{-1}; \mathbb{Z}_p[\rho])$ the specialization of $\mathbb{G}^{(\rho)}_\psi$ by $\kappa_{\mathrm{cyc}}^{k-2}\phi$ becomes a modular form in $M_k(\Gamma_0(Np^*), \psi\omega^{2-k}\phi; \mathbb{Z}_p[\psi\phi])$.

if ψ' is nontrivial. By these facts, we have

$$\lim_{n\to\infty} T_p^{n!}T_N\prod_{l|N}(T_l^{\varphi(Np)}-(\gamma^{s(l)}l)^{\varphi(Np)})\mathbb{F}' \in \mathcal{S}_{Np^\infty}((\Lambda_{\mathrm{cyc},\psi})_{\mathfrak{J}}). \tag{3.93}$$

We denote by $\mathbb{F}'' \in \mathcal{S}_{Np^\infty}((\Lambda_{\mathrm{cyc},\psi})_{\mathfrak{J}})$ the normalized cuspform associated to the form of (3.93). We have $\mathbb{F}'' \equiv \mathbb{F}' \equiv \mathbb{G}_\psi \bmod \mathfrak{J}^l$. There exists a local domain $\mathbb{I}$ which is finite flat over Λ_{cyc}, an $\mathbb{I}$-adic cuspidal eigenform $\mathbb{F} \in \mathcal{S}_{Np^\infty}(\mathbb{I})$ and a height-one prime ideal $\widetilde{\mathfrak{J}} \subset \mathbb{I}$ over the height-one prime ideal $\mathfrak{J} \subset \Lambda_{\mathrm{cyc},\psi}$ such that $\mathbb{F} \equiv \mathbb{F}''$ modulo $\widetilde{\mathfrak{J}}^l$, which is the desired cuspform $\mathbb{F}$. □

(Step II) Let $\rho_{\mathbb{F}} : G_{\mathbb{Q}} \longrightarrow GL_2(\mathbb{K})$ be the irreducible Galois representation associated to $\mathbb{F}$ and let us denote by $\mathbb{V}_{\mathbb{F}} \cong \mathbb{K}^{\oplus 2}$ the representation space of $\rho_{\mathbb{F}}$. We denote by $\mathbb{I}_{\widetilde{\mathfrak{J}}}$ the discrete valuation ring obtained by localizing $\mathbb{I}$ at the height-one prime $\widetilde{\mathfrak{J}}$[58]. For a character $\widetilde{\rho} : G_{\mathbb{Q}} \longrightarrow (\mathbb{I}_{\widetilde{\mathfrak{J}}})^\times$, we denote by $(\mathbb{I}_{\widetilde{\mathfrak{J}}}/\widetilde{\mathfrak{J}})(\widetilde{\rho})$ a simple $(\mathbb{I}_{\widetilde{\mathfrak{J}}}/\widetilde{\mathfrak{J}})$-module of rank one over $(\mathbb{I}_{\widetilde{\mathfrak{J}}}/\widetilde{\mathfrak{J}})$ on which $G_{\mathbb{Q}}$ acts by $\widetilde{\rho}$ and we call $(\mathbb{I}_{\widetilde{\mathfrak{J}}}/\widetilde{\mathfrak{J}})(\widetilde{\rho})$ a $(\mathbb{I}_{\widetilde{\mathfrak{J}}}/\widetilde{\mathfrak{J}})[G_{\mathbb{Q}}]$-module of type $\widetilde{\rho}$. If we denote by $(\mathbb{V}_{\mathbb{F}})^*$ the $\mathbb{K}$-linear dual of $\mathbb{V}_{\mathbb{F}}$, there exists a finitely generated $\mathbb{I}_{\widetilde{\mathfrak{J}}}$-submodule $\mathbb{T}_{\widetilde{\mathfrak{J}}} \subset (\mathbb{V}_{\mathbb{F}})^*$ which is stable by the action of $G_{\mathbb{Q}}$ satisfying the following properties:

(1) The $\mathbb{I}_{\widetilde{\mathfrak{J}}}$-module $\mathbb{T}_{\widetilde{\mathfrak{J}}}$ is free of rank two.
(2) We have a $G_{\mathbb{Q}}$-stable extension:

$$0 \longrightarrow A \longrightarrow \mathbb{T}_{\widetilde{\mathfrak{J}}}/\widetilde{\mathfrak{J}}^l\mathbb{T}_{\widetilde{\mathfrak{J}}} \longrightarrow B \longrightarrow 0 \tag{3.94}$$

satisfying the following conditions:
 (i) We have $A \cong B \cong \mathbb{I}_{\widetilde{\mathfrak{J}}}/\widetilde{\mathfrak{J}}^l$ as $\mathbb{I}_{\widetilde{\mathfrak{J}}}$-modules.
 (ii) Every simple subquotient of A (resp. B) as an $\mathbb{I}_{\widetilde{\mathfrak{J}}}[G_{\mathbb{Q}}]$-module is of type $\mathbf{1}$ (resp. type $\widetilde{\eta}$) where $\widetilde{\eta} := \widetilde{\kappa}\kappa_{\mathrm{cyc}}\psi_{\mathbb{F}}\omega$.

(3) The module $\mathbb{T}_{\widetilde{\mathfrak{J}}}$ has no quotient of type $\mathbf{1}$ as an $\mathbb{I}_{\widetilde{\mathfrak{J}}}[G_{\mathbb{Q}}]$-module.

PROOF. Let us state a brief outline of the proof. The existence of a $G_{\mathbb{Q}}$-stable submodule $\mathbb{T}_{\widetilde{\mathfrak{J}}} \subset (\mathbb{V}_{\mathbb{F}})^*$ which satisfies the assertion (1) follows from the continuity of the representation $\rho_{\mathbb{F}}$ which was explained earlier. For any prime ℓ not dividing Np, we have the following congruence modulo $\widetilde{\mathfrak{J}}$:

$$\begin{aligned}\text{the trace of } \mathrm{Frob}_\ell \text{ acting on } \mathbb{T}_{\widetilde{\mathfrak{J}}}/\widetilde{\mathfrak{J}}\mathbb{T}_{\widetilde{\mathfrak{J}}} &\equiv A_\ell(\mathbb{F}) \\ &\equiv A_\ell(\mathbb{G}_\psi) \equiv \text{the trace of } \mathrm{Frob}_\ell \text{ acting on} (\mathbb{I}_{\widetilde{\mathfrak{J}}}/\widetilde{\mathfrak{J}})(\mathbf{1}) \oplus (\mathbb{I}_{\widetilde{\mathfrak{J}}}/\widetilde{\mathfrak{J}})(\widetilde{\eta}).\end{aligned}$$

Hence, the semi-simplification $(\mathbb{T}_{\widetilde{\mathfrak{J}}}/\widetilde{\mathfrak{J}}\mathbb{T}_{\widetilde{\mathfrak{J}}})^{\mathrm{ss}}$ of the $\mathbb{I}_{\widetilde{\mathfrak{J}}}[G_{\mathbb{Q}}]$-module $\mathbb{T}_{\widetilde{\mathfrak{J}}}/\widetilde{\mathfrak{J}}\mathbb{T}_{\widetilde{\mathfrak{J}}}$ is isomorphic to $(\mathbb{I}_{\widetilde{\mathfrak{J}}}/\widetilde{\mathfrak{J}})(\mathbf{1}) \oplus (\mathbb{I}_{\widetilde{\mathfrak{J}}}/\widetilde{\mathfrak{J}})(\widetilde{\eta})$ by the Chebotarev density theorem. By an argument similar to Step 2a of the Ribet's proof, we can replace the $\mathbb{I}_{\widetilde{\mathfrak{J}}}$-lattice $\mathbb{T}_{\widetilde{\mathfrak{J}}}$ by another lattice so that the assertion (2) is satisfied. Since $\rho_{\mathbb{F}}$ is irreducible, by the argument of Ribet's lemma similar to the one developed in Step 2b of the proof of Ribet, we can replace the lattice $\mathbb{T}_{\widetilde{\mathfrak{J}}}$ by another lattice satisfying the property (3) without spoiling the property (2). □

[58]When $\mathbb{I}$ is not normal, we need to replace $\mathbb{I}$ by its integral closure in $\mathbb{K}$ and $\widetilde{\mathfrak{J}}$ by a height-one prime which is over $\widetilde{\mathfrak{J}}$.

(Step III) The sequence (3.94) is split as an extension of $G_{\mathbb{Q}_p^{\mathrm{ur}}}$-modules where $\mathbb{Q}_p^{\mathrm{ur}}$ is the maximal unramified extension of $\mathbb{Q}_p$.

PROOF. Recall the following result due to Wiles (see [**122**, Thm. 2.2] and [**123**, Thm. 2]) which is a generalization of Step 3 of the proof of Ribet [59].

THEOREM 3.78. *Let $\mathbb{V}_{\mathbb{F}} \cong \mathbb{K}^{\oplus 2}$ be the representation space of the representation $\rho_{\mathbb{F}}$ of $G_{\mathbb{Q}}$. Then the restriction $\rho_{\mathbb{F}}|_{G_{\mathbb{Q}_p}}$ to $G_{\mathbb{Q}_p}$ has the following extension:*

$$0 \longrightarrow \mathbb{K}(\widetilde{\alpha}) \longrightarrow \mathbb{V}_{\mathbb{F}} \longrightarrow \mathbb{K}(\widetilde{\alpha}^{-1}\widetilde{\eta}^{-1}) \longrightarrow 0.$$

Here, $\widetilde{\alpha} : G_{\mathbb{Q}_p} \longrightarrow \mathbb{I}^{\times}$ is an unramified character such that $\widetilde{\alpha}(\mathrm{Frob}_p) = A_p(\mathbb{F})$ where Frob_p denotes the geometric Frobenius element.

Recall that $\mathbb{T}_{\widetilde{\mathfrak{J}}}$ is a lattice of $(\mathbb{V}_{\mathbb{F}})^*$ and note that we have a surjection $(\mathbb{V}_{\mathbb{F}})^* \twoheadrightarrow \mathbb{K}(\widetilde{\alpha}^{-1})$ by taking the $\mathbb{K}$-linear dual of the exact sequence given in Theorem 3.78. We define a $\mathbb{I}_{\widetilde{\mathfrak{J}}}[G_{\mathbb{Q}_p}]$-module $\mathrm{Fil}^{+}\mathbb{T}_{\widetilde{\mathfrak{J}}}$ to be

$$\mathrm{Fil}^{+}\mathbb{T}_{\widetilde{\mathfrak{J}}} := \mathrm{Ker}[\mathbb{T}_{\widetilde{\mathfrak{J}}} \hookrightarrow (\mathbb{V}_{\mathbb{F}})^* \twoheadrightarrow \mathbb{K}(\widetilde{\alpha}^{-1})].$$

Let us set $\mathrm{Fil}^{-}\mathbb{T}_{\widetilde{\mathfrak{J}}} := \mathbb{T}_{\widetilde{\mathfrak{J}}}/\mathrm{Fil}^{+}\mathbb{T}_{\widetilde{\mathfrak{J}}}$. Then $\mathrm{Fil}^{-}\mathbb{T}_{\widetilde{\mathfrak{J}}}$ is isomorphic to $\mathrm{Im}[\mathbb{T}_{\widetilde{\mathfrak{J}}} \hookrightarrow (\mathbb{V}_{\mathbb{F}})^* \twoheadrightarrow \mathbb{K}(\widetilde{\alpha}^{-1})]$. Especially, the module $\mathrm{Fil}^{-}\mathbb{T}_{\widetilde{\mathfrak{J}}}$ is nonzero. By applying the right exact functor $\otimes_{\mathbb{I}_{\widetilde{\mathfrak{J}}}} \mathbb{I}_{\widetilde{\mathfrak{J}}}/\widetilde{\mathfrak{J}}$ to the short exact sequence:

$$0 \longrightarrow \mathrm{Fil}^{+}\mathbb{T}_{\widetilde{\mathfrak{J}}} \longrightarrow \mathbb{T}_{\widetilde{\mathfrak{J}}} \longrightarrow \mathrm{Fil}^{-}\mathbb{T}_{\widetilde{\mathfrak{J}}} \longrightarrow 0,$$

we obtain the exact sequence:

$$\mathrm{Fil}^{+}\mathbb{T}_{\widetilde{\mathfrak{J}}}/\widetilde{\mathfrak{J}}\mathrm{Fil}^{+}\mathbb{T}_{\widetilde{\mathfrak{J}}} \longrightarrow \mathbb{T}_{\widetilde{\mathfrak{J}}}/\widetilde{\mathfrak{J}}\mathbb{T}_{\widetilde{\mathfrak{J}}} \longrightarrow \mathrm{Fil}^{-}\mathbb{T}_{\widetilde{\mathfrak{J}}}/\widetilde{\mathfrak{J}}\mathrm{Fil}^{-}\mathbb{T}_{\widetilde{\mathfrak{J}}} \longrightarrow 0. \tag{3.95}$$

Since $\mathbb{T}_{\widetilde{\mathfrak{J}}}$ is a free module of rank two over the discrete valuation ring $\mathbb{I}_{\widetilde{\mathfrak{J}}}$, the middle term $\mathbb{T}_{\widetilde{\mathfrak{J}}}/\widetilde{\mathfrak{J}}\mathbb{T}_{\widetilde{\mathfrak{J}}}$ of (3.95) is an $\mathbb{I}_{\widetilde{\mathfrak{J}}}/\widetilde{\mathfrak{J}}$-vector space of dimension two. Hence, the dimension of the $\mathbb{I}_{\widetilde{\mathfrak{J}}}/\widetilde{\mathfrak{J}}$-vector space $\mathrm{Fil}^{-}\mathbb{T}_{\widetilde{\mathfrak{J}}}/\widetilde{\mathfrak{J}}\mathrm{Fil}^{-}\mathbb{T}_{\widetilde{\mathfrak{J}}}$ is equal to or smaller than two. Since $\mathrm{Fil}^{-}\mathbb{T}_{\widetilde{\mathfrak{J}}}$ is nonzero, the dimension of $\mathrm{Fil}^{-}\mathbb{T}_{\widetilde{\mathfrak{J}}}/\widetilde{\mathfrak{J}}\mathrm{Fil}^{-}\mathbb{T}_{\widetilde{\mathfrak{J}}}$ is nonzero. If the dimension of $\mathrm{Fil}^{-}\mathbb{T}_{\widetilde{\mathfrak{J}}}/\widetilde{\mathfrak{J}}\mathrm{Fil}^{-}\mathbb{T}_{\widetilde{\mathfrak{J}}}$ is two, we have $\mathbb{T}_{\widetilde{\mathfrak{J}}}/\widetilde{\mathfrak{J}}\mathbb{T}_{\widetilde{\mathfrak{J}}} = \mathrm{Fil}^{-}\mathbb{T}_{\widetilde{\mathfrak{J}}}/\widetilde{\mathfrak{J}}\mathrm{Fil}^{-}\mathbb{T}_{\widetilde{\mathfrak{J}}}$, which implies that the action of $G_{\mathbb{Q}_p}$ on $\mathbb{T}_{\widetilde{\mathfrak{J}}}$ is unramified. Since this contradicts to the fact that the action of $G_{\mathbb{Q}_p}$ on $\mathbb{T}_{\widetilde{\mathfrak{J}}}$ is ramified, we proved that the dimension of $\mathrm{Fil}^{-}\mathbb{T}_{\widetilde{\mathfrak{J}}}/\widetilde{\mathfrak{J}}\mathrm{Fil}^{-}\mathbb{T}_{\widetilde{\mathfrak{J}}}$ is one. By Nakayama's lemma, $\mathrm{Fil}^{-}\mathbb{T}_{\widetilde{\mathfrak{J}}}$ is a cyclic $\mathbb{I}_{\widetilde{\mathfrak{J}}}$-module. Since the $\mathbb{I}_{\widetilde{\mathfrak{J}}}$-module $\mathrm{Fil}^{-}\mathbb{T}_{\widetilde{\mathfrak{J}}}$ is torsion-free, it is free of rank one over $\mathbb{I}_{\widetilde{\mathfrak{J}}}$. The free-ness of $\mathrm{Fil}^{-}\mathbb{T}_{\widetilde{\mathfrak{J}}}$ implies that the first map of the exact sequence (3.95) is injective. Hence, the dimension of $\mathrm{Fil}^{+}\mathbb{T}_{\widetilde{\mathfrak{J}}}/\widetilde{\mathfrak{J}}\mathrm{Fil}^{+}\mathbb{T}_{\widetilde{\mathfrak{J}}}$ is one and this implies that $\mathrm{Fil}^{+}\mathbb{T}_{\widetilde{\mathfrak{J}}}$ is free of rank one over $\mathbb{I}_{\widetilde{\mathfrak{J}}}$ by the same argument as above. We thus obtained the following short exact sequence of $\mathbb{I}_{\widetilde{\mathfrak{J}}}[G_{\mathbb{Q}_p}]$-modules

$$0 \longrightarrow (\mathbb{I}_{\widetilde{\mathfrak{J}}}/\widetilde{\mathfrak{J}}^{l})(\widetilde{\eta}\widetilde{\alpha}) \longrightarrow \mathbb{T}_{\widetilde{\mathfrak{J}}}/\widetilde{\mathfrak{J}}^{l}\mathbb{T}_{\widetilde{\mathfrak{J}}} \longrightarrow (\mathbb{I}_{\widetilde{\mathfrak{J}}}/\widetilde{\mathfrak{J}}^{l})(\widetilde{\alpha}^{-1}) \longrightarrow 0. \tag{3.96}$$

We finish the proof of Step III by (3.94) and (3.96). □

---

[59] The proof of Wiles generalizes the mod p finite flat group scheme $J_1(p)[p]$ associated to the Jacobian variety $J_1(p)$ of the modular curves $X_1(p)$ to the family of the mod p^n finite flat group schemes $J_1(Np^n)[p^n]$ associated to the Jacobian varieties $J_1(Np^n)$ of the modular curves $X_1(Np^n)$ for varying n. He obtain a generalization of the decomposition (3.86) for these finite flat group schemes.

(Step IV) Recall that the Galois cohomology group of degree one classifies extension classes of Galois modules. According to this basic fact, each extension class (3.94) determines an element of the following group:

$$\begin{aligned} H^1(\mathbb{Q}, \mathrm{Hom}_{\mathbb{I}_{\widetilde{\mathfrak{J}}}}(B,A)) &\cong H^1(\mathbb{Q}, (\mathbb{I}_{\widetilde{\mathfrak{J}}}/\widetilde{\mathfrak{J}}^l)(\widetilde{\eta}^{-1})) \\ &\cong H^1(\mathbb{Q}, (\Lambda_{\mathrm{cyc},\psi}/\mathfrak{J}^l)(\widetilde{\eta}^{-1})) \otimes_{\Lambda_{\mathrm{cyc},\psi}} \mathbb{I}. \end{aligned}$$

By Step III, an extension class of (3.94) is trivial as extension classes of $G_{\mathbb{Q}_p^{\mathrm{ur}}}$-modules. That is, the class restricted to

$$H^1(\mathbb{Q}_p^{\mathrm{ur}}, \mathrm{Hom}_{\mathbb{I}_{\widetilde{\mathfrak{J}}}}(B,A)) \cong H^1(\mathbb{Q}_p^{\mathrm{ur}}, (\Lambda_{\mathrm{cyc},\psi}/\mathfrak{J}^l)(\widetilde{\eta}^{-1})) \otimes_{\Lambda_{\mathrm{cyc},\psi}} \mathbb{I}$$

is zero.

For any prime ℓ not dividing the Np^∞, the Galois representation $\rho_{\mathbb{F}}$ is unramified at ℓ by the general property (5) of Hida theory explained before Step I. Hence, the extension class (3.94) restricted to

$$H^1(\mathbb{Q}_\ell^{\mathrm{ur}}, \mathrm{Hom}_{\mathbb{I}_{\widetilde{\mathfrak{J}}}}(B,A)) \cong H^1(\mathbb{Q}_\ell^{\mathrm{ur}}, (\Lambda_{\mathrm{cyc},\psi}/\mathfrak{J}^l)(\widetilde{\eta}^{-1})) \otimes_{\Lambda_{\mathrm{cyc},\psi}} \mathbb{I}$$

is zero for any prime ℓ not dividing the level Np^∞.

For primes ℓ dividing N, the $\widetilde{\mathfrak{J}}$-module

$$\mathrm{Im}[H^1(\mathbb{Q}, \mathrm{Hom}_{\mathbb{I}_{\widetilde{\mathfrak{J}}}}(B,A)) \longrightarrow H^1(\mathbb{Q}_\ell^{\mathrm{ur}}, \mathrm{Hom}_{\mathbb{I}_{\widetilde{\mathfrak{J}}}}(B,A))]$$

is zero if the height-one prime $\widetilde{\mathfrak{J}} \subset \mathbb{I}$ does not divide (p).

By the locally unramified property obtained above, for each height-one prime $\mathfrak{J} \subset \Lambda_{\mathrm{cyc},\psi}$ which does not divide (p), the extension class obtained through Step II and Step III induces a $\Lambda_{\mathrm{cyc},\psi}$-linear map

$$(X_{K^{\mathrm{cyc}}_{\omega\psi^{-1},\infty}})_{\omega\psi^{-1}} \otimes \kappa_{\mathrm{cyc}}^{-1} \longrightarrow \Lambda_{\mathrm{cyc},\psi}/\iota(\mathfrak{J})^l$$

whose cokernel is a pseudo-null $\Lambda_{\mathrm{cyc},\psi}$-module. By the definition of l given in (3.92), we have

$$\mathrm{ord}_{\iota(\mathfrak{J})} \mathrm{Tw}_{\kappa_{\mathrm{cyc}}^{-1}}(L_p(\psi)) \leq \mathrm{ord}_{\iota(\mathfrak{J})} \left(\mathrm{char}_{\Lambda_{\mathrm{cyc},\psi}}(X_{K^{\mathrm{cyc}}_{\omega\psi^{-1},\infty}})_{\omega\psi^{-1}} \otimes \kappa_{\mathrm{cyc}}^{-1}\right)$$

for any height-one prime $\mathfrak{J} \subset \Lambda_{\mathrm{cyc},\psi}$ which does not divide (p). Recall that, we have $\mathrm{ord}_{\iota(\mathfrak{J})} \mathrm{Tw}_{\kappa_{\mathrm{cyc}}^{-1}}(L_p(\psi)) = 0$ for height-one primes $\mathfrak{J} \subset \Lambda_{\mathrm{cyc},\psi}$ dividing (p) by the theorem of Ferrero–Washington (Theorem 3.60). Thus, we proved

$$\mathrm{char}_{\Lambda_{\mathrm{cyc},\psi}}(X_{K^{\mathrm{cyc}}_{\omega\psi^{-1},\infty}})_{\omega\psi^{-1}} \subset (L_p(\psi)) \tag{3.97}$$

for any even Dirichlet character $\psi \neq \mathbf{1}$ of conductor Np^e ($e = 0$ or $e = 1$). By the principle of the analytic class number formula (Theorem 3.74), (3.97) becomes an equality for each ψ and this completes the proof of Theorem 3.66.

3.3.3. Proof of the Iwasawa main conjecture by Euler system. As we remarked at the end of §3.3, there exists another proof of the cyclotomic Iwasawa main conjecture for class groups by means of the method of Euler systems. Historically, the method of Euler systems grew up from a technique invented by Kolyvagin as well as an idea of Thaine. In general, this method bounds the size of a Selmer group from above. Below, we will explain the proof by the method of Euler systems.

There is not so much knowledge which is newly required in this section. However, we have to be familiar with the theories on extensions of global fields and local fields more deeply. For example, the Chebotarev density theorem was already used in the previous section to characterize the semi-simplification of a given Galois representation, but we use the Chebotarev density theorem in a more essential manner in this section. We refer the reader to [**80**, Cap. VII, §13] for the density theorem. Also, we assume that the reader is familiar with the generalities on the structure of local fields and the basic of ramifications of the extension of local fields as found in Part one and Part two of [**108**].

Let us fix a natural number N satisfying $(N,p) = 1$ and an even Dirichlet character $\psi \neq \mathbf{1}$ of conductor Np^e ($e = 0$ or $e = 1$). We set $K = K_\psi \subset \mathbb{Q}(\mu_{Np})$ throughout §3.3.3.

PROPOSITION 3.79. *We have the following* $\mathrm{Gal}(K_\infty^{\mathrm{cyc}}/\mathbb{Q})$*-equivariant exact sequence:*

$$(3.98)\quad 0 \longrightarrow \varprojlim_n (\mathfrak{r}_{K_n^{\mathrm{cyc}}})^\times \otimes_{\mathbb{Z}} \mathbb{Z}_p \longrightarrow \varprojlim_n U_1(\mathfrak{r}_{K_n^{\mathrm{cyc}}} \otimes_{\mathbb{Z}} \mathbb{Z}_p) \longrightarrow \mathfrak{X}_{K_\infty^{\mathrm{cyc}}} \longrightarrow X_{K_\infty^{\mathrm{cyc}}} \longrightarrow 0,$$

where $U_1(\mathfrak{r}_{K_n^{\mathrm{cyc}}} \otimes_{\mathbb{Z}} \mathbb{Z}_p)$ *is the group of principal units (i.e. the subgroup of the group of units of the semi-local ring* $\mathfrak{r}_{K_n^{\mathrm{cyc}}} \otimes_{\mathbb{Z}} \mathbb{Z}_p$ *which consists of units congruent to* 1 *modulo every maximal ideal of* $\mathfrak{r}_{K_n^{\mathrm{cyc}}} \otimes_{\mathbb{Z}} \mathbb{Z}_p$*)*[60].

PROOF. By the class field theory, we have the following exact sequence:

$$(3.99)\quad 0 \longrightarrow (\mathfrak{r}_{K_n^{\mathrm{cyc}}})^\times \otimes_{\mathbb{Z}} \mathbb{Z}_p \longrightarrow U_1(\mathfrak{r}_{K_n^{\mathrm{cyc}}} \otimes_{\mathbb{Z}} \mathbb{Z}_p) \longrightarrow \mathrm{Gal}(M_n^{\mathrm{cyc}}/K_n^{\mathrm{cyc}}) \longrightarrow \mathrm{Gal}(L_n^{\mathrm{cyc}}/K_n^{\mathrm{cyc}}) \longrightarrow 0.$$

We do not give the proof of this here, but (3.99) is obtained by combining (2.7) and the injectivity of the first map $(\mathfrak{r}_{K_n^{\mathrm{cyc}}})^\times \otimes_{\mathbb{Z}} \mathbb{Z}_p \longrightarrow U_1(\mathfrak{r}_{K_n^{\mathrm{cyc}}} \otimes_{\mathbb{Z}} \mathbb{Z}_p)$ due to Brumer's theorem (a special case of Leopoldt conjecture). Note that K_n^{cyc} is an abelian extension of $\mathbb{Q}$, which is a sufficient condition for Leopoldt conjecture to be true (see Remark 2.8 (1)). By definition, we have $\mathfrak{X}_{K_\infty^{\mathrm{cyc}}} = \varprojlim_n \mathrm{Gal}(M_n^{\mathrm{cyc}}/K_n^{\mathrm{cyc}})$ and $X_{K_\infty^{\mathrm{cyc}}} = \varprojlim_n \mathrm{Gal}(L_n^{\mathrm{cyc}}/K_n^{\mathrm{cyc}})$. Since $\mathrm{Gal}(M_n^{\mathrm{cyc}}/K_n^{\mathrm{cyc}})$ and $\mathrm{Gal}(L_n^{\mathrm{cyc}}/K_n^{\mathrm{cyc}})$ are finite groups, we can check the Mittag-Leffler condition, which assures that the sequence (3.98) obtained by taking the projective limit of (3.99) remains to be exact. This completes the proof. □

Let us define the following set[61]

$$\mathfrak{R}_\psi = \left\{ r \in \mathbb{Z}_{\geq 1} \,\middle|\, r \text{ is square-free and relatively prime to } 2Np \right\}.$$

DEFINITION 3.80. Let K be a finite algebraic number field. We denote by $\{z_{n,r}\}_{(n,r)\in\mathbb{Z}_{\geq 1}\times\mathfrak{R}_\psi}$ a set which consists of $z_{n,r} \in \left((\mathfrak{r}_{K_n^{\mathrm{cyc}}(\mu_r)})^\times \otimes_{\mathbb{Z}} \mathbb{Z}_p\right)_\psi$ parametrized by $(n,r) \in \mathbb{Z}_{\geq 1} \times \mathfrak{R}_\psi$. We call such a set $\{z_{n,r}\}_{(n,r)\in\mathbb{Z}_{\geq 1}\times\mathfrak{R}_\psi}$ an **Euler system** (of

[60]In other wards, $U_1(\mathfrak{r}_{K_n^{\mathrm{cyc}}} \otimes_{\mathbb{Z}} \mathbb{Z}_p)$ is the (pro-)p-Sylow subgroup of the profinite group $(\mathfrak{r}_{K_n^{\mathrm{cyc}}} \otimes_{\mathbb{Z}} \mathbb{Z}_p)^\times$.

[61]There appears no ψ explicitly in the definition of $\mathfrak{R}_\psi$ though the notation depends on ψ. However, the set $\mathfrak{R}_\psi$ depends implicitly on ψ since N is the prime-to-p part of the conductor of ψ.

units) if it satisfies the following conditions[62]:

(1) $\mathrm{Nr}_{K_{n+1}^{\mathrm{cyc}}(\mu_r)/K_n^{\mathrm{cyc}}(\mu_r)}(z_{n+1,r}) = z_{n,r}$.

(2) $\mathrm{Nr}_{K_n^{\mathrm{cyc}}(\mu_{r\ell})/K_n^{\mathrm{cyc}}(\mu_r)}(z_{n,r\ell}) = (\mathrm{Frob}_\ell^{-1} - 1) z_{n,r}$ for every prime $\ell \in \mathfrak{R}_\psi$ such that $(\ell, r) = 1$.

Here, $\mathrm{Frob}_\ell^{-1} \in \mathrm{Gal}(K_n^{\mathrm{cyc}}(\mu_r)/\mathbb{Q})$ is an ℓ-th power (arithmetic) Frobenius element[63] and $\mathrm{Nr}_{K_n^{\mathrm{cyc}}(\mu_{r\ell})/K_n^{\mathrm{cyc}}(\mu_r)}$ denotes the norm map.

The following proposition gives a nontrivial example of Euler system of units called an Euler system of cyclotomic units.

PROPOSITION 3.81. *Let $N = 1$ (resp. $N > 1$) and let K be a subfield of $\mathbb{Q}(\mu_{Np})$. For any nonnegative integer n and for any $r \in \mathfrak{R}_\psi$, we set*

$$z'_{n,r} = \frac{\zeta_{p^{n+1}}^a \zeta_r^{\sigma^{-n}} - 1}{\zeta_{p^{n+1}} \zeta_r^{\sigma^{-n}} - 1} \in \mathbb{Z}[\mu_{p^{n+1}r}]^\times$$

$$\left(\text{resp. } z'_{n,r} = \zeta_N^{\sigma^{-n}} \zeta_{p^{n+1}} \zeta_r^{\sigma^{-n}} - 1 \in \mathbb{Z}[\mu_{Np^{n+1}r}]^\times\right)$$

and we define $z_{n,r}$ to be the ψ-part of the projection to pro-p part of the element $\mathrm{Nr}_{\mathbb{Q}(\mu_{Np^{n+1}r})/K_n^{\mathrm{cyc}}(\mu_r)}(z'_{n,r})$. Then the set $\{z_{n,r}\}_{(n,r) \in \mathbb{Z}_{\geq 1} \times \mathfrak{R}_\psi}$ is an Euler system in the sense of Definition 3.80.

We note that, for any nonnegative integer n, we have

$$\mathrm{loc}_p(z_{n,1}) = \left(N_{\mathbb{Q}(\mu_{p^n})/K_n^{\mathrm{cyc}} \cap \mathbb{Q}(\mu_{p^n})}(\widehat{u_n(a)})\right)_\psi$$

$$\left(\text{resp. } \mathrm{loc}_p(z_{n,1}) = \left(N_{\mathbb{Q}(\mu_{Np^n})/K_n^{\mathrm{cyc}} \cap \mathbb{Q}(\mu_{Np^n})}(\widehat{u_{(N),n}})\right)_\psi\right)$$

where loc_p is the natural map from the group of global units $(\mathfrak{r}_{K_n^{\mathrm{cyc}}})^\times$ to the group of local units $(\mathfrak{r}_{K_n^{\mathrm{cyc}}} \otimes_{\mathbb{Z}} \mathbb{Z}_p)^\times$ and $\widehat{u_n(a)}$ (resp. $\widehat{u_{(N),n}}$) is the pro-p part of local cyclotomic units which appeared in Lemma 3.48.

PROOF. The argument for the case $N = 1$ and the argument for the case $N > 1$ is quite similar. Hence, we give the proof only for the case $N > 1$. In order to check the norm condition for $z_{n,r}$, it suffices to check the norm condition for $z'_{n,r}$.

Let us take any nonnegative integer n. Then we have

$$\begin{aligned} \mathrm{Nr}_{\mathbb{Q}(\mu_{Np^{n+2}r})/\mathbb{Q}(\mu_{Np^{n+1}r})}(z'_{n+1,r}) &= \prod_{i=1}^{p-1} \left(\zeta_N^{\sigma^{-(n+1)}} \zeta_{p^{n+2}} \zeta_r^{\sigma^{-(n+1)}} \zeta_p^i - 1\right) \\ &= (\zeta_N^{\sigma^{-(n+1)}} \zeta_{p^{n+2}} \zeta_r^{\sigma^{-(n+1)}})^p - 1. \end{aligned}$$

[62]Of course, it is not Euler who invented Euler systems. The name "Euler system" comes from the fact that the action of the ℓ-th power Frobenius element appearing in the condition (2) is similar to the ℓ-Euler factor of the zeta function (In fact, it is really related to the Euler factor of the zeta-function).

[63]In arithmetic geometry, it is sometimes important to distinguish arithmetic Frobenius elements and geometric Frobenius elements which are opposite to each other. In later chapters of this book, it is rather ℓ-th power geometric Frobenius elements which play important roles (see a footnote to Theorem A.1 for the explanation of geometric Frobenius elements). Since we denote an ℓ-th power geometric Frobenius element by Frob_ℓ later, we denote an ℓ-th power arithmetic Frobenius element by Frob_ℓ^{-1} here

This gives

$$\mathrm{Nr}_{\mathbb{Q}(\mu_{Np^{n+2}r})/\mathbb{Q}(\mu_{Np^{n+1}r})}(z'_{n+1,r}) = z'_{n,r}. \tag{3.100}$$

Let us take an element $r\ell \in \mathfrak{R}_\psi$ such that ℓ is a prime. Then we have

$$\begin{aligned}\mathrm{Nr}_{\mathbb{Q}(\mu_{Np^{n+1}r\ell})/\mathbb{Q}(\mu_{Np^{n+1}r})}(z'_{n,r\ell}) &= \prod_{i=1}^{\ell-1}\left(\zeta_N^{\sigma^{-n}}\zeta_{p^{n+1}}\zeta_{r\ell}^{\sigma^{-n}}\zeta_\ell^i - 1\right)\\ &= \frac{\prod_{i=0}^{\ell-1}\left(\zeta_N^{\sigma^{-n}}\zeta_{p^{n+1}}\zeta_{r\ell}^{\sigma^{-n}}\zeta_\ell^i - 1\right)}{\zeta_N^{\sigma^{-n}}\zeta_{p^{n+1}}\zeta_{r\ell}^{\sigma^{-n}} - 1}\\ &= \frac{\prod_{i=0}^{\ell-1}\left(\zeta_N^{\sigma^{-n}}\zeta_{p^{n+1}}\zeta_{r\ell}^{\sigma^{-n}}\zeta_\ell^i - 1\right)}{z'_{n,r}}.\end{aligned}$$

The numerator of the last term is equal to

$$(\zeta_N^{\sigma^{-n}}\zeta_{p^{n+1}}\zeta_{r\ell}^{\sigma^{-n}})^\ell - 1 = \mathrm{Frob}_\ell^{-1}((\zeta_N)^{\sigma^{-n}}\zeta_{p^{n+1}}\zeta_r^{\sigma^{-n}} - 1) = \mathrm{Frob}_\ell^{-1}(z'_{n,r}).$$

Thus we obtained

$$\mathrm{Nr}_{\mathbb{Q}(\mu_{Np^{n+1}r\ell})/\mathbb{Q}(\mu_{Np^{n+1}r})}(z'_{n,r\ell}) = (\mathrm{Frob}_\ell^{-1} - 1)(z'_{n,r}). \tag{3.101}$$

Note that, $K_n^{\mathrm{cyc}}(\mu_r)$ is a subfield of $\mathbb{Q}(\mu_{Np^{n+1}r})$ since K is a subfield of $\mathbb{Q}(\mu_{Np})$. By defining $z_{n,r}$ to be the ψ-part of the projection to pro-p part of the element $\mathrm{Nr}_{\mathbb{Q}(\mu_{Np^{n+1}r})/K_n^{\mathrm{cyc}}(\mu_r)}(z'_{n,r})$, the result (3.100) (resp. (3.101)) implies te relation (1) (resp. (2)) of Definition 3.80. This completes the proof. □

Rubin and Greither proved the main result of the theory of the Euler system of units as follows (see Remark 3.68 for historical comments on the proof of the cyclotomic Iwasawa main conjecture for ideal class groups):

THEOREM 3.82 (Rubin ([**69**], [**101**]), Greither [**34**]). *Let ψ be an even Dirichlet character such that $\psi \neq \mathbf{1}$. Then, for an Euler system of units $\{z_{n,r}\}_{(n,r)\in\mathbb{Z}_{\geq 1}\times\mathfrak{R}_\psi}$ in the sense of Definition* 3.80*, there exists a nonnegative integer α such that we have:*

$$\begin{aligned}&\mathrm{char}_{\Lambda_{\mathrm{cyc},\psi}}(X_{K_\infty^{\mathrm{cyc}}})_\psi \\ &\quad\supset (p^\alpha)\mathrm{char}_{\Lambda_{\mathrm{cyc},\psi}}\left(\varprojlim_n\left((\mathfrak{r}_{K_n^{\mathrm{cyc}}})^\times\otimes_{\mathbb{Z}}\mathbb{Z}_p\right)\Big/\varprojlim_n z_{n,1}\Lambda_{\mathrm{cyc},\psi}\right)_\psi.\end{aligned} \tag{3.102}$$

Further, when the order of ψ is prime to p, we can take α to be zero.

REMARK 3.83. Let $\{z_{n,r}\}_{(n,r)\in\mathbb{Z}_{\geq 1}\times\mathfrak{R}_\psi}$ be an Euler system of units as defined in Definition 3.80. If we divide the exact sequence obtained in Proposition 3.79 by submodules generated by the first layers of Euler system and take the ψ-covariant quotient, we obtain the following exact sequence:

$$(3.103)\quad 0 \longrightarrow \left(\varprojlim_n \left((\mathfrak{r}_{K_n^{\mathrm{cyc}}})^\times \otimes_{\mathbb{Z}} \mathbb{Z}_p \right) \Big/ \varprojlim_n z_{n,1}\Lambda_{\mathrm{cyc},\psi} \right)_\psi \otimes_{\mathbb{Z}_p} \mathbb{Q}_p$$
$$\longrightarrow \left(\varprojlim_n U_1(\mathfrak{r}_{K_n^{\mathrm{cyc}}} \otimes_{\mathbb{Z}} \mathbb{Z}_p) \Big/ \varprojlim_n \mathrm{loc}_p(z_{n,1})\Lambda_{\mathrm{cyc},\psi} \right)_\psi \otimes_{\mathbb{Z}_p} \mathbb{Q}_p$$
$$\longrightarrow \left(\mathfrak{X}_{K_\infty^{\mathrm{cyc}}}\right)_\psi \otimes_{\mathbb{Z}_p} \mathbb{Q}_p \longrightarrow \left(X_{K_\infty^{\mathrm{cyc}}}\right)_\psi \otimes_{\mathbb{Z}_p} \mathbb{Q}_p \longrightarrow 0,$$

where the last term $\left(X_{K_\infty^{\mathrm{cyc}}}\right)_\psi \otimes_{\mathbb{Z}_p} \mathbb{Q}_p$ is a torsion $\Lambda_{\mathrm{cyc},\psi} \otimes_{\mathbb{Z}_p} \mathbb{Q}_p$-module thanks to Theorem 3.4 and the third term $\left(\mathfrak{X}_{K_\infty^{\mathrm{cyc}}}\right)_\psi \otimes_{\mathbb{Z}_p} \mathbb{Q}_p$ is a torsion $\Lambda_{\mathrm{cyc},\psi} \otimes_{\mathbb{Z}_p} \mathbb{Q}_p$-module thanks to Theorem 3.69 and the assumption that ψ is even. The first two terms of (3.103) are torsion $\Lambda_{\mathrm{cyc},\psi} \otimes_{\mathbb{Z}_p} \mathbb{Q}_p$-modules if and only if $\varprojlim_n z_{n,1}$ is a nontorsion element of $\Lambda_{\mathrm{cyc},\psi}$.

(1) When we take the Euler system $\{z_{n,r}\}_{(n,r)\in\mathbb{Z}_{\geq 1}\times\mathfrak{R}_\psi}$ in (3.103) to be the Euler system of cyclotomic units $\{z_{n,r}^{\mathrm{cyc},\psi}\}_{(n,r)\in\mathbb{Z}_{\geq 1}\times\mathfrak{R}_\psi}$ obtained in Proposition 3.81, the first two terms of (3.103) are torsion $\Lambda_{\mathrm{cyc},\psi} \otimes_{\mathbb{Z}_p} \mathbb{Q}_p$-modules.
(2) When the order of the character ψ is prime to p, the sequence (3.103) is exact even without taking $\otimes_{\mathbb{Z}_p}\mathbb{Q}_p$[64].
(3) By the Ferrero–Washington theorem (Theorem 3.60) and the principle of the analytic class number formula (Theorem 3.74), the μ-invariants of all four terms in (3.103) before taking $\otimes_{\mathbb{Z}_p}\mathbb{Q}_p$ are zero.

Before proving Theorem 3.82, we will prove that Theorem 3.82 implies (+)-type Iwasawa main conjecture (Theorem 3.67).

PROOF OF THEOREM 3.82⇒THEOREM 3.67. If we prove

$$(3.104)\qquad \mathrm{char}_{\Lambda_{\mathrm{cyc},\psi}} \left((\mathfrak{X}_{K_\infty^{\mathrm{cyc}}})_\psi \right)^\bullet \otimes \kappa_{\mathrm{cyc}} \supset (L_p(\psi))$$

for all even Dirichlet characters $\psi \neq \mathbf{1}$ whose conductor divide Np, it follows by the principle of the analytic class number formula (Theorem 3.74), that we have $\mathrm{char}_{\Lambda_{\mathrm{cyc},\psi}} \left((\mathfrak{X}_{K_\infty^{\mathrm{cyc}}})_\psi \right)^\bullet \otimes \kappa_{\mathrm{cyc}} = (L_p(\psi))$ for all even Dirichlet characters $\psi \neq \mathbf{1}$ whose conductor divide Np[65]. By Remark 3.83 (3), the proof of (3.104) is reduced to showing

$$(3.105)\quad \mathrm{length}_{(\Lambda_{\mathrm{cyc},\psi})_{\mathfrak{I}}} \left(\left(\left((\mathfrak{X}_{K_\infty^{\mathrm{cyc}}})_\psi \right)^\bullet \otimes \kappa_{\mathrm{cyc}} \right) \otimes_{\Lambda_{\mathrm{cyc},\psi}} (\Lambda_{\mathrm{cyc},\psi})_{\mathfrak{I}} \right) \leq \mathrm{ord}_{\mathfrak{I}}(L_p(\psi))$$

for every height one prime $\mathfrak{I}$ of $\Lambda_{\mathrm{cyc},\psi}$ which does not divide (p). By the construction of p-adic L-function à la Coleman using cyclotomic units (see (3.2.4)), we have

$$(3.106)\qquad \mathrm{ord}_{\mathfrak{I}}(L_p(\psi)) = \mathrm{length}_{(\Lambda_{\mathrm{cyc},\psi})_{\mathfrak{I}}} \left((\mathcal{U}/\mathcal{C})_\psi^\bullet \otimes \kappa_{\mathrm{cyc}} \right) \otimes_{\Lambda_{\mathrm{cyc},\psi}} (\Lambda_{\mathrm{cyc},\psi})_{\mathfrak{I}},$$

where $\mathcal{U} = \varprojlim_n U_1(\mathfrak{r}_{K_n^{\mathrm{cyc}}} \otimes_{\mathbb{Z}} \mathbb{Z}_p)$ and $\mathcal{C} = \varprojlim_n \mathrm{loc}_p(z_{n,1}^{\mathrm{cyc},\psi})\Lambda_{\mathrm{cyc},\psi}$. Note that the localization functor $\otimes_{\Lambda_{\mathrm{cyc},\psi}\otimes_{\mathbb{Z}_p}\mathbb{Q}_p} (\Lambda_{\mathrm{cyc},\psi})_{\mathfrak{I}}$ preserves the exactness of (3.103) and that the alternating some of the lengths of terms in a short exact sequence of

[64]See comments given after Definition 3.64 for some facts on the ψ-part and the ψ-quotient for ψ whose order is prime to p.

[65]By the discussion given in Remark 3.33 (3) and by Stickelberger's theorem (Theorem 3.94), which will be presented later, Theorem 3.67 holds trivially true for $\psi = \mathbf{1}$. Hence, we can exclude the case $\psi = \mathbf{1}$ in this proof.

torsion $(\Lambda_{\mathrm{cyc},\psi})_{\mathfrak{J}}$-modules must be zero. By these facts and by (3.106), the validity of (3.105) is equivalent to the validity of the following inequality

$$\begin{aligned}(3.107)\quad &\mathrm{length}_{(\Lambda_{\mathrm{cyc},\psi})_{\mathfrak{J}}}(X_{K_\infty^{\mathrm{cyc}}})_\psi \otimes_{\Lambda_{\mathrm{cyc},\psi}} (\Lambda_{\mathrm{cyc},\psi})_{\mathfrak{J}} \\ &\leq \mathrm{length}_{(\Lambda_{\mathrm{cyc},\psi})_{\mathfrak{J}}} \left(\varprojlim_n \left((\mathfrak{r}_{K_n^{\mathrm{cyc}}})^\times \otimes_{\mathbb{Z}} \mathbb{Z}_p \right) \Big/ \varprojlim_n z_{n,1}^{\mathrm{cyc},\psi} \Lambda_{\mathrm{cyc},\psi} \right)_\psi \otimes_{\Lambda_{\mathrm{cyc},\psi}} (\Lambda_{\mathrm{cyc},\psi})_{\mathfrak{J}}\end{aligned}$$

for every height one prime $\mathfrak{J}$ of $\Lambda_{\mathrm{cyc},\psi}$ which does not divide (p). Since (3.107)) follows immediately by applying Theorem 3.82 to the Euler system $\{z_{n,r}\} = \{z_{n,r}^{\mathrm{cyc},\psi}\}$, we complete the proof. □

Let us prepare some notation and some results which will be necessary for the proof of Theorem 3.82. From now on, we fix a natural number m. For each $r \in \mathfrak{R}_\psi$, we consider the following condition (CD_m) :

(CD_m) Every prime factor ℓ of r is completely decomposed in the extension $K_m^{\mathrm{cyc}}(\mu_p)/\mathbb{Q}$.

If a natural number r satisfies the condition (CD_m), every prime factor ℓ of r is completely decomposed in the extension $\mathbb{Q}(\mu_{p^{m+1}})/\mathbb{Q}$ since $\mathbb{Q}(\mu_{p^{m+1}})$ is a subfield of $K_m^{\mathrm{cyc}}(\mu_p)$. Hence, every prime factor ℓ of r satisfies $\ell \equiv 1 \bmod p^{m+1}$ if a natural number r satisfies the condition (CD_m).

For every natural number n, we consider the following isomorphism

$$\begin{aligned}(3.108)\quad \mathrm{res}_{n,r,m} : H^1(K_n^{\mathrm{cyc}}, \mu_{p^{m+1}}) &= (K_n^{\mathrm{cyc}})^\times/((K_n^{\mathrm{cyc}})^\times)^{p^{m+1}} \\ &\xrightarrow{\sim} \left((K_n^{\mathrm{cyc}}(\mu_r))^\times/((K_n^{\mathrm{cyc}}(\mu_r))^\times)^{p^{m+1}} \right)^{\mathrm{Gal}(K_n^{\mathrm{cyc}}(\mu_r)/K_n^{\mathrm{cyc}})},\end{aligned}$$

which is induced by the restriction map of the Galois cohomologies with coefficient in $\mu_{p^{m+1}}$ from $G_{K_n^{\mathrm{cyc}}}$ to its subgroup $G_{K_n^{\mathrm{cyc}}(\mu_r)}$. The restriction map as above is not necessarily injective for a general finite algebraic number field K. However, when K is totally real and p is odd, we can show that the kernel and the cokernel of $\mathrm{res}_{n,r,m}$ is zero by the Inflation-Restriction sequence.

For each odd prime number $\ell \nmid Np$, we fix a generator σ_ℓ of the cyclic group $\mathrm{Gal}(K_n^{\mathrm{cyc}}(\mu_\ell)/K_n^{\mathrm{cyc}}) \cong \mathrm{Gal}(\mathbb{Q}(\mu_\ell)/\mathbb{Q})$. For each $r \in \mathfrak{R}_\psi$, we have the following decomposition of the Galois group $\mathrm{Gal}(K_n^{\mathrm{cyc}}(\mu_r)/K_n^{\mathrm{cyc}})$:

$$\mathrm{Gal}(K_n^{\mathrm{cyc}}(\mu_r)/K_n^{\mathrm{cyc}}) \cong \mathrm{Gal}(K_n^{\mathrm{cyc}}(\mu_{\ell_1})/K_n^{\mathrm{cyc}}) \times \cdots \times \mathrm{Gal}(K_n^{\mathrm{cyc}}(\mu_{\ell_s})/K_n^{\mathrm{cyc}})$$

corresponding to the prime decomposition $r = \ell_1 \cdots \ell_s$. According to this decomposition, we define an element $\mathbb{D}_r \in \mathbb{Z}[\mathrm{Gal}(K_n^{\mathrm{cyc}}(\mu_r)/K_n^{\mathrm{cyc}})]$ by $\mathbb{D}_r = \prod_{i=1}^s \mathbb{D}_{\ell_i}$ where $\mathbb{D}_{\ell_i} = \sum_{j=0}^{\ell_i - 1} j\sigma_{\ell_i}^j$ for each i. We call $\mathbb{D}_r$ a **Kolyvagin operator**. Let us denote by $z_{n,r,m} \in (K_n^{\mathrm{cyc}}(\mu_r))^\times/((K_n^{\mathrm{cyc}}(\mu_r))^\times)^{p^{m+1}}$ the image of $z_{n,r} \in (K_n^{\mathrm{cyc}}(\mu_r))^\times \otimes_{\mathbb{Z}} \mathbb{Z}_p$ by the map:

$$(K_n^{\mathrm{cyc}}(\mu_r))^\times \otimes_{\mathbb{Z}} \mathbb{Z}_p \twoheadrightarrow (K_n^{\mathrm{cyc}}(\mu_r))^\times/((K_n^{\mathrm{cyc}}(\mu_r))^\times)^{p^{m+1}}.$$

PROPOSITION 3.84. *When a natural number $r \in \mathfrak{R}_\psi$ satisfies the condition (CD_m), for every natural number n, we have*

$$\mathbb{D}_r(z_{n,r,m}) \in \left((K_n^{\mathrm{cyc}}(\mu_r))^\times/((K_n^{\mathrm{cyc}}(\mu_r))^\times)^{p^{m+1}} \right)^{\mathrm{Gal}(K_n^{\mathrm{cyc}}(\mu_r)/K_n^{\mathrm{cyc}})}.$$

PROOF. When $r = 1$, there is nothing to prove. We will prove the proposition by an induction argument with respect to the number of prime divisors of r. Let $r > 1$ and let ℓ be a prime factor of r. We set $r' = \frac{r}{\ell}$. Then we have

$$\begin{aligned}
&(\sigma_\ell - 1)\mathbb{D}_r(z_{n,r,m}) \\
&= (\sigma_\ell - 1)\mathbb{D}_\ell \mathbb{D}_{r'}(z_{n,r,m}) \\
&= \#\mathrm{Gal}(K_n^{\mathrm{cyc}}(\mu_\ell)/K_n^{\mathrm{cyc}})\mathbb{D}_{r'}(z_{n,r,m}) - \mathrm{Nr}_{K_n^{\mathrm{cyc}}(\mu_{lr'})/K_n^{\mathrm{cyc}}(\mu_{r'})}\mathbb{D}_{r'}(z_{n,r,m}) \\
&= \#\mathrm{Gal}(K_n^{\mathrm{cyc}}(\mu_\ell)/K_n^{\mathrm{cyc}})\mathbb{D}_{r'}(z_{n,r,m}) - (\mathrm{Frob}_\ell^{-1} - 1)\mathbb{D}_{r'}(z_{n,r',m}) \\
&= 0.
\end{aligned}$$

In the above equation, the first equality is nothing but the definition of $\mathbb{D}_r$, the second is easily checked by developing $(\sigma_\ell - 1)\mathbb{D}_l$. By the commutativity of the ring $\mathbb{Z}[\mathrm{Gal}(K_n^{\mathrm{cyc}}(\mu_r)/K_n^{\mathrm{cyc}})]$ and by the property (2) of Definition 3.80, we have

$$\begin{aligned}
\mathrm{Nr}_{K_n^{\mathrm{cyc}}(\mu_{lr'})/K_n^{\mathrm{cyc}}(\mu_{r'})}\mathbb{D}_{r'}(z_{n,r,m}) &= \mathbb{D}_{r'}\mathrm{Nr}_{K_n^{\mathrm{cyc}}(\mu_{\ell r'})/K_n^{\mathrm{cyc}}(\mu_{r'})}(z_{n,r,m}) \\
&= \mathbb{D}_{r'}(\mathrm{Frob}_\ell^{-1} - 1)(z_{n,r',m}) \\
&= (\mathrm{Frob}_\ell^{-1} - 1)\mathbb{D}_{r'}(z_{n,r',m})
\end{aligned}$$

which implies the third equality in the previous equation. By the condition (CD_m), we have $\#\mathrm{Gal}(K_n^{\mathrm{cyc}}(\mu_\ell)/K_n^{\mathrm{cyc}}) = \ell - 1 \equiv 0 \bmod p^{m+1}$. By applying the induction hypothesis to r', we have $(\mathrm{Frob}_\ell^{-1} - 1)\mathbb{D}_{r'}(z_{n,r',m}) = 0$. This implies the fourth equality in the previous equation.

Since σ_ℓ for prime factors ℓ of r generates $\mathrm{Gal}(K_n^{\mathrm{cyc}}(\mu_r)/K_n^{\mathrm{cyc}})$, the fact $(\sigma_\ell - 1)\mathbb{D}_r(z_{n,r,m}) = 0$ for all prime factors ℓ of r implies that $\mathbb{D}_r(z_{n,r,m})$ is invariant by the action of $\mathrm{Gal}(K_n^{\mathrm{cyc}}(\mu_r)/K_n^{\mathrm{cyc}})$. This completes the proof of proposition. □

DEFINITION 3.85. Let us take an element $r \in \mathfrak{R}_\psi$ satisfying the condition (CD_m) and let n be a natural number. Then we define

$$\kappa_{n,r,m} \in (K_n^{\mathrm{cyc}})^\times/((K_n^{\mathrm{cyc}})^\times)^{p^{m+1}}$$

to be the inverse image

$$\kappa_{n,r,m} = \mathrm{res}_{n,r,m}^{-1}(\mathbb{D}_r(z_{n,r,m}))$$

via the isomorphism (3.108). We call $\kappa_{n,r,m}$ a **Kolyvagin derivative** of $z_{n,r,m}$.

We prepare some important maps which play important roles in the proof of Theorem 3.82.

DEFINITION 3.86. Let us take a prime number $\ell \in \mathfrak{R}_\psi$ satisfying the condition (CD_m) and let n be a natural number. For each prime λ of K_n^{cyc} over (ℓ), we denote the completion of K_n^{cyc} at λ by $K_{n,\lambda}^{\mathrm{cyc}}$ and its ring of integers by $\mathfrak{r}_{n,\lambda}$.

(1) For a factor λ of (ℓ) as above, we denote the composite

$$(K_n^{\mathrm{cyc}})^\times/((K_n^{\mathrm{cyc}})^\times)^{p^{m+1}} \longrightarrow ((K_{n,\lambda}^{\mathrm{cyc}})^\times/\mathfrak{r}_{n,\lambda}^\times)\Big/ ((K_{n,\lambda}^{\mathrm{cyc}})^\times/\mathfrak{r}_{n,\lambda}^\times)^{p^{m+1}} \xrightarrow{\sim} \mathbb{Z}/(p^{m+1})$$

by $\mathrm{ord}_{\lambda,m}$[66].

[66] In other words, the map $\mathrm{ord}_{\lambda,m}$ is the map which assigns $\mathrm{ord}_{\lambda,m}(x) = \mathrm{ord}_\lambda(\widetilde{x}) \bmod p^{m+1}$ to each $x \in (K_n^{\mathrm{cyc}})^\times/((K_n^{\mathrm{cyc}})^\times)^{p^{m+1}}$ where $\widetilde{x} \in (K_n^{\mathrm{cyc}})^\times$ is a lift of x and $\mathrm{ord}_\lambda : (K_n^{\mathrm{cyc}})^\times \longrightarrow \mathbb{Z}$ is the normalized additive valuation at λ.

(2) Let $\ell \in \mathfrak{R}_\psi$ be a prime satisfying the condition (CD_m) and let us fix a prime $\widetilde{\ell}$ of K_n^{cyc} over (ℓ). Since the prime (ℓ) is totally decomposed at K_n^{cyc}, there is a unique element $g_\lambda \in \mathrm{Gal}(K_n^{\mathrm{cyc}}/\mathbb{Q})$ such that $\lambda = (\widetilde{\ell})^{g_\lambda}$ for each prime λ of K_n^{cyc} over (ℓ). Hence, we have the following map

$$(K_n^{\mathrm{cyc}})^\times/((K_n^{\mathrm{cyc}})^\times)^{p^{m+1}} \longrightarrow \mathbb{Z}/(p^{m+1})[\mathrm{Gal}(K_n^{\mathrm{cyc}}/\mathbb{Q})]$$
$$x \mapsto \bigoplus_{\lambda|\ell} \mathrm{ord}_{\lambda,m}(x)\cdot g_\lambda.$$

Let us set $\Lambda_{\psi,n,m} := \Lambda_{\mathrm{cyc},\psi}/(\omega_n, p^{m+1})$. By taking the ψ-quotient of the above map, we obtain the map:

$$\mathrm{Div}_{n,m}^{(\ell)} : \left((K_n^{\mathrm{cyc}})^\times/((K_n^{\mathrm{cyc}})^\times)^{p^{m+1}}\right)_\psi \longrightarrow \Lambda_{\psi,n,m}.$$

Here we recall that $\Lambda_{\mathrm{cyc},\psi}/(\omega_n, p^{m+1}) = \left(\mathbb{Z}/(p^{m+1})[\mathrm{Gal}(K_n^{\mathrm{cyc}}/\mathbb{Q})]\right)_\psi$.

DEFINITION 3.87. Let n, m, ℓ, $\widetilde{\ell}$ and g_λ be as in Definition 3.86. Recall that σ_ℓ is a fixed generator of the cyclic group $\mathrm{Gal}(K_n^{\mathrm{cyc}}(\mu_\ell)/K_n^{\mathrm{cyc}})$. Since the extension $K_n^{\mathrm{cyc}}(\mu_\ell)$ is totally ramified at every prime λ of K_n^{cyc} over (ℓ), we denote the unique prime of $K_n^{\mathrm{cyc}}(\mu_\ell)$ over λ by the same symbol λ.

(1) We denote by u_λ the image of $(1-\sigma_\ell)(\varpi_\lambda) = \varpi_\lambda/\varpi_\lambda^{\sigma_\ell} \in \mathfrak{r}_{n,\lambda}^\times$ in

$$(\mathfrak{r}_{K_n^{\mathrm{cyc}}(\mu_\ell)}/\lambda)^\times/((\mathfrak{r}_{K_n^{\mathrm{cyc}}(\mu_\ell)}/\lambda)^\times)^{p^{m+1}} = (\mathfrak{r}_{K_n^{\mathrm{cyc}}}/\lambda)^\times/((\mathfrak{r}_{K_n^{\mathrm{cyc}}}/\lambda)^\times)^{p^{m+1}}$$

where $\varpi_\lambda \in K_n^{\mathrm{cyc}}(\mu_\ell)$ is a uniformizer of the prime λ.

(2) We define a noncanonical homomorphism depending on the choice of σ_ℓ:

$$\varphi_{n,m}^{(\lambda,\sigma_\ell)} : (\mathfrak{r}_{K_n^{\mathrm{cyc}}}/\lambda)^\times/((\mathfrak{r}_{K_n^{\mathrm{cyc}}}/\lambda)^\times)^{p^{m+1}} \longrightarrow \mathbb{Z}/(p^{m+1})$$

to be the unique homomorphism from a multiplicative cyclic group to an additive group such that $\varphi_{n,m}^{(\lambda,\sigma_\ell)}(u_\lambda) = 1$. Here we note that u_λ is a generator of the cyclic group $(\mathfrak{r}_{K_n^{\mathrm{cyc}}}/\lambda)^\times/((\mathfrak{r}_{K_n^{\mathrm{cyc}}}/\lambda)^\times)^{p^{m+1}}$ because the extension $K_n^{\mathrm{cyc}}(\mu_\ell)/K_n^{\mathrm{cyc}}$ is tamely ramified at λ. We also note that we have $(\mathrm{ord}_{\lambda,m}\circ N_\ell)(\varpi_\lambda) = 1$ where $N_\ell : K_n^{\mathrm{cyc}}(\mu_\ell) \longrightarrow K_n^{\mathrm{cyc}}$ is the norm map.

(3) We define a noncanonical homomorphism depending on the choices of σ_ℓ and $\widetilde{\ell}$:

$$\varphi_{n,m}^{(\widetilde{\ell},\sigma_\ell)} : \left((\mathfrak{r}_{K_n^{\mathrm{cyc}}}/(\ell))^\times\right)_\psi \longrightarrow \Lambda_{\psi,n,m} = \left(\mathbb{Z}/(p^{m+1})[\mathrm{Gal}(K_n^{\mathrm{cyc}}/\mathbb{Q})]\right)_\psi$$

to be the ψ-quotient of the map

$$(\mathfrak{r}_{K_n^{\mathrm{cyc}}}/(\ell))^\times \longrightarrow \mathbb{Z}/(p^{m+1})[\mathrm{Gal}(K_n^{\mathrm{cyc}}/\mathbb{Q})]$$

sending $x \in (\mathfrak{r}_{K_n^{\mathrm{cyc}}}/(\ell))^\times$ to $\oplus_{\lambda|\ell}\varphi_{n,m}^{(\lambda,\sigma_\ell)}(x_\lambda)g_\lambda$ where x_λ for each λ is given by

$$\left(\mathfrak{r}_{K_n^{\mathrm{cyc}}}/(\ell)\right)^\times \xrightarrow{\sim} \bigoplus_{\lambda|\ell}(\mathfrak{r}_{K_n^{\mathrm{cyc}}}/\lambda)^\times \twoheadrightarrow \bigoplus_{\lambda|\ell}(\mathfrak{r}_{K_n^{\mathrm{cyc}}}/\lambda)^\times/((\mathfrak{r}_{K_n^{\mathrm{cyc}}}/\lambda)^\times)^{p^{m+1}}$$
$$x \mapsto \bigoplus_{\lambda|\ell} x_\lambda.$$

Since the situation of the above definition is a bit complicated, we will illustrate the relation of some operations at each λ below.

$$\begin{array}{ccccc}
K_n^{\mathrm{cyc}}(\mu_\ell)^\times/(K_n^{\mathrm{cyc}}(\mu_\ell)^\times)^{p^{m+1}} & & & & \\
\downarrow{\scriptstyle 1-\sigma_\ell} & \searrow^{\mathrm{Nr}_{K_n^{\mathrm{cyc}}(\mu_\ell)/K_n^{\mathrm{cyc}}}} & & & \\
(\mathfrak{r}_{K_n^{\mathrm{cyc}}(\mu_\ell)}/\lambda)^\times/((\mathfrak{r}_{K_n^{\mathrm{cyc}}(\mu_\ell)}/\lambda)^\times)^{p^{m+1}} & & (K_n^{\mathrm{cyc}})^\times/((K_n^{\mathrm{cyc}})^\times)^{p^{m+1}} & & \\
\Vert & & & \searrow^{\mathrm{ord}_{\lambda,m}} & \\
(\mathfrak{r}_{K_n^{\mathrm{cyc}}}/\lambda)^\times/((\mathfrak{r}_{K_n^{\mathrm{cyc}}}/\lambda)^\times)^{p^{m+1}} & & \xrightarrow{\varphi_{n,m}^{(\lambda,\sigma_\ell)}} & & \mathbb{Z}/(p^{m+1}).
\end{array}$$

Below, we will prove the main theorem (Theorem 3.82) of the Euler system argument by using two key propositions below (Proposition 3.88 and Proposition 3.90).

PROPOSITION 3.88. *Under the setting of Definition 3.87, we have the following equality in* $\Lambda_{\psi,n,m}$*:*

$$\mathrm{Div}_{n,m}^{(\ell)}(\kappa_{n,r,m}) = \varphi_{n,m}^{(\widetilde{\ell},\sigma_\ell)}(\kappa_{n,r/\ell,m}) \tag{3.109}$$

where we denote the image of $\kappa_{n,r/\ell,m} \in (K_n^{\mathrm{cyc}})^\times/((K_n^{\mathrm{cyc}})^\times)^{p^{m+1}}$ *in* $(\mathfrak{r}_{K_n^{\mathrm{cyc}}}/(\ell))^\times$ *also by the same symbol* $\kappa_{n,r/\ell,m}$ *by abuse of notation.*

REMARK 3.89. We do not have a natural map from $(K_n^{\mathrm{cyc}})^\times/((K_n^{\mathrm{cyc}})^\times)^{p^{m+1}}$ to $(\mathfrak{r}_{K_n^{\mathrm{cyc}}}/(\ell))^\times$. However we have a natural map from the subgroup

$$\left\{ x \in (K_n^{\mathrm{cyc}})^\times/((K_n^{\mathrm{cyc}})^\times)^{p^{m+1}} \;\middle|\; \mathrm{ord}_{\lambda,m}(x) = 0 \text{ for every } \lambda|(\ell) \right\}$$

to $(\mathfrak{r}_{K_n^{\mathrm{cyc}}}/(\ell))^\times$. Thus the image of $\kappa_{n,r/\ell,m}$ in $(\mathfrak{r}_{K_n^{\mathrm{cyc}}}/(\ell))^\times$ discussed in the above proposition makes sense since we have $\mathrm{ord}_{\lambda,m}(\kappa_{n,r/\ell,m}) = 0$ for every $\lambda|(\ell)$.

PROPOSITION 3.90. *Let* m, n *be natural numbers satisfying* $p^m\mathrm{Cl}(K_n^{\mathrm{cyc}})[p^\infty] = 0$*. Suppose we are given the following data:*

(i) an element $c \in \mathrm{Cl}(K_n^{\mathrm{cyc}})[p^m]$,

(ii) a $\mathrm{Gal}(K_n^{\mathrm{cyc}}/\mathbb{Q})$*-stable finite* $\Lambda_{\psi,n,m}$*-submodule* $W \subset \left((K_n^{\mathrm{cyc}})^\times/((K_n^{\mathrm{cyc}})^\times)^{p^{m+1}}\right)_\psi$,

(iii) a $\mathrm{Gal}(K_n^{\mathrm{cyc}}/\mathbb{Q})$*-equivariant* $\Lambda_{\psi,n,m}$*-linear homomorpihsm* $\phi_W : W \to \Lambda_{\psi,n,m}$.

Then there exist a prime number $\ell_c \in \mathfrak{R}_\psi$ *satisfying the condition* (CD_m)*, a prime* $\widetilde{\ell}_c$ *of* K_n^{cyc} *over* ℓ_c *and a generator* σ_{ℓ_c} *of* $\mathrm{Gal}(\mathbb{Q}(\mu_{\ell_c})/\mathbb{Q})$ *such that we have*[67]*:*

$$\begin{cases} [\widetilde{\ell}_c] = c, \\ \mathrm{ord}_{\lambda,m}(w) = 0 \textit{ for any } w \in W \textit{ and for any prime } \lambda \textit{ of } K_n^{\mathrm{cyc}} \textit{ over } \ell_c, \\ \varphi_{n,m}^{(\widetilde{\ell}_c,\sigma_{\ell_c})}(w) = t \cdot \phi_W(w) \ (\textit{with } t \in (\mathbb{Z}/(p^{m+1})[\psi])^\times), \textit{ for any } w \in W \end{cases} \tag{3.110}$$

where $[\widetilde{\ell}_c]$ *denotes the ideal class of* $\widetilde{\ell}_c$*. Further, there exist infinitely many such primes* ℓ_c*.*

[67]Below, we denote the projection of $w \in \left((K_n^{\mathrm{cyc}})^\times/((K_n^{\mathrm{cyc}})^\times)^{p^{m+1}}\right)_\psi$ to $\left((\mathfrak{r}_{K_n^{\mathrm{cyc}}}/(l))^\times\right)_\psi$ by the same symbol w by abuse of notation. We emphasize the same remark as Remark 3.89 here again.

The proof of Proposition 3.88 is based on the norm compatible property of the Euler system $z_{n,r}$ (see Definition 3.80 (2)) and some properties of Kolyvagin operators $\mathbb{D}_r(z_{n,r,m})$. The proof of Proposition 3.90 is based on the Chebotarev density theorem. We omit the proof of the above two propositions and we will concentrate on proving Theorem 3.82 using these two propositions. However, we refer to [**99**, Prop. 2.4] for the proof of Proposition 3.88 and to [**99**, Thm. 3.1] for the proof of Proposition 3.90.

PROOF OF THEOREM 3.82. By Theorem 3.4, $(X_{K^{\mathrm{cyc}}_\infty})_\psi$ is a finitely generated torsion $\Lambda_{\mathrm{cyc},\psi}$-module. Hence, there are finitely many elements $g^{(1)}, \ldots, g^{(s)} \in \Lambda_{\mathrm{cyc},\psi}$ so that we have a $\Lambda_{\mathrm{cyc},\psi}$-linear injection:

$$(\Lambda_{\mathrm{cyc},\psi}/(g^{(1)})) \oplus \cdots \oplus (\Lambda_{\mathrm{cyc},\psi}/(g^{(s)})) \hookrightarrow (X_{K^{\mathrm{cyc}}_\infty})_\psi \tag{3.111}$$

with finite cokernel[68]. By taking the ψ-quotient of the 4-term exact sequence (3.51) in Theorem 3.57, we have the following exact sequence:

$$\begin{aligned}(3.112)\quad 0 \longrightarrow \text{the maximal } p\text{-power torsion part of } &\left(\varprojlim_n U_1(\mathfrak{r}_{K^{\mathrm{cyc}}_n} \otimes_{\mathbb{Z}} \mathbb{Z}_p)\right)_\psi \\ \longrightarrow \left(\varprojlim_n U_1(\mathfrak{r}_{K^{\mathrm{cyc}}_n} \otimes_{\mathbb{Z}} \mathbb{Z}_p)\right)_\psi &\longrightarrow \Lambda_{\mathrm{cyc},\psi} \longrightarrow \text{a finite } p\text{-group} \longrightarrow 0.\end{aligned}$$

On the other hand, by the assumption that ψ is even, the ψ-quotient of the first injection of the 4-term exact sequence (3.98) of Proposition 3.79 gives the following $\Lambda_{\mathrm{cyc},\psi}$-linear map

$$\left(\varprojlim_n (\mathfrak{r}_{K^{\mathrm{cyc}}_n})^\times \otimes_{\mathbb{Z}} \mathbb{Z}_p\right)_\psi \longrightarrow \left(\varprojlim_n U_1(\mathfrak{r}_{K^{\mathrm{cyc}}_n} \otimes_{\mathbb{Z}} \mathbb{Z}_p)\right)_\psi \tag{3.113}$$

whose kernel and cokernel are $\Lambda_{\mathrm{cyc},\psi}$-torsion. By combining (3.112) and (3.113), there exists a nonnegative integer α and an ideal $\mathcal{J} \subset \Lambda_{\mathrm{cyc},\psi}$ of height two such that we have a $\Lambda_{\mathrm{cyc},\psi}$-linear map:

$$\left(\varprojlim_n (\mathfrak{r}_{K^{\mathrm{cyc}}_n})^\times \otimes_{\mathbb{Z}} \mathbb{Z}_p\right)_\psi \longrightarrow \Lambda_{\mathrm{cyc},\psi} \tag{3.114}$$

whose kernel and cokernel are annihilated by (p^α) and by $\mathcal{J}$ respectively[69].

By the structure theorem of Iwasawa modules (Theorem 2.39), we have the following 4-term exact sequence

$$\begin{aligned}(3.115)\quad 0 \longrightarrow \text{the maximal } p\text{-power torsion part of } &\varprojlim_n (\mathfrak{r}_{K^{\mathrm{cyc}}_n})^\times \otimes_{\mathbb{Z}} \mathbb{Z}_p \\ \longrightarrow \varprojlim_n (\mathfrak{r}_{K^{\mathrm{cyc}}_n})^\times \otimes_{\mathbb{Z}} \mathbb{Z}_p &\longrightarrow \Lambda_{\mathrm{cyc},\psi} \longrightarrow \text{a finite group} \longrightarrow 0.\end{aligned}$$

We denote by $h \in \Lambda_{\mathrm{cyc},\psi}$ the image of $\varprojlim_n z_{n,1} \in \left(\varprojlim_n (\mathfrak{r}_{K^{\mathrm{cyc}}_n})^\times \otimes_{\mathbb{Z}} \mathbb{Z}_p\right)_\psi$ by the map (3.114). In the rest of the proof, we fix a pseudo-isomorphism as in (3.111)

[68] Since ψ is an even character, we should have $s = 0$ if we assume Greenberg's conjecture (Conjecture 3.26). Hence, Theorem 3.82 holds trivially true under Greenberg's conjecture (see Proposition 3.93 (2)).

[69] We note that the map (3.114) is injective when the order of the character ψ is prime to p.

and a $\Lambda_{\mathrm{cyc},\psi}$-linear map as in (3.114). Note that we have

$$\mathrm{char}_{\Lambda_{\mathrm{cyc},\psi}}\left(\varprojlim_n(\mathfrak{r}_{K_n^{\mathrm{cyc}}})^\times\otimes_{\mathbb{Z}}\mathbb{Z}_p\right)_\psi \Big/ \varprojlim_n z_{n,1}\Lambda_{\mathrm{cyc},\psi}=(h)$$

by definition. Hence, our goal in this proof is to prove the inclusion

$$\left(\prod_{i=1}^{s} g^{(i)}\right)\supset(p^\alpha h). \tag{3.116}$$

Under this situation, we will control ideal class groups and groups of global units of K_n^{cyc} when the intermediate field K_n^{cyc} varies.

As for the class group, by Step 3 of the proof of Iwasawa's class number formula (Theorem 3.1) given in §3.1.2, there exists a constant n_0 such that the natural maps

$$(X_{K_\infty^{\mathrm{cyc}}})_\psi/\omega_n(X_{K_\infty^{\mathrm{cyc}}})_\psi\longrightarrow(\mathrm{Cl}(K_n^{\mathrm{cyc}})[p^\infty])_\psi \tag{3.117}$$

are surjective and their kernels are finite groups whose orders are uniformly bounded independently of $n\geq n_0$. For each integer i satisfying $1\leq i\leq s$ and for any natural number n, we set

$$A_n^{(i)}:=\Lambda_{\mathrm{cyc},\psi}/(g^{(i)},\omega_n).$$

By applying the functor $\otimes_{\Lambda_{\mathrm{cyc},\psi}}\Lambda_{\mathrm{cyc},\psi}/(\omega_n)$ to (3.111) and by composing with the map (3.117), we have $\Lambda_{\mathrm{cyc},\psi}$-linear homomorphisms

$$\alpha_n:\ A_n^{(1)}\oplus\cdots\oplus A_n^{(s)}\longrightarrow(\mathrm{Cl}(K_n^{\mathrm{cyc}})[p^\infty])_\psi \tag{3.118}$$

whose kernels and cokernels are finite groups whose orders are uniformly bounded independently of n. We choose and fix an ideal $\mathcal{I}\subset\Lambda_{\mathrm{cyc},\psi}$ of height two which annihilates the kernels and the cokernels of (3.118) for every n.

As for the control of the group of global units, we first recall the control of the group of local units which is easier than that of the group of global units. Then we study the difference between local units and global units. Recall that the norm map of local units under the local cyclotomic $\mathbb{Z}_p$-extension induces the following isomorphism:

$$\varprojlim_n U_1(\mathfrak{r}_{K_n^{\mathrm{cyc}}}\otimes_{\mathbb{Z}}\mathbb{Z}_p)\bmod(\omega_n,p^{m+1})\xrightarrow{\sim}U_1(\mathfrak{r}_{K_n^{\mathrm{cyc}}}\otimes_{\mathbb{Z}}\mathbb{Z}_p).$$

Since the group of units of a local field is related to the Galois group of the maximal abelian extension over this local field by the local class field theory, this isomorphism can be proved by controling the Galois group of the maximal abelian when the local field varies. We omit the proof of this isomorphism, but we refer the reader to [**69**, Chap. 5] for example. Let us consider the following commutative diagram

$$\begin{array}{ccccccc}
\left(\varprojlim_{m,n}A_{m,n}\right)_{m,n} & \longrightarrow & \left(\varprojlim_{m,n}B_{m,n}\right)_{m,n} & \longrightarrow & \left(\varprojlim_{m,n}(B_{m,n}/A_{m,n})\right)_{m,n} & \longrightarrow & 0\\
\downarrow & & \downarrow & & \downarrow & & \\
A_{m,n} & \longrightarrow & B_{m,n} & \longrightarrow & B_{m,n}/A_{m,n} & \longrightarrow & 0
\end{array} \tag{3.119}$$

where we set

$$A_{m,n}=(\mathfrak{r}_{K_n^{\mathrm{cyc}}})^\times/((\mathfrak{r}_{K_n^{\mathrm{cyc}}})^\times)^{p^{m+1}},$$
$$B_{m,n}=U_1(\mathfrak{r}_{K_n^{\mathrm{cyc}}}\otimes_{\mathbb{Z}}\mathbb{Z}_p)/U_1(\mathfrak{r}_{K_n^{\mathrm{cyc}}}\otimes_{\mathbb{Z}}\mathbb{Z}_p)^{p^{m+1}}$$

and $(\ )_{m,n}$ on the upper horizontal sequence means taking the functor the functor $\otimes_{\Lambda_{\mathrm{cyc},\psi}} \Lambda_{\mathrm{cyc},\psi}/(\omega_n, p^{m+1})$. By (3.98), the rightmost vertical map fits into the leftmost vertical map of the following diagram:

(3.120)

$$\begin{array}{ccccccc} \left(\varprojlim_{m,n}(B_{m,n}/A_{m,n})\right)_{m,n} & \longrightarrow & \left(\mathfrak{X}_{K_\infty^{\mathrm{cyc}}}\right)_{m,n} & \longrightarrow & \left(X_{K_\infty^{\mathrm{cyc}}}\right)_{m,n} & \longrightarrow & 0 \\ \downarrow & & \downarrow & & \downarrow & & \\ B_{m,n}/A_{m,n} & \longrightarrow & \mathfrak{X}_{K_n^{\mathrm{cyc}}}/(\mathfrak{X}_{K_n^{\mathrm{cyc}}})^{p^{m+1}} & \longrightarrow & X_{K_n^{\mathrm{cyc}}}/(X_{K_n^{\mathrm{cyc}}})^{p^{m+1}} & \longrightarrow & 0 \end{array}$$

where $\mathfrak{X}_{K_n^{\mathrm{cyc}}}$ is the Galois group of the maximal p-ramified p-extension of K_n^{cyc} and $X_{K_n^{\mathrm{cyc}}}$ is the Galois group of the maximal unramified p-extension of K_n^{cyc}.

By the diagram chasing argument using the snake lemma and the diagrams (3.119), (3.120), we check that the kernel and the cokernel of the map

$$\left(\varprojlim_n (\mathfrak{r}_{K_n^{\mathrm{cyc}}})^\times \otimes_{\mathbb{Z}} \mathbb{Z}_p\right)_\psi \ \mathrm{mod}\ (\omega_n, p^{m+1}) \longrightarrow \left((\mathfrak{r}_{K_n^{\mathrm{cyc}}})^\times/((\mathfrak{r}_{K_n^{\mathrm{cyc}}})^\times)^{p^{m+1}}\right)_\psi$$

induced by the norm map of global units under the global cyclotomic $\mathbb{Z}_p$-extension are annihilated universally by an ideal in $\Lambda_{\psi,n,m}$ of height two which does not depend on m, n (see [**69**, Appendix §6] for more detail of the argument here). There exist a nonnegative integer α and an ideal $\mathcal{J} \subset \Lambda_{\mathrm{cyc},\psi}$ of height two such that, for each $\eta \in (p^\alpha)(\gamma-1)\mathcal{J}$, the map

$$\left(\varprojlim_n (\mathfrak{r}_{K_n^{\mathrm{cyc}}})^\times \otimes_{\mathbb{Z}} \mathbb{Z}_p\right)_\psi \ \mathrm{mod}\ (\omega_n, p^{m+1}) \longrightarrow \Lambda_{\psi,n,m} \xrightarrow{\times\eta} \Lambda_{\psi,n,m}$$

induced by (3.117) mod ω_n and the multiplication by η is uniquely extended to a $\Lambda_{\psi,n,m}$-linear map

$$\beta^{(\eta)}_{n,m} : \left((\mathfrak{r}_{K_n^{\mathrm{cyc}}})^\times/((\mathfrak{r}_{K_n^{\mathrm{cyc}}})^\times)^{p^{m+1}}\right)_\psi \longrightarrow \Lambda_{\psi,n,m}. \tag{3.121}$$

Let us fix natural numbers m, n satisfying $(\mathrm{Cl}(K_n^{\mathrm{cyc}})[p^\infty])_\psi = (\mathrm{Cl}(K_n^{\mathrm{cyc}})[p^m])_\psi$. We denote by $g^{(i)}_{n,m}, h_{n,m} \in \Lambda_{\psi,n,m}$ the images of $g^{(i)}, h \in \Lambda_{\mathrm{cyc},\psi}$ via $\Lambda_{\mathrm{cyc},\psi} \twoheadrightarrow \Lambda_{\psi,n,m}$ respectively. For each i satisfying $i = 1, \ldots, s$, we fix a generator c_i of the $\mathbb{Z}_p[\mathrm{Gal}(K_n^{\mathrm{cyc}}/\mathbb{Q})]$-module $A_n^{(i)}$ which appears in (3.118). We set $c_{s+1} = 0$.

(Step 1) Let us choose an arbitrary element $\eta_1 \in (p^\alpha)(\gamma-1)\mathcal{J}$ and apply Proposition 3.90 to c_1, $W_1 = \left((\mathfrak{r}_{K_n^{\mathrm{cyc}}})^\times/((\mathfrak{r}_{K_n^{\mathrm{cyc}}})^\times)^{p^{m+1}}\right)_\psi$ and $\phi_{W_1} = \beta^{(\eta_1)}_{n,m}$. Then, there exist a prime $\ell_1 \in \mathfrak{R}_\psi$ satisfying the condition (CD_m), a prime $\widetilde{\ell}_1$ of K_n^{cyc} over ℓ_1 and an element $t_1 \in (\mathbb{Z}/(p^{m+1})[\psi])^\times$ such that we have

$$\begin{cases} [\ell_1] - c_1, \\ \varphi^{(\widetilde{\ell}_1,\sigma_{\ell_1})}_{n,m}(\kappa_{n,1,m}) = t_1\eta_1 h_{n,m}. \end{cases} \tag{3.122}$$

(Step 2) We want to show that there exist $\ell_2, \ldots, \ell_{s+1}$ satisfying the condition (CD_m), primes $\widetilde{\ell}_2, \ldots, \widetilde{\ell}_{s+1}$ of K_n^{cyc} over $\ell_2, \ldots, \ell_{s+1}$ and elements $t_2, \ldots, t_{s+1} \in (\mathbb{Z}/(p^{m+1})[\psi])^\times$ such that we have

$$\begin{cases} [\widetilde{\ell}_k] = c_k, \\ g^{(k-1)}_{n,m}\varphi^{(\widetilde{\ell}_k,\sigma_{\ell_k})}_{n,m}(\kappa_{n,\ell_1\cdots\ell_k,m}) = t_k\eta_k\varphi^{(\widetilde{\ell}_{k-1},\sigma_{\ell_{k-1}})}_{n,m}(\kappa_{n,\ell_1\cdots\ell_{k-1},m}), \end{cases} \tag{3.123}$$

for every natural number k satisfying $2 \leq k \leq s+1$a where η_k is an arbitrary element in $\mathcal{I}$ for each k.

By induction, we assume that (3.123) holds true for $k = i-1$ and we will prove it for $k = i$.

By construction, the element

$$\kappa_{n,\ell_1\cdots\ell_i,m} \in \left((K_n^{\mathrm{cyc}})^\times/((K_n^{\mathrm{cyc}})^\times)^{p^{m+1}}\right)_\psi = H^1(K_n^{\mathrm{cyc}}, \mu_{p^{m+1}})_\psi$$

is unramified outside primes dividing $\ell_1\cdots\ell_i 2Np$. That is, $\kappa_{n,\ell_1\cdots\ell_i,m}$ is contained in $H^1(K^{\mathrm{cyc}}_{n,\Sigma_{\ell_1\cdots\ell_i}}/K_n^{\mathrm{cyc}}, \mu_{p^{m+1}})_\psi \subset H^1(K_n^{\mathrm{cyc}}, \mu_{p^{m+1}})_\psi$ where $\Sigma_{\ell_1\cdots\ell_i}$ is the set of primes of K_n^{cyc} dividing $\ell_1\cdots\ell_i 2Np$ and $K^{\mathrm{cyc}}_{n,\Sigma_{\ell_1\cdots\ell_i}}$ is the largest Galois extension of K_n^{cyc} unramified outside the primes over $\Sigma_{\ell_1\cdots\ell_i}$.

By the global duality theorem of Galois cohomology, we have the following lemma[70]:

LEMMA 3.91. *Let F be a finite algebraic number field, Σ a finite set of finite primes of F and m a natural number. We consider the localization map of Galois cohomology at Σ as follows:*

$$\mathrm{loc}_\Sigma(\mu_{p^{m+1}}) : H^1(F, \mu_{p^{m+1}}) \longrightarrow \bigoplus_{v\in\Sigma} H^1(F_v, \mu_{p^{m+1}}),$$

$$\mathrm{loc}_\Sigma(\mathbb{Z}/(p^{m+1})) : H^1(F, \mathbb{Z}/(p^{m+1})) \longrightarrow \bigoplus_{v\in\Sigma} H^1(F_v, \mathbb{Z}/(p^{m+1})).$$

Then, for any

$$x \in H^1(F_\Sigma/F, \mu_{p^{m+1}})$$

and for any

$$y \in \mathrm{Cl}(F)[p^{m+1}] \subset H^1(F, \mathbb{Z}/(p^{m+1})),$$

the local duality pairing of their localizations at Σ:

$$\bigoplus_{v\in\Sigma} (H^1(F_v, \mu_{p^{m+1}}) \times H^1(F_v, \mathbb{Z}/(p^{m+1}))) \longrightarrow \bigoplus_{v\in\Sigma} H^2(F_v, \mu_{p^{m+1}}) \longrightarrow \mathbb{Z}/(p^{m+1}),$$

$$\bigoplus_{v\in\Sigma}(\mathrm{loc}_v(x), \mathrm{loc}_v(y)) \mapsto \bigoplus_{v\in\Sigma} \mathrm{loc}_v(x) \cup \mathrm{loc}_v(y) \longrightarrow \sum_{v\in\Sigma} \mathrm{loc}_v(x) \cup \mathrm{loc}_v(y)$$

is zero where $\cup$ denotes the cup product. Here, in the last map, we regard $\mathrm{loc}_v(x) \cup \mathrm{loc}_v(y)$ as an element of $\mathbb{Z}/(p^{m+1})$ via the canonical isomorphism $H^2(F_v, \mu_{p^{m+1}}) \cong \mathbb{Z}/(p^{m+1})$.

Now, we denote by $B_{i-1} \subset \mathrm{Cl}(K_n^{\mathrm{cyc}})[p^m]$ the subgroup generated by the primes of $\mathfrak{r}_{K_n^{\mathrm{cyc}}}$ over $\ell_1, \ldots, \ell_{i-1}$. By Lemma 3.91, $\mathrm{ord}_{\lambda,m}(\kappa_{n,\ell_1\cdots\ell_i,m}) \in \mathbb{Z}/(p^{m+1})$ annihilates the image of λ in $\mathrm{Cl}(K_n^{\mathrm{cyc}})[p^m]/B_{i-1}$ for any prime λ of K_n^{cyc} over ℓ_i. Hence, for any $\eta_i \in \mathcal{I}$, the fixed generator $g^{(i)}_{n,m} \in \Lambda_{\psi,n,m}$ of the annihilator ideal of c_i divides $\eta_i \mathrm{Div}^{(\ell_i)}_{n,m}(\kappa_{n,\ell_1\cdots\ell_i,m}) \in \Lambda_{\psi,n,m}$. This implies that there exists a $\Lambda_{\psi,n,m}$-linear homomorpihsm $\phi^{(\eta_i)}_{W_i} : W_i \longrightarrow \Lambda_{\psi,n,m}$ such that we have

$$g^{(i)}_{n,m}\phi^{(\eta_i)}_{W_i}(\kappa_{n,\ell_1\cdots\ell_i,m}) = \eta_i \mathrm{Div}^{(\ell_i)}_{n,m}(\kappa_{n,\ell_1\cdots\ell_i,m}) \tag{3.124}$$

where W_i is the $\mathrm{Gal}(K_n^{\mathrm{cyc}}/\mathbb{Q})$-stable $\Lambda_{\psi,n,m}$-submodule of

$$\left((K_n^{\mathrm{cyc}})^\times/((K_n^{\mathrm{cyc}})^\times)^{p^{m+1}}\right)_\psi$$

[70]For the global duality theorem, we refer the reader to [**109**, Chap. II, §6] for example.

generated by $\kappa_{n,\ell_1\cdots\ell_{i-1},m}$. By Proposition 3.88, we have

$$\mathrm{Div}^{(\ell_i)}_{n,m}(\kappa_{n,\ell_1\cdots\ell_i,m}) = \varphi^{(\widetilde{\ell}_{i-1},\sigma_{\ell_{i-1}})}_{n,m}(\kappa_{n,\ell_1\cdots\ell_{i-1},m}). \tag{3.125}$$

By combining (3.124) and (3.125), we obtain

$$g^{(i)}_{n,m}\phi^{(\eta_i)}_{W_i}(\kappa_{n,\ell_1\cdots\ell_i,m}) = \eta_i\varphi^{(\widetilde{\ell}_{i-1},\sigma_{\ell_{i-1}})}_{n,m}(\kappa_{n,\ell_1\cdots\ell_{i-1},m}).$$

If we apply Proposition 3.90 to the data $c = c_i$, $W = W_i$ and $\phi_W = \phi^{(\eta_i)}_{W_i}$, we prove that (3.123) holds for $k = i$.
(Step 3) By putting Step 1 and Step 2 together, we obtain the following equality in $\Lambda_{\psi,n,m}$:

$$g^{(1)}_{n,m}\cdots g^{(s)}_{n,m}\varphi^{(\widetilde{\ell}_{s+1},\sigma_{\ell_{s+1}})}_{n,m}(\kappa_{n,\ell_1\cdots\ell_{s+1},m}) = (t_1\cdots t_{s+1})(\eta_1\cdots\eta_{s+1})h_{n,m}. \tag{3.126}$$

Note that the choice of $\eta_1 \in (p^\alpha)\mathcal{J}$ and $\eta_2,\dots,\eta_{s+1} \in \mathcal{I}$ is independent of m and n. Hence, we have $g^{(1)}\cdots g^{(s)}|(\eta_1\cdots\eta_{s+1})h$ for any $\eta_1 \in (p^\alpha)\mathcal{J}$ and for any $\eta_2,\dots,\eta_{s+1} \in \mathcal{I}$. Since the ideals $\mathcal{I}$ and $\mathcal{J}$ are of height two, we can choose another set of elements $\eta'_1 \in \mathcal{J}$, $\eta'_2,\dots,\eta'_{s+1} \in \mathcal{I}$ so that the common factor of $\eta_1\cdots\eta_{s+1}$ and $\eta'_1\cdots\eta'_{s+1}$ is contained in $(p^\alpha)(\gamma-1)$. Since the principal ideal $(\gamma-1)$ does not divide $\mathrm{char}_{\Lambda_{\mathrm{cyc},\psi}}(X_{K^{\mathrm{cyc}}_\infty})_\psi$, the validity of $g^{(1)}\cdots g^{(s)}|(\eta_1\cdots\eta_{s+1})h$ and $g^{(1)}\cdots g^{(s)}|(\eta'_1\cdots\eta'_{s+1})h$ implies (3.116), which completes the proof. □

REMARK 3.92. (1) Here, we discussed the Euler system argument over a cyclotomic Iwasawa algebra. However we could have started from the Euler system argument in a simpler situation over a discrete valuation ring to study the structure of the ideal class group over a fixed finite algebraic number field. In such a situation, the Euler system argument is less complicated and it is easier to understand the essential of the Euler system argument since we do not have to consider the action the Galois group for K^{cyc}_n/K. For the reader who wishes to see better the essential of the Euler system theory starting from the case over a discrete valuation ring, we refer to the book [**101**].

(2) We can reformulate the Euler system theory according to the idea of "Iwasawa theory of Galois deformation" as is presented later in Theorem 8.29 of §8.2 of the second volume of this book. Theorem 8.29 can be regarded as a generalization of the Euler system theory for cyclotomic deformations of various motives which was already established by Kato [**56**], Rubin [**101**], Perrin-Riou [**91**]. We recall that the proof in preceding work [**56**], [**101**] and [**91**] contains the complexity of dealing with the action the Galois group for K^{cyc}_n/K which was discussed above. Thus Theorem 8.29 obtained as a generalization of the Euler system which can be applied to Galois deformations such as Hida deformations gives also a much simplified another proof when we consider the classical setting of cyclotomic deformations of various motives.

It is a bit surprising that the philosophy and the technique of "Iwasawa theory of Galois deformation" which is the main theme of the part III of the book helps us to simplify the proof of the original Euler system theory in the setting of "the cyclotomic Iwasawa theory" which is discussed in the part I and the part II of the book.

Before the complete proof of the cyclotomic Iwasawa main conjecture for ideal class groups was established, it was already known that the cyclotomic Iwasawa main conjecture for ideal class groups follows from other conjectures Since these results motivated and supported this conjecture historically, it is important to recall these results below.

The case $K = \mathbb{Q}(\mu_p)$ is the most typical and the most elementary example. For this K, $(\zeta_p - 1)$ is the only prime of K which is over p and the prime $(\zeta_p - 1)$ is totally ramified in the cyclotomic $\mathbb{Z}_p$-extension $K_\infty^{\mathrm{cyc}}/K$. As we saw in the proof of Iwasawa's class number formula (Theorem 3.1), a lot of things become technically much easier under this situation. It was known that the cyclotomic Iwasawa main conjecture for ideal class groups for $K = \mathbb{Q}(\mu_p)$ follows from Vandiver's conjecture (Conjecture 3.25) or it is valid under some milder conditions. Iwasawa himself studied the case $K = \mathbb{Q}(\mu_p)$ deeply (see [**49**] for example).

As for the correlation between conjectures with which we can deduce the cyclotomic Iwasawa main conjecture for ideal class groups follows from some other conjectures, we have the following result.

PROPOSITION 3.93. *Let us take a primitive Dirichlet character ψ satisfying the same assumption as that of Theorem* 3.66. *Then we have the following implication (without assuming the results of Mazur–Wiles and Rubin explained above):*

(3.127) *Multiplicity one conjecture (Conjecture* 3.23*)*

$\Longrightarrow$ *$(-)$-type Iwasawa main conjecture (Conjecture* 3.66*)*

(3.128) *Greenberg's conjecture (Conjecture*3.26*) for $K = K_\psi$*

$\Longrightarrow$ *$(+)$-type Iwasawa main conjecture (Conjecture* 3.67*) for ψ*

Before giving the proof of this proposition, we recall that Stickelberger elements which appeared in §3.2.3 act on fractional ideals of abelian fields and their ideal classes are annihilated by Stickelberger elements. This well-known result is called Stickelberger's theorem (see [**117**, Thm. 6.10] for example). This type of Stickelberger's theorem combined with Iwasawa's construction of p-adic L-function using Stickelberger elements (see §3.2.3) implies the following theorem which we also call Stickelberger's theorem:

THEOREM 3.94 (Stickelberger's theorem). *Let us take a primitive Dirichlet character ψ satisfying the same assumption as that of Theorem* 3.66. *Then any $x \in (X_{K_\infty^{\mathrm{cyc}}})_{\omega\psi^{-1}}$ is annihilated by the principal ideal $L_p(\psi)\Lambda_{\mathrm{cyc},\psi} \cap \Lambda_{\mathrm{cyc},\psi}$*[71].

PROOF OF PROPOSITION 3.93. Let ψ be a nontrivial even character and let us assume Conjecture 3.23 for $\eta = \omega\psi^{-1}$. By Stickelberger's theorem for ψ, we obtain $(L_p(\psi)) \subset \mathrm{char}_{\Lambda_{\mathrm{cyc},\psi}}(X_{K_\infty^{\mathrm{cyc}}})_{\omega\psi^{-1}}$. Since this inclusion holds true for any nontrivial even character ψ whose conductor divides Np^e, the principle of the analytic class number formula (Theorem 3.74) implies that $(-)$-type Iwasawa main conjecture (Conjecture 3.66) is true for any nontrivial even character ψ whose conductor divides Np^e, which completes the proof of the first implication.

[71]Stickelberger's theorem as stated here is obtained by taking the projective limit of the statement of the original Stickelberger's theorem explained above for each intermediate field of K^{cyc}/K.

Next, we will prove the second implication. Since $K = K_\psi$ is a totally real number field, the Greenberg's conjecture for K implies that $X_{K_\infty^{\mathrm{cyc}}}$ is a pseudo-null Λ_{cyc}-module. Especially, $X_{K_\infty^{\mathrm{cyc}}}$ is a finite abelian group. Since the last term $(X_{K_\infty^{\mathrm{cyc}}})_\psi \otimes_{\mathbb{Z}_p} \mathbb{Q}_p$ of the exact sequence (3.103) is zero, Remark 3.83 (3) implies

$$\mathrm{char}_{\Lambda_{\mathrm{cyc},\psi}} \left(\varprojlim_n U_1(\mathfrak{r}_{K_n^{\mathrm{cyc}}} \otimes_{\mathbb{Z}} \mathbb{Z}_p) \Big/ \varprojlim_n \mathrm{loc}_p(z_{n,1})\Lambda_{\mathrm{cyc},\psi} \right)_\psi \subset \mathrm{char}_{\Lambda_{\mathrm{cyc},\psi}} (\mathfrak{X}_{K_\infty^{\mathrm{cyc}}})_\psi.$$

On the other hand, by (3.52), we have

$$\mathrm{char}_{\Lambda_{\mathrm{cyc},\psi}} \left(\varprojlim_n U_1(\mathfrak{r}_{K_n^{\mathrm{cyc}}} \otimes_{\mathbb{Z}} \mathbb{Z}_p) \Big/ \varprojlim_n \mathrm{loc}_p(z_{n,1})\Lambda_{\mathrm{cyc},\psi} \right)_\psi = (L_p(\psi)).$$

We thus obtain

$$(L_p(\psi)) \subset (\mathrm{char}_{\Lambda_{\mathrm{cyc},\psi}} (\mathfrak{X}_{K_\infty^{\mathrm{cyc}}})_\psi)^\bullet$$

for any nontrivial even character of $\mathrm{Gal}(\mathbb{Q}(\mu_{Np})/\mathbb{Q})$ and the principle of the analytic class number formula (Theorem 3.74) implies that (+)-type Iwasawa main conjecture (Conjecture 3.67)) is true for any nontrivial even character ψ whose conductor divides Np^e, which completes the proof of the second implication. □

3.4. Iwasawa main conjecture for class groups over general base fields

Let F be a finite algebraic number field. Let K/F be a finite abelian extension and ψ a character of the Galois group $\mathrm{Gal}(K/F)$. Recall that, for $F = \mathbb{Q}$, we formulated the cyclotomic Iwasawa theory for ideal class groups for any ψ and p and we proved it.

Now, we would like to ask if we can formulate and prove the Iwasawa main conjecture when F is a general finite algebraic number field. Specifically, we ask the following questions:

(1) Under appropriate assumptions, does there exist an analytic p-adic L-function which interpolates special values of related L-functions?
(2) If such a p-adic L-function exists, can we expect the Iwasawa main conjecture which relates the p-adic L-function to the characteristic ideal of certain ideal group?

3.4.1. Case of abelian extensions over totally real fields.

Most of results for Iwasawa theory stated earlier for abelian extensions of $\mathbb{Q}$ are generalized to abelian extensions of a totally real finite algebraic number field. Let us recall the following result:

THEOREM 3.95 (Siegel-Klingen). *Let F be a totally real finite algebraic number field and η a finite order character of the absolute Galois group G_F. Let us regard η as a Hecke character of finite order by class field theory and denote the L-function of the Hecke character η by $L(\eta, s)$ (see Definition* A.11 *for $L(\eta, s)$). Then, we have $L(\eta, 1-r) \in \mathbb{Q}[\eta]$ for any negative integer $1-r$.*

The Siegel-Klingen theorem above holds true for finite order characters of the absolute Galois group G_F for any number fields F. In Theorem 3.95, we restricted ourselves to the case where F is totally real since we consider only the totally real situation in this section. For the proof of the Siegel-Klingen theorem, we refer the reader to [**80**, Chap. VII, §9] for example. For a totally real finite algebraic number field F, we have $L(\eta, 1-r) \neq 0$ if and only if $(-1)^{1-r} = \eta(c)$ where $c \in G_F$ is the complex conjugate element. This fact follows from the functional equation of $L(\eta, s)$. For this fact, we refer the reader to [**80**, Chap. VII, §12] as well as the proof of Theorem 3.100 given later.

We set $\Gamma_{\mathrm{cyc},F} = \mathrm{Gal}(F_\infty^{\mathrm{cyc}}/F)$[72]. A totally real generalization of the construction of the p-adic L-function given in Theorem 3.30 is as follows:

THEOREM 3.96 (Barsky [**3**], Cassou-Noguès [**11**], Deligne-Ribet [**23**]). *Let F be a finite algebraic number field. Let ψ be a finite order character of G_F such that the fix field K_ψ of $\mathrm{Ker}(\psi)$ is totally real and that K_ψ satisfies $K_\psi \cap F_\infty^{\mathrm{cyc}} = F$. Then, there exists a unique element*

$$L_p(F,\psi) \in \begin{cases} \mathbb{Z}_p[\psi][[\Gamma_{\mathrm{cyc},F}]] & \text{when } \psi \neq \mathbf{1}, \\ \dfrac{1}{\gamma - \kappa_{\mathrm{cyc}}(\gamma)} \mathbb{Z}_p[[\Gamma_{\mathrm{cyc},F}]] & \text{when } \psi = \mathbf{1}, \end{cases}$$

which satisfies the following equality in $\overline{\mathbb{Q}}_p$:

$$\kappa_{\mathrm{cyc}}^{1-r} \phi(L_p(F,\psi)) = \prod_{\mathfrak{p}} \left(1 - \frac{(\psi\phi^{-1}\omega^{r-1})(\mathfrak{p})}{N(\mathfrak{p})^{1-r}}\right) \cdot L(\psi\phi^{-1}\omega^{r-1}, 1-r)$$

for any integer $r \geq 1$ and for any finite order character ϕ of $\Gamma_{\mathrm{cyc},F}$ [73].

REMARK 3.97. The construction by Barsky and the construction by Cassou-Noguès use both a theory of Shintani and they are quite similar. On the other and, the construction by Deligne-Ribet uses a compactification of a Hilbert modular variety and is totally different from the others. We refer the reader to [**18**], [**60**], [**95**] for surveys on these constructions.

Wiles [**124**] proved the cyclotomic Iwasawa main conjecture for class groups over totally real fields. The result is as follows:

THEOREM 3.98. *($(-)$-type Iwasawa main conjecture) Let F be a totally real finite algebraic number field. Let ψ be a finite order character of G_F such that the fix field K_ψ of $\mathrm{Ker}(\psi)$ is totally real and that K_ψ satisfies $K_\psi \cap F_\infty^{\mathrm{cyc}} = F$. Then the following equality holds:*

$$\mathrm{char}_{\mathbb{Z}_p[\psi][[\Gamma_{\mathrm{cyc},F}]]}(X_{K^{\mathrm{cyc}}_{\omega\psi^{-1},\infty}})_{\omega\psi^{-1}} = \begin{cases} (L_p(F,\psi)) & \text{when } \psi \neq \mathbf{1}, \\ (\gamma - \kappa_{\mathrm{cyc}}(\gamma))(L_p(F,\mathbf{1})) & \text{when } \psi = \mathbf{1}. \end{cases}$$

As in the case over $\mathbb{Q}$, we can also formulate $(+)$-type Iwasawa main conjecture in the general totally real situation and we can prove the equivalence with $(-)$-type Iwasawa main conjecture by the principle of Kummer duality. Also, there are some other results which are formulated and proved in the same way in the general totally real situation. We will not touch such results. On the other hand, there are some problems which are solved over $\mathbb{Q}$ but their totally real generalizations are difficult to prove. Knowing some problems which can not be worked out in a parallel manner seems more interesting. Here are some of important examples among such problems:

(1) As a totally real analogue of the Ferrero–Washington theorem (Theorem 3.60), it is conjectured that the μ-invariant of the p-adic L-function $L_p(F,\psi)$ is zero. This conjecture is still unsolved.

---

[72] Note that the natural map $\Gamma_{\mathrm{cyc},F} \longrightarrow \Gamma_{\mathrm{cyc},\mathbb{Q}} = \Gamma_{\mathrm{cyc}}$ is isomorphic if $F \cap \mathbb{Q}_\infty = \mathbb{Q}$.

[73] When $\psi = \mathbf{1}$, we exclude the case $r = 1$ and $\phi = \mathbf{1}$.

(2) Since the Leopoldt conjecture is still unsolved for abelian extensions of general totally real fields, the argument of the four-term exact sequence of Proposition 3.79 is not available in the general totally real situation.
(3) We can prove the cyclotomic Iwasawa main conjecture for class groups over totally real fields (Theorem 3.98) by the method of the Eisenstein ideal. Though the proof is technically more complicated, the main idea of the proof is similar by replacing elliptic modular forms by Hilbert modular forms. On the other hand, another proof of the cyclotomic Iwasawa main conjecture for class groups over totally real fields by the Euler system method is not known yet. The reason comes from the difficulty to find an Euler system over a general totally real field which generalizes the Euler system of cyclotomic units discussed in §3.3.3.

REMARK 3.99. Let F be a totally real finite algebraic number field and let $\widetilde{F}$ be the composite of abelian extensions of p-power degree over F which are unramified outside p. When $F = \mathbb{Q}$, we have $\widetilde{F} = \mathbb{Q}_\infty$.

In general, $\widetilde{F}$ is a finite extension of F_∞^{cyc} if the Leopoldt conjecture is true for (F, p), but $\widetilde{F}$ might not be equal to F_∞^{cyc} even when the Leopoldt conjecture is true for (F, p). We can construct a p-adic L-function $\widetilde{L}_p(F, \psi) \in \mathbb{Z}_p[\psi][[\mathrm{Gal}(\widetilde{F}/F)]]$ such that its image via $\mathbb{Z}_p[\psi][[\mathrm{Gal}(\widetilde{F}/F)]] \twoheadrightarrow \mathbb{Z}_p[\psi][[\Gamma_{\mathrm{cyc},F}]]$ coincides with the p-adic L-function $L_p(F, \psi) \in \mathbb{Z}_p[\psi][[\Gamma_{\mathrm{cyc},F}]]$ given in Theorem 3.96. On the algebraic side, we have an analogue of the $\mathbb{Z}_p[\psi][[\Gamma_{\mathrm{cyc},F}]]$-module $(X_{K^{\mathrm{cyc}}_{\omega\psi^{-1},\infty}})_{\omega\psi^{-1}}$ over $\mathbb{Z}_p[\psi][[\mathrm{Gal}(\widetilde{F}/F)]]$. It is interesting to try to generalize the Iwasawa main conjecture over $\mathbb{Z}_p[\psi][[\Gamma_{\mathrm{cyc},F}]]$ to the Iwasawa main conjecture over $\mathbb{Z}_p[\psi][[\mathrm{Gal}(\widetilde{F}/F)]]$ (see [**97**] for example).

3.4.2. Case of abelian extensions over CM-fields.

Next, we discuss the case over a finite algebraic number field which is not totally real. Let us recall the following theorem.

THEOREM 3.100. *Let F be a finite algebraic number field which is not totally real. For any finite order character of G_F and for any negative integer $1-k$, we have $L(\eta, 1-k) = 0$.*

PROOF. Recall that $r_1(F)$ (resp. $r_2(F)$) is the number of real embeddings of F (resp. the number of conjugate pairs of complex embeddings of F. We set

$$\Lambda(\eta, s) = \left(|D_F| N_{F/\mathbb{Q}}(C(\eta))\right)^{s/2} \Gamma_{\mathbb{R}}(s)^{r_1(F)} \Gamma_{\mathbb{C}}(s)^{r_2(F)} L(\eta, s)$$

where we define $\Gamma_{\mathbb{R}}(s) = \pi^{-s/2}\Gamma(s/2)$, $\Gamma_{\mathbb{C}}(s) = 2(2\pi)^{-s}\Gamma(s)$ and we denote by $C(\eta)$ the conductor of η. Then there exists a constant $W(\eta)$ with which we have the following functional equation[74]:

$$\Lambda(\eta, s) = W(\eta)\Lambda(\overline{\eta}, 1-s). \tag{3.129}$$

We have $\frac{\Gamma_{\mathbb{R}}(s)}{\Gamma_{\mathbb{R}}(1-s)} = \cos(\pi s/2)\Gamma_{\mathbb{C}}(s)$, $\frac{\Gamma_{\mathbb{C}}(s)}{\Gamma_{\mathbb{C}}(1-s)} = \frac{1}{2}\sin(\pi s)\Gamma_{\mathbb{C}}(s)^2$. Recall that the Gamma function $\Gamma(s)$ has no zeros and it has only poles of order 1 at integers equal to or smaller than 0. Hence, we have $L(\eta, 1-k) = 0$ for every negative integer $1-k$ when $r_2(F) > 0$. □

[74]See [**80**, Chap. VII] for the proof of the functional equation.

Let F be a finite algebraic number field F which is not totally real and ψ a finite order character of G_F. An analogue of the p-adic L-function given in Theorem 3.96 interpolating $L(\psi\phi^{-1}\omega^{r-1}, 1-r)$ for integers $r \geq 1$ and for finite order characters ϕ of $\Gamma_{\mathrm{cyc},F}$ must be identically zero by the above theorem. Since we have such a problem on the analytic side, we can not expect a natural analogue of the cyclotomic Iwasawa theory for ideal class groups in this case.

However, if F contains a CM-field, we might have a fairly reasonable setting of Iwasawa theory by considering the composite of all $\mathbb{Z}_p$-extensions $\widetilde{F}_\infty$ (see Theorem 2.6 for $\widetilde{F}_\infty$) in place of the cyclotomic $\mathbb{Z}_p$-extension of F.

Recall that Weil's theorem insists that a finite algebraic number field F contains a CM-field if and only if there exists a Hecke character η on F which is not of the form $\eta = N_F^r\psi$ with N_F the norm character, ψ a finite order Hecke character and r an integer (see Theorem A.8 for the precise statement). By the definition of the norm character, we have the equality $L(N_F^r\psi, s) = L(\psi, r+s)$. When F is not totally real, we have $L(\psi, j) = 0$ for every negative integer j as explained in Theorem 3.100 and it is difficult to draw some meaningful arithmetic invariants to interpolate from these special values. However, if F contains a CM-field[75], the values $L(\eta, j)$ of algebraic Hecke characters η which are not of the form $\eta = N_F^r\psi$ give interesting nonzero arithmetic invariants. An important algebraicity theorem for the ratio between $L(\eta, j)$ and a Deligne period related to η is known and we can expect a nonzero p-adic L-function which interpolates some of these values $L(\eta, j)$ over the space of continuous characters of $\mathrm{Gal}(\widetilde{F}_\infty/F)$[76]. In fact, when F is a CM-field, we have a p-adic L-function as is presented in Theorem 3.104 below and we can discuss the Iwasawa main conjecture using this p-adic L-function (see Conjecture 3.107).

There are not so many known results when F contains a CM-field but F itself is not a CM-field. Hence, we will assume that F is a CM-field in the rest of §3.4.2 and we present some known results or conjectures under this situation. We denote by $\widetilde{\Gamma}_F$ the Galois group $\mathrm{Gal}(\widetilde{F}_\infty/F)$ where $\widetilde{F}_\infty$ is the composite of all $\mathbb{Z}_p$-extensions of F and we will discuss Iwasawa theory over $\mathcal{O}[[\widetilde{\Gamma}_F]]$.

We prepare some notation.

DEFINITION 3.101. Let F be a CM finite algebraic number field.

(1) Let J_F be the set of embeddings of F into $\overline{\mathbb{Q}}$. Then a subset Σ of J_F is called a **CM type** of F if we have $\Sigma \bigcup \overline{\Sigma} = J_F$ and $\Sigma \bigcap \overline{\Sigma} = \emptyset$ with $\overline{\Sigma}$ the complex conjugate of Σ.

(2) A CM type Σ of F is called a **p-ordinary CM type** if the p-adic valuation on F induced from the embedding $\iota_p \circ \iota_\infty^{-1} \circ \sigma : F \hookrightarrow \overline{\mathbb{Q}}_p$ and the p-adic valuation on F induced from the embedding $\iota_p \circ \iota_\infty^{-1} \circ \tau : F \hookrightarrow \overline{\mathbb{Q}}_p$ are different for any $\sigma \in \Sigma$ and for any $\tau \in \overline{\Sigma}$.

REMARK 3.102. Let F be a CM finite algebraic number field F^+ the maximal totally real subfield of F.

(1) There exists a p-ordinary CM type of F if and only if all primes of F^+ over p split at F/F^+.

---

[75]Even when F is a non totally real finite algebraic number field which does not contain any CM-fields we have a formulation of [**90**] through the p-adic Beilinson conjecture.

[76]Later in §6.1 of the second volume of this book, we discuss a condition called the critical condition for the algebraicity of η and we give a Hodge theoretic criterion for the critical condition.

(2) Suppose that all primes of F^+ over p split at F/F^+ and denote the number of primes of F^+ over p by s. Then the number of p-ordinary CM types Σ of F is equal to 2^s.

We fix a complex torus $\mathbb{C}^\Sigma/\mathfrak{A}$ where $\mathfrak{A}$ is a fractional ideal of F which is prime to p. Let us fix an element $\delta \in F$ which satisfies the following two conditions:

(i) We have $\overline{\delta} = -\delta$ and we have $\mathrm{Im}(\sigma(\delta)) > 0$ for any $\sigma \in \Sigma$.

(ii) The exterior product $\langle u, v\rangle = \frac{u^c v - v^c u}{2\delta}$ induces the isomorphism $\mathfrak{r}_F \wedge_{\mathfrak{r}_{F^+}} \mathfrak{r}_F \cong d_{F^+}^{-1}\mathfrak{c}^{-1}$ where d_{F^+} is the absolute different of F^+ and $\mathfrak{c}$ is a fractional ideal of F^+ which is prime to p.

By the choice of an element δ as above, $\mathbb{C}^\Sigma/\mathfrak{A}$ is equipped with a polarization, hence there exists an abelian variety $B(\mathfrak{A})_{\mathbb{C}}$ over $\mathbb{C}$ of dimension $d = \frac{[F:\mathbb{Q}]}{2}$ such that $B(\mathfrak{A})_{\mathbb{C}}(\mathbb{C}) \cong \mathbb{C}^\Sigma/\mathfrak{A}$ (see [**78**, Chap. I] for example). By the theory of complex multiplication, there exists an abelian variety $B(\mathfrak{A})_{\overline{\mathbb{Q}}}$ over $\overline{\mathbb{Q}}$ satisfying the following conditions (see [**111**, Chap. III] for example):

(1) We have $B(\mathfrak{A})_{\overline{\mathbb{Q}}} \times_{\mathrm{Spec}(\overline{\mathbb{Q}})} \mathrm{Spec}(\mathbb{C}) \cong B(\mathfrak{A})_{\mathbb{C}}$.

(2) The abelian variety $B(\mathfrak{A})_{\overline{\mathbb{Q}}}$ has good reduction over $\overline{\mathbb{Z}}_{(p)}$ where $\overline{\mathbb{Z}}_{(p)}$ is the localization of the ring of integers $\overline{\mathbb{Z}}$ of $\overline{\mathbb{Q}}$ at the maximal ideal induced by the inverse image of the maximal ideal of $\overline{\mathbb{Q}}_p$ by the embedding ι_p fixed in (3.76).

Let $B(\mathfrak{A})_{\overline{\mathbb{Z}}_{(p)}}$ be a smooth projective model of $B(\mathfrak{A})_{\overline{\mathbb{Q}}}$ and $\Omega^1_{B(\mathfrak{A})_{\overline{\mathbb{Z}}_{(p)}}}$ the space of regular one-forms on $B(\mathfrak{A})_{\overline{\mathbb{Z}}_{(p)}}$. Since $\mathfrak{r}_F$ acts on $B(\mathfrak{A})_{\overline{\mathbb{Z}}_{(p)}}$, $\Omega^1_{B(\mathfrak{A})_{\overline{\mathbb{Z}}_{(p)}}}$ is an $\mathfrak{r}_F \otimes_{\mathbb{Z}} \overline{\mathbb{Z}}_{(p)}$-module. In fact, $\Omega^1_{B(\mathfrak{A})_{\overline{\mathbb{Z}}_{(p)}}}$ is a free $\mathfrak{r}_F \otimes_{\mathbb{Z}} \overline{\mathbb{Z}}_{(p)}$-module of rank one. We choose and fix an $\mathfrak{r}_F \otimes_{\mathbb{Z}} \overline{\mathbb{Z}}_{(p)}$-basis $\omega(\mathfrak{A})$ of $\Omega^1_{B(\mathfrak{A})_{\overline{\mathbb{Z}}_{(p)}}}$. The assumption that all primes of F^+ over p split at F/F^+ implies that the abelian variety $B(\mathfrak{A})_{\overline{\mathbb{Q}}}$ is p-ordinary type at every prime of $\overline{\mathbb{Q}}$ over p. Let $\mathbb{C}_p$ be the completion of $\overline{\mathbb{Q}}_p$ and $\mathcal{O}_{\mathbb{C}_p}$ the ring of integers of $\mathbb{C}_p$. The choice of an element δ satisfying the conditions (i) and (ii) above gives an isomorphism $i_p : \widehat{(\mathbb{G}_m)_{\mathcal{O}_{\mathbb{C}_p}}} \otimes_{\mathbb{Z}} d_{F^+}^{-1} \xrightarrow{\sim} \widehat{B(\mathfrak{A})_{\mathcal{O}_{\mathbb{C}_p}}}$ where $\widehat{B(\mathfrak{A})_{\mathcal{O}_{\mathbb{C}_p}}}$ is the formal group over $\mathcal{O}_{\mathbb{C}_p}$ associated to $B(\mathfrak{A})_{\overline{\mathbb{Z}}_{(p)}}$ [77]. We set $\omega_{\mathrm{can}}(\mathfrak{A}) = (i_p)_*\left(\frac{dt}{t} \otimes 1\right)$ where $\frac{dt}{t} \otimes 1$ is the standard regular one-form on $\widehat{(\mathbb{G}_m)_{\mathcal{O}_{\mathbb{C}_p}}} \otimes_{\mathbb{Z}} d_{F^+}^{-1}$. We define an element $C_p^{\mathrm{CM}} \in (\mathfrak{r}_F \otimes_{\mathbb{Z}} \mathcal{O}_{\mathbb{C}_p})^\times$ to be

$$C_p^{\mathrm{CM}} = \omega(\mathfrak{A})/\omega_{\mathrm{can}}(\mathfrak{A}). \tag{3.130}$$

Let $p_\infty : \mathbb{C}^\Sigma \twoheadrightarrow B(\mathfrak{A})_{\mathbb{C}}$ be the complex uniformization of $B(\mathfrak{A})_{\mathbb{C}} = \mathbb{C}^\Sigma/\mathfrak{A}$ and we set $\omega_{\mathrm{trans}}(\mathfrak{A}) = (p_\infty)_*(\wedge_{\sigma\in\Sigma} dz_\sigma)$. We define an element $C_\infty^{\mathrm{CM}} \in (F \otimes_{\mathbb{Q}} \mathbb{C})^\times$ to be

$$C_\infty^{\mathrm{CM}} = \omega(\mathfrak{A})/\omega_{\mathrm{trans}}(\mathfrak{A}). \tag{3.131}$$

DEFINITION 3.103. For each $\sigma \in \Sigma$, taking the σ-component of the element $C_p^{\mathrm{CM}} \in (\mathfrak{r}_F \otimes_{\mathbb{Z}} \mathcal{O}_{\mathbb{C}_p})^\times$ (resp. $C_\infty^{\mathrm{CM}} \in (F \otimes_{\mathbb{Q}} \mathbb{C})^\times$) given above, we obtain $C_p^{\mathrm{CM}}(\sigma) \in \mathcal{O}_{\mathbb{C}_p}^\times$ (resp. $C_\infty^{\mathrm{CM}}(\sigma) \in \mathbb{C}^\times$). For $n = \sum_{\sigma\in\Sigma} n_\sigma \sigma \in \mathbb{Z}[\Sigma]$, we define $(C_p^{\mathrm{CM}})^n \in \mathcal{O}_{\mathbb{C}_p}$ (resp. $(C_\infty^{\mathrm{CM}})^n \in \mathbb{C}$) to be $\prod_{\sigma\in\Sigma}(C_p^{\mathrm{CM}}(\sigma))^{n_\sigma}$ (resp. $\prod_{\sigma\in\Sigma}(C_\infty^{\mathrm{CM}}(\sigma))^{n_\sigma}$). We call this element a **p-adic CM period** (resp. a **complex CM period**).

[77] See [**59**, §5.1, 5.8] for the definition of i_p and its dependence on δ.

We set $\Sigma_p = \{\iota_p \circ \iota_\infty^{-1} \circ \sigma : F \hookrightarrow \overline{\mathbb{Q}}_p | \sigma \in \Sigma\}$ and we denote by S_F the set of primes of F over p which are induced by elements of Σ_p. Recall that $\widetilde{F}_\infty$ is the composite of all $\mathbb{Z}_p$-extensions of F and $\widetilde{\Gamma}_F$ is defined to be $\mathrm{Gal}(\widetilde{F}_\infty/F)$.

THEOREM 3.104 (Katz [**59**], Hida-Tilouine [**43**]). *Let F be a CM finite algebraic number field. Suppose that all primes of the maximal totally real subfield F^+ of F over p split at F/F^+ and we choose and fix a p-ordinary CM type Σ of F. We also fix an element $\delta \in F$ which satisfies the conditions (i) and (ii) above. Let ψ be a finite order character of G_F such that the fix field K_ψ of $\mathrm{Ker}(\psi)$ satisfies $K_\psi \cap F_\infty^{\mathrm{cyc}} = F$.*

Then there exists a unique element $L_p(F,\Sigma,\psi) \in \mathcal{O}_{\mathbb{C}_p}[[\widetilde{\Gamma}_F]]$ which satisfies the following equality in $\overline{\mathbb{Q}}_p$:

$$
\begin{aligned}
(3.132)\quad \frac{\eta^{\mathrm{gal}}(L_p(F,\Sigma,\psi))}{(C_p^{\mathrm{CM}})^{\{m+2d(\sigma)\}}} &= [\mathfrak{r}_F^\times : \mathfrak{r}_{F^+}^\times] \frac{(-1)^{md}\prod_{\sigma\in\Sigma}(\pi^{d(\sigma)}\Gamma(m+d(\sigma)))}{\sqrt{|D_{F^+}|}\prod_{\sigma\in\Sigma}\mathrm{Im}(\sigma(\delta))^{d(\sigma)}}\tau(\eta\psi) \\
&\times \prod_{\mathfrak{p}\in S_F}(1-\eta\psi(\overline{\mathfrak{p}}))(1-(\eta\psi)^*(\overline{\mathfrak{p}}))\cdot \frac{L(\eta\psi,0)}{(C_\infty^{\mathrm{CM}})^{\{m+2d(\sigma)\}}}
\end{aligned}
$$

for any algebraic Hecke character η on F satisfying the following two conditions:

(i) There exist $m \in \mathbb{Z}_{>0}$ and $\{d(\sigma) \in \mathbb{Z}_{\geq 0}\}_{\sigma\in\Sigma}$ (or $m \in \mathbb{Z}_{\leq 0}$ and $\{d(\sigma) \in \mathbb{Z} \mid m+d(\sigma)-1 \geq 0\}_{\sigma\in\Sigma}$) such that the weight of η is equal to $-m\sum_{\sigma\in\Sigma}\sigma - \sum_{\sigma\in\Sigma}d(\sigma)(\sigma-\overline{\sigma})$ [78].

(ii) The Galois character η^{gal} associated to η factors through the quotient $\mathrm{Gal}(\widetilde{F}_\infty/F)$ of G_F [79].

Here the symbol $\{m+2d(\sigma)\}$ which appears in the both sides of (3.132) *denotes $\sum_{\sigma\in\Sigma}(m+2d(\sigma))\sigma \in \mathbb{Z}[\Sigma]$, $\overline{\sigma}$ denotes the complex conjugate of σ and D_{F^+} is the absolute discriminant of F^+. Also, $(\eta\psi)^*$ denotes the algebraic Hecke character $(\eta\psi)^*(x) = \eta\psi(\overline{x})^{-1}N_F(x)^{-1}$ and $\tau(\eta\psi)$ denotes the generalized Gauss sum associated to an algebraic Hecke character $\eta\psi$ [80].*

REMARK 3.105.
(1) Condition (i) of Theorem 3.104 is equivalent to "the critical condition" studied by Shimura and Deligne for special values of automorphic L-functions or Hasse–Weil L-functions. Under the critical condition, we conjecture that the ratio of the special value and a suitable complex period is an algebraic number (see [**21**] for the precise formulation of the conjecture). In fact, this conjecture is proved by Damerell when F is an imaginary quadratic field and by Katz when F is a general CM field. We remark that the right-hand side of the interpolation formula (3.132) is an element of $\overline{\mathbb{Q}}$ hence it naturally makes sense as an element of $\mathbb{C}_p$.
(2) The idea of the construction of the p-adic L-function $L_p(F,\Sigma,\psi)$ in Theorem 3.104 is to specialize a p-adic family of Eisenstein series on the algebraic group $GL(2)$ over the totally real field F^+ at CM points.

[78] See Definition A.6 for the definition of the weight of an algebraic Hecke character.

[79] See Proposition A.10 for the definition of the Galois character associated to an algebraic Hecke character.

[80] Here we omit the precise definition of the generalized Gauss sum $\tau(\eta\psi)$, but we refer the reader to Definition 6.46 for the definition of the generalized Gauss sum when F is an imaginary quadratic field.

We explain the idea through an example of the most elementary case where F is an imaginary quadratic field F (i.e. $F^+ = \mathbb{Q}$). Let us present the ring of integers of F as $\mathbb{Z} + \alpha\mathbb{Z}$. If we substitute $z = \alpha$ in the real analytic Eisenstein series:

$$E_{k,s}(z)|_{s=j} = \sum_{\substack{(m,n)\in\mathbb{Z}^2\\(m,n)\neq(0,0)}} \frac{1}{(mz+n)^k|mz+n|^{2j}},$$

the value is equal to the special value $L(\eta_k N_F^j, 0)$ where η_k is an algebraic Hecke character of weight $k\sigma$ such that $\eta_k((a)) = a^k$. Hence, if we consider a p-adic family of real analytic Eisenstein series and we evaluate the family at a CM point, we obtain a p-adic L-function which interpolates $L(\eta_k N_F^j, 0)$ in which η_k and j vary[81].

(3) Recall that the Kubota–Leopoldt p-adic L-function obtained at Theorem 3.30 has another construction by Coleman map using a norm-compatible system of cyclotomic units (see §3.2.4). When F is an imaginary quadratic field, the CM p-adic L-function $L_p(F, \Sigma, \psi)$ has another construction by Coleman map using a norm-compatible system of elliptic units. After a pioneering work of Coates–Wiles [**14**] relating a system of elliptic units to L-values, this construction was first done by Yager ([**128**], [**129**]) when the class number of F is one. Then de Shalit[**25**] generalized this construction to the case where the class number of F is not necessarily one.

REMARK 3.106. Assume that F is an imaginary quadratic field and let $\mathfrak{p}$ be the prime of F over p which is induced by the fixed p-ordinary CM-type of F. We denote the unique $\mathbb{Z}_p$-extension of F which is ramified only over $\mathfrak{p}$ by $F_\infty^{(\mathfrak{p})}$ and the Galois group $\mathrm{Gal}(F_\infty^{(\mathfrak{p})}/F)$ by $\Gamma_\mathfrak{p}$. Gillard [**31**] and Schneps [**103**] applied the method of Sinnott which was explained after Ferrero–Washington's theorem (Theorem 3.60) as another proof and they proved that the image of the two-variable CM p-adic L-function $L_p(F, \Sigma, \psi) \in \mathcal{O}[[\widetilde{\Gamma}_F]]$ by the quotient map to a one-variable Iwasawa algebra $\mathcal{O}[[\widetilde{\Gamma}_F]] \twoheadrightarrow \mathcal{O}[[\Gamma_\mathfrak{p}]]$ is not divisible by a uniformizer ϖ of $\mathcal{O}$ (that is, the Iwasawa μ-invariant of the image in $\mathcal{O}[[\Gamma_\mathfrak{p}]]$ is zero). As a corollary, it follows immediately that the two-variable CM p-adic L-function $L_p(F, \Sigma, \psi) \in \mathcal{O}[[\widetilde{\Gamma}_F]]$ is not divisible by ϖ. Thus an analogue of Ferrero–Washington theorem holds true for the image in $\mathcal{O}[[\Gamma_\mathfrak{p}]]$. However, the analogue of Ferrero–Washington theorem is not known for the cyclotomic p-adic L-functions associated to algebraic Hecke characters obtained by specializing $L_p(F, \Sigma, \psi) \in \mathcal{O}[[\widetilde{\Gamma}_F]]$

When F is a general CM finite algebraic number field, Hida [**42**] proves that $L_p(F, \Sigma, \psi) \in \mathcal{O}[[\widetilde{\Gamma}_F]]$ is not divisible by ϖ. Hida proves this by showing that the μ-invariant of the image in the anticyclotomic Iwasawa algebra is zero in place of considering the image in the $\mathfrak{p}$-ramified Iwasawa algebra as in Gillard [**31**] and Schneps [**103**].

[81]The explanation here is only a rough sketch of the idea and is not precise enough. If we make the meaning of "evaluating at a CM point" precise, complex periods and p-adic periods come into the description of the interpolation property.

Let K be a finite abelian extension of F such that $K \cap \widetilde{F}_\infty = F$. We denote the composite $K\widetilde{F}_\infty$ by $\widetilde{K}_\infty^{\mathrm{CM}}$ and the maximal abelian p-power extension of $\widetilde{K}_\infty^{\mathrm{CM}}$ which is unramified outside S_F by M^Σ. The Galois group $X_K^\Sigma = \mathrm{Gal}(M^\Sigma/\widetilde{K}_\infty^{\mathrm{CM}})$ is a compact $\mathbb{Z}_p$-module with conjugate action of $\widetilde{\Gamma}_F = \mathrm{Gal}(\widetilde{K}_\infty^{\mathrm{CM}}/K)$, hence is regarded as a compact $\mathbb{Z}_p[[\widetilde{\Gamma}_F]]$-module. It is known that X_K^Σ is a finitely generated torsion $\mathbb{Z}_p[[\widetilde{\Gamma}_F]]$-module (see [**44**, Thm. 1.2.2]).

CONJECTURE 3.107 (Iwasawa Main Conjecture). *Under the same setting as Theorem 3.104, we have the following equality of ideals of $\mathcal{O}_{\mathbb{C}_p}[[\widetilde{\Gamma}_F]]$:*

$$\mathrm{char}_{\mathcal{O}_{\mathbb{C}_p}[[\widetilde{\Gamma}_F]]}\left((X_{K_\psi}^\Sigma)_\psi \otimes_{\mathbb{Z}_p[\psi][[\widetilde{\Gamma}_F]]} \mathcal{O}_{\mathbb{C}_p}[[\widetilde{\Gamma}_F]]\right) = (L_p(F,\Sigma,\psi)). \tag{3.133}$$

When F is an imaginary quadratic field, we have elliptic units on abelian extensions of F. Under the assumption that p is decomposed in F and $p \nmid w_H \#\mathrm{Cl}(F)|\psi|$, Rubin [**100**, Thm. 4.1] proved

$$\begin{aligned}(3.134)\quad & \mathrm{char}_{\mathbb{Z}_p[\psi][[\widetilde{\Gamma}_F]]}(X_{K_\psi}^\Sigma)_\psi \\ & \supset \mathrm{char}_{\mathbb{Z}_p[\psi][[\widetilde{\Gamma}_F]]}(\text{local units over } \widetilde{K}_{\psi,\infty}^{\mathrm{CM}})_\psi \,/(\text{elliptic units})_\psi\end{aligned}$$

using the Euler system of elliptic units as we discussed in the argument using the Euler system of cyclotomic units in §3.3.3. Here, w_H is the number of roots of unity contained in the Hilbert class field H of F and $|\psi|$ is the order of ψ.

On the other hand, as we explained in Remark 3.105 (3), Yager and de Shalit proved that

$$\mathrm{char}_{\mathbb{Z}_p[\psi][[\widetilde{\Gamma}_F]]}(\text{local units over } \widetilde{K}_{\psi,\infty}^{\mathrm{CM}})_\psi \,/(\text{elliptic units})_\psi = (L_p(F,\Sigma,\psi)). \tag{3.135}$$

Hence, when all technical assumptions required in (3.134), (3.135) are satisfied, we obtain

$$\mathrm{char}_{\mathcal{O}_{\mathbb{C}_p}[[\widetilde{\Gamma}_F]]}\left((X_{K_\psi}^\Sigma)_\psi \otimes_{\mathbb{Z}_p[\psi][[\widetilde{\Gamma}_F]]} \mathcal{O}_{\mathbb{C}_p}[[\widetilde{\Gamma}_F]]\right) \supset (L_p(F,\Sigma,\psi)). \tag{3.136}$$

For a fixed abelian extension K of F, we can prove the principle of the analytic class number formula as follows:
(3.137)

$$\prod_{K_\psi = K} \mathrm{char}_{\mathcal{O}_{\mathbb{C}_p}[[\widetilde{\Gamma}_F]]}\left((X_{K_\psi}^\Sigma)_\psi \otimes_{\mathbb{Z}_p[\psi][[\widetilde{\Gamma}_F]]} \mathcal{O}_{\mathbb{C}_p}[[\widetilde{\Gamma}_F]]\right) = \prod_{K_\psi = K} (L_p(F,\Sigma,\psi)),$$

which is an analogue of Theorem 3.74 for $F = \mathbb{Q}$. By combining (3.136) and (3.137), we obtain the following theorem:

THEOREM 3.108 (Rubin). *Let F be an imaginary quadratic field of class number one, p an odd prime number which is decomposed in F. If we have $p \nmid w_H$ and the order of ψ is prime to p, the Iwasawa main conjecture (Conjecture 3.107) holds true.*

REMARK 3.109. (1) For general CM-fields F which are not imaginary quadratic fields, analogues of elliptic units are not found yet. Hence, a generalization of Theorem 3.108 by the method of Euler systems is an open problem. On the other hand, there are some progress by the method using the congruence of p-adic families of modular forms as follows.

(a) We can decompose $\widetilde{\Gamma}_F$ as $\widetilde{\Gamma}_F \cong (\widetilde{\Gamma}_F)^+ \times (\widetilde{\Gamma}_F)^-$ according to the action of complex conjugate. We note that, when the Leopoldt conjecture is true, we have $(\widetilde{\Gamma}_F)^+ \cong \mathbb{Z}_p$ and $(\widetilde{\Gamma}_F)^+$ is nothhing but $\Gamma_{\mathrm{cyc},F}$. The group $(\widetilde{\Gamma}_F)^-$ is isomorphic to $\mathbb{Z}_p^d$ where d is a positive integer equal or larger than $\frac{[F:\mathbb{Q}]}{2}$ and we have $d = \frac{[F:\mathbb{Q}]}{2}$ when the Leopoldt conjecture is true. We call the extension of F whose Galois group is $(\widetilde{\Gamma}_F)^-$ the anti-cyclotomic $\mathbb{Z}_p^d$-extension of F. Under certain technical conditions, Mazur–Tilouine [**74**] proved the inclusion

(3.138)
$$\text{(char. ideal of the Selmer group)} \subset \text{(principal ideal of the } p\text{-adic } L\text{-function)}$$

in $\mathscr{O}_{\mathbb{C}_p}[[(\widetilde{\Gamma}_F)^-]]$ by the method using the congruence module over the Hecke algebra which controls the congruence between modular forms. We remark that this inclusion is the opposite to the one obtained by the method of Euler systems. This method was generalized by Hida–Tilouine ([**43**], [**44**]) to prove (3.138) for F general CM-fields F. It seems that the condition on ψ is a bit delicate in this modular method and the validity of one of two inequalities does not imply the validity of the desired equality as in Rubin's case discussed above. Hida [**40**] deduces the equality of the Iwasawa main conjecture assuming $R = T$ theorem obtained in the preprint [**30**] by Taylar-Wiles system argument.

(b) Hsieh [**45**] applies the method ofthe Eisenstein ideal for modular forms on the algebraic group $\mathrm{U}(2,1)$, which is a higher-dimensional analogue of the method of Mazur–Wiles for $\mathrm{GL}(2)$ over $\mathbb{Q}$. He studies the congruence between a cuspform and an Eisenstein series in the Hida family of $\mathrm{U}(2,1)$ for the inclusion of type (3.138) in $\mathscr{O}_{\mathbb{C}_p}[[\widetilde{\Gamma}_F]]$ under certain conditions.

(2) The same type of generalization discussed in Remark 3.99 when the base field F was totally real should be studied. That is, we can study the Iwasawa main conjecture over $\mathscr{O}_{\mathbb{C}_p}[[\mathrm{Gal}(\widetilde{F}/F)]]$ where $\widetilde{F}$ is a finite extension of $\widetilde{F}_\infty$ which is defined to be the composite of abelian extensions of p-power degree over F which are unramified outside p.

3.5. Problems and perspective beyond the Iwasawa main conjecture

Through this chapter until the last section, we developed the Iwasawa theory by focusing on the Iwasawa main conjecture. Hence, we cut some of topics which are not directly linked to the Iwasawa main conjecture. In this section, we discuss some of these topics. We recall some well-known conjectures and problems. Also we will try to propose some important directions and plans of research. Some of these propositions will make sense also in the settings of Part II and Part III of this book. We hope that they contain some seeds for the future research of Iwasawa theory.

3.5.1. Problems and perspective on ideal class groups (algebraic aspect). In this subsection, we concentrate on the side of ideal group. We will present some open problems and some conjectures as well as related known results.

(1) There are divisions of opinion among the researchers on the validity of Vandiver's conjecture (Conjecture 3.25) stated earlier. In any case, we expect to progress either in the sense that the conjecture is solved partially or a counter example is found. As a partial result, Kurihara [**66**] proved $\mathrm{Cl}(\mathbb{Q}(\mu_p))[p]^{\omega^{p-3}} = 0$ for any prime p. Also, Soulé [**114**] proved $\mathrm{Cl}(\mathbb{Q}(\mu_p))[p]^{\omega^{p-n}} = 0$ when the prime number p is sufficiently large compared to n through the calculation of higher K-groups of the ring of integers $\mathbb{Z}$.

(2) As for the Greenberg's conjecture for totally real finite algebraic number fields K (Conjecture 3.26), only known results are essentially for the case where K is a real quadratic field. On the other hand, a generalization of the conjecture to the case where K is not necessarily totally real is known. We denote by $\widetilde{K}_\infty$ the composite of all $\mathbb{Z}_p$-extensions of K and we set $\widetilde{\Gamma}_K = \mathrm{Gal}(\widetilde{K}_\infty/K)$ [82]. We denote by $\widetilde{L}_\infty$ the composite of everywhere unramified abelian extensions of p-power degree over K_∞ and we set $X_{\widetilde{K}_\infty} = \mathrm{Gal}(\widetilde{L}_\infty/\widetilde{K}_\infty)$. Then the **general Greenberg's conjecture** asserts that $X_{\widetilde{K}_\infty}$ is a pseudo-null $\mathbb{Z}_p[[\widetilde{\Gamma}_K]]$-module.

(3) As we have seen in the proof of the Iwasawa main conjecture of §3.3.2, modular forms and Hecke algebras for the algebraic group GL(2) often play important roles to study the ideal group which is the Selmer group for the algebraic group GL(1) by the method of the Eisenstein ideal. For example, by Harder–Pink, Kurihara, it was known that the structure of the ψ-part of ideal groups for a Dirichlet character ψ is related to the structure of a local component of the Hecke algebra which is associated to an Eisenstein series for ψ. More recently, Ohta (see [**85**], [**86**] for example) discovered that the validity of Conjecture 3.21 (the semi-simplicity conjecture) and Conjecture 3.23 (the multiplicity one conjecture) for ideal class groups is more or less equivalent to the validity of the Gorenstein property and the principality of the Eisenstein ideal of the Hecke algebra for modular forms on the algebraic group GL(2) respectively. After Ohta, Sharifi [**110**] formulated a refined conjecture for the structure of ideal group by means of modular forms on the algebraic group GL(2).

We expect that some of the problems on the algebraic side discussed above are studied and developed in a suitable manner also in the context of the generalized Iwasawa theory discussed after the next chapter.

3.5.2. Problems and perspective on p-adic L-functions (analytic aspect). As we have seen in Theorem 3.30, the Kubota–Leopoldt p-adic L-function $L_p(\psi) \in \Lambda_{\mathrm{cyc},\psi}$ was characterized by the interpolation property of values at arithmetic characters $\kappa^j_{\mathrm{cyc}}\phi$ with $j \leq 0$. Recall also that the value of $L_p(\psi)$ at $\kappa^j_{\mathrm{cyc}}\phi$ with $j = 1$ was given by the Leopoldt formula (Theorem 3.62). What can we say about the value of $L_p(\psi)$ at $\kappa^j_{\mathrm{cyc}}\phi$ with $j \geq 2$?

(1) Recall that Deligne conjecture (see [**21**]), the Beilinson conjecture (see [**132**]) and the Tamagawa number conjecture of Bloch–Kato (see [**5**]) are known for the geometric meaning of the special values for complex

[82] See Theorem 2.6 (2) and Conjecture 2.7 for the $\mathbb{Z}_p$-rank of $\mathrm{Gal}(\widetilde{K}_\infty/K)$.

L-functions such as Dirichlet L-functions $L(\eta, s)$ or more general Hasse–Weil L-functions. It seems interesting to consider p-adic analoques of known results or known conjectures on the special values for complex L-functions.

(2) For example, it is conjectured that special values of Riemann's zeta function $\zeta(s)$ at integers $j \geq 2$ are all transcendental numbers. For even j's, the ratio of $\zeta(j)$ and π^j are nonzero rational numbers, hence the conjecture holds to be true. However, very few results are known for odd j's. Essentially, we only know that Apéry proved that $\zeta(3)$ is an irrational number and that Rivoal proved that there are infinitely many odd positive integers j such that $\zeta(j)$ are irrational numbers. As for the p-adic analogue of these results, the only known results for the Kubota–Leopoldt p-adic L-function $L(\eta, s)$ are the irrationality of special values at κ_{cyc}^j for $j = 2, 3$ and $p = 2, 3$ proved by Beukers and Calegari.

(3) As for special values of the Kubota–Leopoldt p-adic L-function $L(\eta, s)$ at $\kappa_{\text{cyc}}^j \phi$, these values are nonzero algebraic numbers for integers $j \leq 0$ according to the interpolation property of Theorem 3.30. Since these values are expected to be irrational for positive integers j, they must be always nonzero. However the nonvanishing of values for positive integers j is not yet proved.

Below, we discuss a p-adic analogue of the Beilinson conjecture mentioned in (1) above in a bit more detailed manner.

In general, assuming that the Hasse–Weil L-function $L(\mathbb{M}, s)$ associated to a motive $\mathbb{M}$ is well-defined and meromorphically continued to the whole $\mathbb{C}$-plane, we have some conjectures on the geometric meaning of values $L(\mathbb{M}, j) = L(\mathbb{M}(j), 0)$ at integers j. These conjectures are formulated differently depending on the classification of the integer j. If the Tate-twist $\mathbb{M}(j)$ is critical in the sense of Deligne (see Remark 5.4), we have the Deligne conjecture for $L(\mathbb{M}, j)$. If the Tate-twist $\mathbb{M}(j)$ is noncritical, we have the Beilinson conjecture for $L(\mathbb{M}, j)$. Now we consider the case where $\mathbb{M}$ is a Dirichlet motive associated to a Dirichlet character η. In this case, we have $L(\mathbb{M}, s) = L(\eta, s)$ and we note that

$$\text{“}\mathbb{M}(j) \text{ is critical”} \iff \text{“}j \geq 1 \text{ and } \eta(-1) = (-1)^j\text{”} \text{ or } \text{“}j \leq 0 \text{ and } \eta(-1) \neq (-1)^j\text{”}.$$

For an integer $j \geq 1$, we have the following functional equation[83]:

$$L(\eta, j) = \frac{\tau(\eta)}{(\sqrt{-1})^{p(\eta)}\sqrt{M}} \left(\frac{M}{\pi}\right)^{\frac{1-2j}{2}} \lim_{s \to j} \frac{\Gamma(\frac{1-s+p(\eta)}{2})}{\Gamma(\frac{s+p(\eta)}{2})} L(\overline{\eta}, 1-s), \tag{3.139}$$

where M is the conductor of η and $p(\eta) \in \{0, 1\}$ is defined by $\eta(-1) = (-1)^{p(\eta)}$. Hence, it suffices to study the values at integers $j \geq 1$.

The period integral plays an important role for the Deligne conjecture and the regulator map on K-groups plays an important role for the Beilinson conjecture.

[83] The function $\Gamma(s)$ in the right-hand side has a pole of order one at every negative integer j, but the function $L(\overline{\eta}, s)$ has a zero of order one at such an integer j. Hence, a pole and a zero cancel in the right-hand side.

The following facts are known for the regulator map:

THEOREM 3.110. *The following statements hold for a finite algebraic number field F.*

(1) *For every nonnegative integer m, the m-th K-group $K_m(\mathfrak{r}_F)$ of the ring of integers $\mathfrak{r}_F$ is a finitely generated abelian group. For $m \geq 2$, we have*[84]*:*

$$\mathrm{rank}_{\mathbb{Z}} K_m(\mathfrak{r}_F) = \begin{cases} 0 & m \equiv 0 \ mod\ 2, \\ r_1(F) + r_2(F) & m \equiv 1 \ mod\ 4, \\ r_2(F) & m \equiv 3 \ mod\ 4. \end{cases}$$

(2) *For any natural number $j \geq 1$, the $\mathbb{R}$-vector space $\prod_{\iota : F \hookrightarrow \mathbb{C}} (2\pi\sqrt{-1})^{j-1}\mathbb{R}$ is of dimension $[F : \mathbb{Q}]$. The complex conjugate acts on*

$$\prod_{\iota : F \hookrightarrow \mathbb{C}} (2\pi\sqrt{-1})^{j-1}\mathbb{R}$$

by the action on $(2\pi\sqrt{-1})^{j-1}$ and by the twist of $\iota : F \hookrightarrow \mathbb{C}$ composing with the conjugate $\mathbb{C} \longrightarrow \mathbb{C}$. We denote by $\left[\prod_{\iota : F \hookrightarrow \mathbb{C}} (2\pi\sqrt{-1})^{j-1}\mathbb{R}\right]^+$ the fixed part by the action of complex conjugate. For each $j \geq 1$, Borel defined a ***regulator map:***

$$\mathrm{reg}_{F,j,\infty} : K_{2j-1}(\mathfrak{r}_F) \longrightarrow \left[\prod_{\iota : F \hookrightarrow \mathbb{C}} (2\pi\sqrt{-1})^{j-1}\mathbb{R} \right]^+ \tag{3.140}$$

whose kernel coincides with the torsion part of $K_{2j-1}(\mathfrak{r}_F)$.

When $j = 1$, the above regulator map on $K_1(\mathfrak{r}_F) = \mathfrak{r}_F^\times$ is identified with the classical Dirichlet regulator map on the group of units. Hence, $\mathrm{reg}_{F,1,\infty}(K_1(\mathfrak{r}_F))$ spans a lattice of the hyperplane of $[\prod_{\iota : F \hookrightarrow \mathbb{C}} \mathbb{R}]^+$ consisting of elements whose trace are 0.

When $j > 1$, we have $\mathrm{rank}_{\mathbb{Z}} K_{2j-1}(\mathfrak{r}_F) = \dim_{\mathbb{R}} \left[\prod_{\iota : F \hookrightarrow \mathbb{C}} (2\pi\sqrt{-1})^{j-1}\mathbb{R}\right]^+$ by Theorem 3.110 (1). Hence, by Theorem 3.110 (2), $\mathrm{reg}_{F,j,\infty}(K_{2j-1}(\mathfrak{r}_F))$ spans a lattice of $\left[\prod_{\iota : F \hookrightarrow \mathbb{C}} (2\pi\sqrt{-1})^{j-1}\mathbb{R}\right]^+$.

DEFINITION 3.111. When $j = 1$, we call the volume of a fundamental domain of the lattice spanned by $\mathrm{reg}_{F,1,\infty}(K_1(\mathfrak{r}_F))$ in a hyperplane of $[\prod_{\iota : F \hookrightarrow \mathbb{C}} \mathbb{R}]^+$ the **regulator** and we denote it by $\mathrm{Reg}_{F,1,\infty}$.

When $j > 1$, we call the volume of a fundamental domain of the lattice spanned by $\mathrm{reg}_{F,j,\infty}(K_{2j-1}(\mathfrak{r}_F))$ in $\left[\prod_{\iota : F \hookrightarrow \mathbb{C}} (2\pi\sqrt{-1})^{j-1}\mathbb{R}\right]^+$ the **regulator** and we denote it by $\mathrm{Reg}_{F,j,\infty}$.

Let η be a Dirichlet character. By regarding η as a character of $\mathrm{Gal}(F/\mathbb{Q})$ for an abelian extension F of $\mathbb{Q}$, we can define the η-part $\mathrm{reg}_{\eta,j,\infty}$ of $\mathrm{reg}_{F,j,\infty}$ with respect to the action of $\mathrm{Gal}(F/\mathbb{Q})$ on the regulator map $\mathrm{reg}_{F,j,\infty}$. We denote by $\mathrm{Reg}_{\eta,j,\infty}$ the volume of a fundamental domain of the lattice spanned by $\mathrm{reg}_{\eta,j,\infty}(K_{2j-1}(\mathfrak{r}_F))$ in the η-part of $\left[\prod_{\iota : F \hookrightarrow \mathbb{C}} (2\pi\sqrt{-1})^{j-1}\mathbb{R}\right]^+$.

The following results are known for special values of Dirichlet L-functions.

[84] The rank and the order of the torsion part of $K_m(\mathfrak{r}_F)$ are important invariants which are related to special values of the zeta function of F. We refer the reader to [**119**] for known results and some examples on this topic.

THEOREM 3.112. *Let η be a Dirichlet character. Then the following assertions hold true for $j \geq 1$.*

(1) *When $\eta(-1) = (-1)^j$, we have $L(\eta, j) \in \overline{\mathbb{Q}}^\times \cdot \pi^j$.*
(2) *When $\eta(-1) \neq (-1)^j$, we have $L(\eta, j) \in \overline{\mathbb{Q}}^\times \cdot \mathrm{Reg}_{\eta,j,\infty}$.*

REMARK 3.113. (1) The assertion (1) of Theorem 3.112 is a special case of the Deligne conjecture and the assertion (2) of Theorem 3.112 is a special case of the Beilinson conjecture.
(2) The assertion (1) of Theorem 3.112 follows from Theorem 3.36 and the functional equation (3.139). The assertion (2) of Theorem 3.112 for $j = 1$ is equivalent to the classical Dirichlet's class number formula (Theorem 3.63). In fact, we have $K_1(\mathfrak{r}_F) = \mathfrak{r}_F^\times$ for $j = 1$ and we have $\mathrm{reg}_{F,1,\infty}(u) = \prod_{\iota : F \hookrightarrow \mathbb{C}} \log|\iota(u)|_{\mathbb{C}}$ for $u \in \mathfrak{r}_F^\times$. The assertion (2) of Theorem 3.112 for $j > 1$ was proved by Borel [**6**] by calculating $\mathrm{Reg}_{\eta,j,\infty}$.

Recall that a character $\kappa_{\mathrm{cyc}}^j\phi$ with a finite character ϕ and an integer $j \in \mathbb{Z}$ for the p-adic L-function $L_p(\psi)$ plays a role of an integer point for the complex L-function. Below, we give an important remark on p-adic analogues of the Deligne conjecture and the Beilinson conjecture explained above.

REMARK 3.114. For the Kubota–Leopoldt p-adic L-function $L_p(\psi)$, $\kappa_{\mathrm{cyc}}^j\phi$'s with $j \leq 0$ correspond to critical motives and $\kappa_{\mathrm{cyc}}^j\phi$'s with $j \geq 1$ correspond to noncritical motives. This is different from the situation of Dirichlet L-functions $L(\eta, s)$ where $L(\eta, j)$ being critical or noncritical depends on the parity of j. The inconsistency comes from the fact that the interpolation of the p-adic L-function is related to special values of Dirichlet L-functions twisted by ω^{j-1} for each j.

By the above remark, the interpolation property of the Kubota–Leopoldt p-adic L-function $L_p(\psi)$ at $\kappa_{\mathrm{cyc}}^j\phi$ with $j \leq 0$ given in Theorem 3.30 is a p-adic analogue of the Deligne conjecture. Further, we expect a p-adic analogue of the Beilinson conjecture for values of the Kubota–Leopoldt p-adic L-function $L_p(\psi)$ at $\kappa_{\mathrm{cyc}}^j\phi$ with $j \geq 1$. In the p-adic case, we can define a p-adic regulator map $\mathrm{reg}_{F,j,p} : K_{2j-1}(\mathfrak{r}_F) \longrightarrow \prod_{\iota : F \hookrightarrow \overline{\mathbb{Q}}_p} \overline{\mathbb{Q}}_p$ and a p-adic regulator $\mathrm{Reg}_{F,j,p} \in \overline{\mathbb{Q}}_p$ (see [**35**], [**36**]), which are natural analogues of a complex regulator map and a complex regulator due to Borel and Beilinson. Let η be a Dirichlet character. By regarding η as a character of $\mathrm{Gal}(F/\mathbb{Q})$ for an abelian extension F of $\mathbb{Q}$, we can define $\mathrm{Reg}_{\eta,j,p} \in \overline{\mathbb{Q}}_p$ similarly as $\mathrm{Reg}_{\eta,j,\infty} \in \mathbb{C}$.

We have the following result by Coleman and Gros (see [**4**, Prop. 4.17], [**17**], [**35**] and [**36**]):

THEOREM 3.115 (p-adic Beilinson conjecture). *For any integer $j \geq 1$ and for any finite character ϕ on Γ_{cyc}, we have*

$$\kappa_{\mathrm{cyc}}^j\phi(L_p(\psi)) = \left(1 - \frac{(\psi\omega^{j-1}\phi^{-1})(p)}{p^j}\right) \cdot \mathrm{Reg}_{\psi\omega^{j-1}\phi^{-1},j,p} \cdot \frac{L(\psi\omega^{j-1}\phi^{-1}, j)}{\mathrm{Reg}_{\psi\omega^{j-1}\phi^{-1},j,\infty}}.$$

REMARK 3.116. (1) In Chapter 5 of the second volume of this book, we will present the p-adic Beilinson conjecture in the setting of cyclotomic Iwasawa theory for motives due to Perrin-Riou [**90**, Chap. 4] (see also a survey article by Colmez [**19**]).

(2) The classical Beilinson conjecture over $\mathbb{C}$ allows an ambiguity of multiplication by elements of $\overline{\mathbb{Q}}^\times$ as in Theorem 3.112 (2), which comes from the ambiguity of the choice of bases on the source and the target space of the regulator map. However, we have no such ambiguity in the p-adic Beilinson conjecture because we have both a complex regulator and a p-adic regulator ($\mathrm{Reg}_{\psi\omega^{j-1}\phi^{-1},j,\infty}$ and $\mathrm{Reg}_{\psi\omega^{j-1}\phi^{-1},j,p}$ in the above theorem) and the ambiguity of the choice of bases cancels by taking the ratio of these regulators. Hence, the equality of the p-adic Beilinson conjecture has no ambiguity coming from the choice of bases.

REMARK 3.117. (1) Special values $L(\eta, j)$ at positive integers j with $\eta(-1) \neq (-1)^j$ which appeared in Theorem 3.112 (2) are deeply related to the polylogarithm $\mathrm{Li}_j(z) = \sum_{n=1}^{\infty} \frac{z^n}{n^j}$ (cf. Zagier conjecture)

(2) In the proof of the p-adic Beilinson conjecture by [**17**], [**35**], [**36**], a p-adic analogue of the polylogarithm $\mathrm{Li}_j(z) = \sum_{n=1}^{\infty} \frac{z^n}{n^j}$ plays an essential role and the proof consists of two steps (A) and (B) as follows:
 (A) Coleman [**17**] established a theory of p-adic integration of dimension one called "theory of Coleman integration". As an application, he defined a j-th p-adic polylogarithm $\mathrm{Li}_j^{(p)}(x)$ on $\overline{\mathbb{Q}}_p^\times$ under the same setting as Definition 3.61[85]. The function $\mathrm{Li}_j^{(p)}(x)$ is an extention of the function $\mathrm{Li}_j^{(p)}(1+z) = \sum_{n=1}^{\infty} \frac{(-1)^{n-1}z^n}{n^k}$ on $1+\overline{P}$ to $\overline{\mathbb{Q}}_p^\times$ and it coincides with $\log_p(x)$ when $j = 1$. Then, in Chap. VII of [**17**], he proved that $\kappa^j_{\mathrm{cyc}}\phi(L_p(\psi))$ is a $\overline{\mathbb{Q}}$-linear combination of special values of $\mathrm{Li}_j^{(p)}(x)$ at cyclotomic units x.
 (B) Gros [**35**], [**36**] constructed a p-adic regulator map $\mathrm{reg}_{F,j,p}$ from the K-group K_{2j-1} of a cyclotomic field F to its syntomic cohomology. Then he expresses special values of $\mathrm{Li}_j^{(p)}$ at cyclotomic units x by the image of Soulé elements in the K-group K_{2j-1} of a cyclotomic field F via this p-adic regulator map.

(3) We refer the reader to [**4**] for numerical examples of Theorem 3.115 as well as a generalization of the p-adic Beilinson conjecture to Artin L-functions.

REMARK 3.118. Since regulator maps $\mathrm{reg}_{F,j,\infty}$ and $\mathrm{reg}_{\eta,j,\infty}$ by Borel are isomorphisms of $\mathbb{R}$-vector spaces (see Theorem 3.110), regulators $\mathrm{Reg}_{F,j,\infty}$ and $\mathrm{Reg}_{\eta,j,\infty}$ are nonzero real numbers.

On the other hand, p-adic regulator maps $\mathrm{reg}_{F,j,p}$ and $\mathrm{reg}_{F,j,p}$ are not known to be injective. Though we expect that p-adic regulators $\mathrm{Reg}_{F,j,p}$ and $\mathrm{Reg}_{\eta,j,p}$ are nonzero, it is not proved yet. "Nondegeneracy of p-adic regulator maps" is an important issue that we should attack in future.

3.5.3. Some other problems and perspective. Up to now, we discussed potential problems on cyclotomic Iwasawa theory of ideal class groups for the algebraic aspect and for the analytic aspect in §3.5.1 and in §3.5.2 respectively. In this section, we will discuss some other problems and perspective which were not classified in the algebraic aspect nor in the analytic aspect.

---

[85] In fact, Coleman [**17**] defined $\mathrm{Li}_j^{(p)}(x)$ as a function over $\mathbb{C}_p^\times = (\widehat{\overline{\mathbb{Q}}_p})^\times$.

(1) When the base field F is $\mathbb{Q}$ or an imaginaly quadratic field, we have an Euler system of cyclotomic units or an Euler system of elliptic units and we can prove the Iwasawa main conjecture for an abelian extension of F by the method of Euler systems. Will it be possible to construct a nice system of units (or an Euler system, in other words) on abelian extensions of F when the base field F is a general totally real field or a general CM field of finite degree? **Search of new Euler systems** is an important issue.

(2) Let F be a totally real finite algebraic number field. Recall that the cyclotomic Iwasawa main conjecture for ideal class groups was formulated for each finite character ψ of the absolute Galois group G_F[86] The Iwasawa main conjecture was formulated as an equality between the characteristic ideal of $(X_{K_\infty^{\mathrm{cyc}}})_{\omega\psi^{-1}}$ and the principal ideal generated by $L_p(\psi)$ on the Iwasawa algebra $\Lambda_{\mathrm{cyc},\psi}$. Now we consider a finite abelian extension K/F. If the character $\omega\psi^{-1}$ factors through the quotient $G_F \twoheadrightarrow \mathrm{Gal}(K/F)$, $(X_{K_\infty^{\mathrm{cyc}}})_{\omega\psi^{-1}}$ is the specialization of $\mathbb{Z}_p[\mathrm{Gal}(K/F)][[\Gamma_{\mathrm{cyc}}]]$-module $X_{K_\infty^{\mathrm{cyc}}}$ by $\omega\psi^{-1}$ and $L_p(\psi)$ is the specialization of a p-adic L-function $L_p(K)$ in the total quotient ring of $\mathbb{Z}_p[\mathrm{Gal}(K/F)][[\Gamma_{\mathrm{cyc}}]]$. Thus we expect an Iwasawa main conjecture which compares the algebraic object $X_{K_\infty^{\mathrm{cyc}}}$ and the analytic object $L_p(K)$ called an **equivariant Iwasawa main conjecture** over $\mathbb{Z}_p[\mathrm{Gal}(K/F)][[\Gamma_{\mathrm{cyc}}]]$. We remark that there is no natural notion of the characteristic ideal for $\mathbb{Z}_p[\mathrm{Gal}(K/F)][[\Gamma_{\mathrm{cyc}}]]$-modules. Hence, we need to find some ideas and to develop some foundation to overcome the absence of the characteristic ideal.

(3) Let G be a p-adic Lie group which has a surjection $G \twoheadrightarrow \Gamma_{\mathrm{cyc}}$ and L a Galois extension of K such that $\mathrm{Gal}(L/K) \cong G$. Coates proposed an Iwasawa theory over the noncommutative Iwasawa algebra $\mathbb{Z}_p[[G]] = \varprojlim \mathbb{Z}_p[G/U]$ (U runs through open normal subgroups of G).

Noncommutative generalizations over $\mathbb{Z}_p[[G]]$ of cyclotomic Iwasawa theory for ideal class groups studied in this chapter as well as cyclotomic Iwasawa theory for p-adic Galois representations (see Chapter 5 and Chapter 6) are studied by a lot of researchers. This theory is called a **noncommutative Iwasawa theory**.

On the algebraic side, one of essential problems in the noncommutative setting is that we do not have the "structure theorem of Iwasawa modules" and the theory of characteristic ideal does not work well[87]. However, Venjakob and Kato made some breakthroughs by using K-theory and they found some important algebraic invariants replacing the characteristic ideal. On the analytic side, one of the issues is that the analytic continuation and the algebraicity of special values of an L-function twisted by a nonabelian Artin representation is not well-established yet. We refer the reader to **[12]** for the formulation of the noncommutative Iwasawa main conjecture for p-adic Galois representations[88].

---

[86]See Theorem 3.66 and Theorem 3.66 when $F = \mathbb{Q}$. See Theorem 3.98 when F is a general totally real finite algebraic number field.

[87]For example, there exist pseudo-null modules whose characteristic ideals are nontrivial.

[88]The article **[12]** formulates the noncommutative Iwasawa main conjecture only for elliptic curves, but the formulation should be similar for general p-adic Galois representations.

In the case of the noncommutative Iwasawa main conjecture for ideal class groups, we already obtained remarkable progress. At the early stage, Kato, Ritter–Weiss, Hara, Kakde contributed by proving some partial results. After deeper understanding on the noncommutative Iwasawa theory through these results, Burns and Kato discovered a new idea to reduce the noncommutative Iwasawa theory to the commutative Iwasawa theory. Based on these progress, Kakde [**55**] and Ritter–Weiss [**98**] proved independently the noncommutative Iwasawa main conjecture for ideal class groups, which is a noncommutative generalization of Theorem 3.66, Theorem 3.66 and Theorem 3.98. We expect that, in near future, these work will be extended to prove the noncommutative Iwasawa main conjecture for p-adic Galois representations.

(4) The Iwasawa main conjecture is an equality between the characteristic ideal of the algebraic Iwasawa module $(X_{K^{\mathrm{cyc}}_{\omega\psi^{-1},\infty}})_{\omega\psi^{-1}}$ and the principal ideal generated by the analytic p-adic L-function. Note that the characteristic ideal of $(X_{K^{\mathrm{cyc}}_{\omega\psi^{-1},\infty}})_{\omega\psi^{-1}}$ is the reflexive closure of the 0-th Fitting ideal of $(X_{K^{\mathrm{cyc}}_{\omega\psi^{-1},\infty}})_{\omega\psi^{-1}}$. Since we can recover the characteristic ideal from the 0-th Fitting ideal, the 0-th Fitting ideal is an invariant with deeper information compared to the characteristic ideal. Further, we also have the i-th Fitting ideal of $(X_{K^{\mathrm{cyc}}_{\omega\psi^{-1},\infty}})_{\omega\psi^{-1}}$ for $i > 0$ and it is important to study these refined algebraic invariants comparing with some analytic invariants. In this direction, **Iwasawa theory for higher Fitting ideals** is studied actively by Kurihara et al.

(5) Recall that the zeta value $\zeta(m)$ for $m > 1$ is given by an infinite sum $\sum_{n\geq 1}\frac{1}{n^m}$. The multiple zeta value is a multiple version of the zeta value given by an infinite sum $\sum_{n_1>n_2>\ldots>n_s>0}\frac{1}{n_1^{m_1}n_2^{m_2}\cdots n_s^{m_s}}$. What is the p-adic L-function interpolating these multiple zeta values? What is the corresponding algebraic object? It will be interesting to investigate **multiple Iwasawa theory** related to multiple zeta values.

(6) It is also interesting to study **Iwasawa theory of function fields** for function fields of one variable of characteristic $p > 0$. Witte [**127**] studies the case over function fields of characteristic $p > 0$ for ℓ-adic Lie extensions with $\ell \neq p$ and proposes a formulation of Iwasawa theory of function fields in this case. The case over function fields of characteristic $p > 0$ for p-adic Lie extensions is more difficult and we expect a good formulation surrounding Selmer groups and p-adic L-functions.

Bookguide

BookguideBookguide This book is intended as a first textbook on the generalized Iwasawa theory for motives and Galois representations. However, as for the cyclotomic Iwasawa theory of ideal class groups discussed on the first volume, we can find some other textbooks. Since existing textbooks might give different viewpoints, for the reader's convenience we have listed some of them here. The books covering both the algebraic side and the analytic side of the cyclotomic Iwasawa theory of ideal class groups is as follows:

(1) An elementary textbook for the cyclotomic Iwasawa theory of ideal class groups which is read by most of people might be [**117**] by Washington. Though we do not have to all the detail of [**117**] to study the generalized Iwasawa theory for motives and Galois representations, [**117**] is traditionally the most standard introductory book to start to study Iwasawa theory.

(2) The textbook [**69**] by Lang is also aimed for the cyclotomic Iwasawa theory of ideal class groups. The book [**69**] is also interesting since it covers some of topics which are not contained in [**117**].

We will also list some books which are elementary enough but focused on more special subjects of the cyclotomic Iwasawa theory of ideal class groups compared to the above two books.

(1) The textbook [**51**] by Iwasawa is specialized on the Kubota–Leopoldt p-adic L-function explained in §3.2.3 of this book. This book explains several constructions of the Kubota–Leopoldt p-adic L-function. Especially Iwasawa's construction of the Kubota–Leopoldt p-adic L-function (see §3.2.3) is explained in detail. On the very end of the book, the algebraic aspect of the Iwasawa theory is discussed a little bit.

(2) The textbook [**13**] by Coates–Sujatha explains the construction of the Kubota–Leopoldt p-adic L-function from the norm compatible sequence of cyclotomic units à la Coleman (see §3.2.4) as well as the proof of the Iwasawa main conjecture by using the Euler system of cyclotomic units (see §3.3.3). In [**13**], they do not prove the Iwasawa main conjecture in full generality but prove it only over the base field $\mathbb{Q}(\mu_{p^\infty})$, which makes the proof much simpler and much more accessible to the reader.

(3) The proceeding [**135**] of the workshop held at Indian Institute of Technology, Guwahati in September 2008 explains the proof of the Iwasawa main conjecture by using the Eisenstein ideal (see §3.3.2). The book also covers Ribet's proof of an unramified p-extension of $\mathbb{Q}(\mu_p)$ given in (3.75) before discussing the proof of the Iwasawa main conjecture by Mazur–Wiles and Wiles.

(4) We can find the collected works of Iwasawa [**134**]. When we study certain theory, it is sometimes very inspiring to look into and study the articles of the origin written by the founder of the theory. Thus [**134**] is highly recommended.

APPENDIX A

As an appendix to Chapter 3, we summarize basic facts on modular forms and Hecke characters.

A.1. Modular forms and associated Galois representations

First, we recall p-adic Galois representations associated to modular forms.

THEOREM A.1 (Deligne, Shimura, Ribet). *Let $f \in S_k(\Gamma_1(M))$ be a normalized cuspidal eigenform of weight $k \geq 2$ and $\mathcal{K}$ a finite extension of $\mathbb{Q}_p$ which contains all coefficients of the q-expansion of f. Then, there exists a $\mathcal{K}$-vector space V_f of dimension 2 equipped with a continuous $\mathcal{K}$-linear Galois representation $\rho_f : G_{\mathbb{Q}} \longrightarrow \mathrm{Aut}_{\mathcal{K}} V_f \cong GL_2(\mathcal{K})$ satisfying the following conditions:*

(i) The representation on V_f is unramified over primes of $\mathbb{Q}$ not dividing pM.
(ii) For a geometric Frobenius[1] $\mathrm{Frob}_\ell \in G_{\mathbb{Q}}$ *of any prime number ℓ not dividing pM, we have* $\mathrm{Tr}(\rho_f(\mathrm{Frob}_\ell)) = a_\ell(f)$.
(iii) The representation of $G_{\mathbb{Q}}$ on V_f is irreducible.

By the Chebotarev density theorem[2]*, the representation V_f is uniquely characterized up to isomorphism by the above properties.*

Below, we summarize the history of the above theorem.

REMARK A.2. (1) For a normalized cuspidal eigenform f of weight $k \geq 2$ and level $\Gamma_1(M)$, Deligne [**20**] constructed the p-adic Galois representation V_f by cutting out the f-isotipic component of the action of the Hecke algebra on the parabolic part of the etale cohomology $H^1_{\text{ét}}(Y_1(M)_{\overline{\mathbb{Q}}}, \mathcal{L}_{k-2}(\mathcal{K}))$ on a local system $\mathcal{L}_{k-2}(\mathcal{K})$ over a modular curve $Y_1(M)$.

(2) In the construction of V_f given by the etale cohomology as explained above, the property (i) follows from the smooth base change theorem of the etale cohomology theory. The property (ii) is the deepest among three properties and its proof consists of the "congruence relation" which relates the ℓ-Hecke correspondence on the mod ℓ reduction of a modular curve to the algebraic correspondence induced by the geometric Frobenius at ℓ. The property (iii) is due to Ribet [**94**, Prop. 4.4].

---

[1] A geometric Frobenius is a lift of the inverse element of the arithmetic Frobenius $x \mapsto x^p$ in $G_{\mathbb{F}_p}$ to the decomposition subgroup at p of $G_{\mathbb{Q}}$. The decomposition group at p is only unique up to conjugation in $G_{\mathbb{Q}}$. Hence, a geometric Frobenius is only unique up to conjugation. However $\mathrm{Tr}(\rho_f(\mathrm{Frob}_\ell))$ is well-defined since the trace of a matrix depends only on its conjugacy class.

[2] We refer the reader to [**107**, I.2.3] for the precise statement of the Chebotarev density theorem.

(3) Shimura [111] reduced the proof of the above theorem for normalized cuspidal eigenforms of general weight $k \geq 2$ to the proof of for normalized cuspidal eigenforms of weight $k = 2$ by using the method of group cohomology and the congruence modulo p-power. Since Shimura constructs the p-adic Galois representation V_f by taking the limit of modulo p-power reductions of p-adic Galois representations for cuspidal eigenforms of weight 2, we lose the information on the complex absolute values of eigenvalues of Frobenius elements at unramified primes in Shimura's construction. Thus, though the construction of V_f by Deligne [20] implies the proof of the Ramanujan conjecture, the construction of V_f by Shimura [111] does not imply the proof of the Ramanujan conjecture.

(4) The construction of V_f due to Deligne [20] uses an etale cohomology group of a local system. Hence, the construction of [20] is not motivic. Later, Scholl [104] gave the construction of V_f by using an etale cohomology of Kuga-Sato variety, which implies for example that V_f is cristalline at p when the level of f is prime to p.

Below, we remark that there are two different ways for the construction of the p-adic Galois representation V_f of a modular form f. We have to be careful not to confuse these two.

REMARK A.3. For a normalized cuspidal eigenform $f \in S_k(\Gamma_1(M))$ of weight $k \geq 2$, we have a cohomological Galois representation and a homological Galois representation. The homological Galois representation is the linear dual of the cohomological Galois representation. Theorem A.1 is stated with the cohomological formulation. If we state it with the homological formulation, we have to replace the condition Theorem A.1(ii) using a geometric Frobenius element by the following condition using an arithmetic Frobenius element

(ii)$'$ For an arithmetic Frobenius $\mathrm{Frob}_\ell^{-1} \in G_{\mathbb{Q}}$ of any prime number ℓ not dividing pM, we have $\mathrm{Tr}(\rho_f(\mathrm{Frob}_\ell^{-1})) = a_\ell(f)$.

There are some articles written with the homological formulation (Wiles's articles [122], [123], [124] for example) and some articles written with the cohomological formulation (Deligne's article [20] for example). When we uses several different articles, we have to be careful not to mix different normalizations.

Let us consider the case of a normalized cuspidal eigenform f of weight 2 and level $\Gamma_1(M)$ in order to explain the cohomological formulation and the homological formulation a bit more. In this case, we can canonically associate an abelian variety B_f to f, which is a quotient of the Jacobian variety of modular curve $X_1(N)$ (see [112]). Then the p-adic Galois representation for f of the cohomological formulation is equal to $H^1_{\text{ét}}(B_f \underset{\mathrm{Spec}(\mathbb{Q})}{\times} \mathrm{Spec}(\overline{\mathbb{Q}}), \mathbb{Q}_p)$ and the p-adic Galois representation for f of the homological formulation is equal to $T_p B_f \otimes_{\mathbb{Z}_p} \mathbb{Q}_p$.

THEOREM A.4 (Deligne, Mazur–Wiles). *Let $f \in S_k(\Gamma_1(M))$ be a normalized cuspidal eigenform of weight $k \geq 2$, $\mathcal{K}$ a finite extension of $\mathbb{Q}_p$ which contains all coefficients of the q-expansion of f. Assume that f is p-stabilized*[3] *and that $a_p(f)$ is a p-adic unit. Then the restriction $\rho_f|_{G_{\mathbb{Q}_p}}$ to the decomposition group $G_{\mathbb{Q}_p}$ of*

[3]We say that f is p-stabilized if M is divisible by p and f is eigenvector of the Hecke operator at p.

$\rho_f : G_{\mathbb{Q}} \longrightarrow \mathrm{Aut}_{\mathcal{K}}(V_f) \cong GL_2(\mathcal{K})$ is equivalent to a representation of the form

$$\rho_f(g) = \begin{pmatrix} \alpha(g) & * \\ 0 & \beta(g) \end{pmatrix} \qquad (g \in G_{\mathbb{Q}_p}) \tag{A.1}$$

where α is an unramified character of $G_{\mathbb{Q}_p}$ such that $\alpha(\mathrm{Frob}_p) = a_p(f)$.

REMARK A.5. (1) A mod p variant of the above result was proved in an unpublised letter from Deligne to Serre in 1974 (see [**37**]). Also, Mazur–Wiles [**75**, Chap. 3 §2] and Wiles [**122**, Theorem 2.2] proved the above result.

The proof by Deligne depends on the technique of the vanishing cycles and the proof by Mazur–Wiles and Wiles depends on the study of Néron models over p-adic fields of the Jacobian varieties of modular curves. In the latter method, only the p-adic Galois representations for cuspforms of weight 2 can be studied by using the Jacobian varieties of modular curves. However, Shimura's technique of the congruence modulo p-power of the group cohomology (see Remark A.2 (3)) or Hida theory which is a refinement of Shimura's technique permits us to prove the desired statement for the case of cuspforms of weight $k > 2$ by reducing to the case of cuspforms of weight $k = 2$.

There is the third proof different from the above two proofs which is based on the theorem of Katz-Messing [**62**]. Recall that Scholl [**104**] constructs a motif of $f \in S_k(\Gamma_1(M))$ (we will give a summary on the motif in the appendix of the second volume) by using a Kuga-Sato variety of weight k and level M. Since a Kuga-Sato variety is known to be a smooth projective variety, the characteristic polynomial of a crystalline Frobenius operator acting on its i-th crystalline cohomology coincides with the characteristic polynomial of a p-Frobenius element acting on its i-th ℓ-adic etale cohomology for every natural number i. This implies that the Newton polygon coincides with the Hodge polygon for $\rho_f|_{G_{\mathbb{Q}_p}}$. Since a crystalline p-adic Galois representation whose Newton polygon and Hodge polygon coincide is known to be p-ordinary[4] by [**88**]. This implies Theorem A.4 when N_f is prime to p since the p-adic Galois representation V_f is crystalline in this case. By using Hida theory, any p-stabilized normalized cuspidal eigenform f with $a_p(f)$ a p-unit but N_f not necessarily prime to p is p-adically approximated by a sequence of p-stabilized normalized cuspidal eigenforms $\{f_i\}$ such that $a_p(f_i)$ are p-units and N_{f_i} are prime to p[5]. Thus the method using the theorem of Katz-Messing covers forms f with N_f divisible by p.

(2) We recall that Theorem A.4 is stated with the cohomological formulation as explained in Remark A.3. If we take the homological notation as in the papers of Mazur–Wiles and Wiles, the corresponding variant of Theorem A.4 is different from the actual version of Theorem A.4. In both of these cases, the images of $\rho_f|_{G_{\mathbb{Q}_p}}$ are contained in the upper triangular subgroup of $GL_2(\mathcal{K})$ after changing the $\mathcal{K}$-basis of V_f if necessary. However, in

[4] We have not yet defined the notion of p-ordinary, but it roughly means a representation of the form (A.1).

[5] Note that the weights k_i of f_i are not the same as the weight k of f. However, k_i converges p-adically to k.

the homological variant of Theorem A.4, the restriction $\rho_f|_{G_{\mathbb{Q}_p}}$ to the decomposition group $G_{\mathbb{Q}_p}$ is equivalent to a representation of the form

$$\rho_f(g) = \begin{pmatrix} \beta'(g) & * \\ 0 & \alpha'(g) \end{pmatrix} \qquad (g \in G_{\mathbb{Q}_p}) \tag{A.2}$$

where α' is an unramified character of $G_{\mathbb{Q}_p}$ such that $\alpha(\mathrm{Frob}_p^{-1}) = a_p(f)$.

A.2. Algebraic Hecke characters and associated Galois characters

DEFINITION A.6. (1) Let F be a finite algebraic number field, $\mathfrak{f}$ an integral ideal of F. We denote by $I_{\mathfrak{f}}$ the group of fractional ideals of F which are prime to $\mathfrak{f}$. Let J_F be the set of embeddings of F into $\overline{\mathbb{Q}}$. Then a group homomorphism $\eta : I_{\mathfrak{f}} \longrightarrow \overline{\mathbb{Q}}^{\times}$ is called an **algebraic Hecke character** [6] of weight $\sum_{\sigma \in J_F} n_\sigma \sigma \in \mathbb{Z}[J_F]$, level $\mathfrak{f}$ if η satisfies

$$\eta((\alpha)) = \prod_{\sigma \in J_F} \sigma(\alpha)^{n_\sigma}$$

for any totally positive elements $\alpha \in F^{\times}$ such that $\mathrm{ord}_{\mathfrak{p}}(\alpha - 1) \geq \mathrm{ord}_{\mathfrak{p}}\mathfrak{f}$ holds true for every prime ideal $\mathfrak{p}$ dividing $\mathfrak{f}$.

(2) An algebraic Hecke character of level $\mathfrak{f}$ can be naturally regarded as an algebraic Hecke character of level $\mathfrak{f}'$ for any ideal $\mathfrak{f}'$ satisfying $\mathfrak{f} \supset \mathfrak{f}'$. For an algebraic Hecke character η of level $\mathfrak{f}$, the largest ideal $\mathfrak{f}_0$ satisfying $\mathfrak{f}_0 \supset \mathfrak{f}$ such that there exists an algebraic Hecke character η_0 of level $\mathfrak{f}_0$ and η is naturally induced from η_0 is called a **conductor** of $\mathfrak{f}$.

EXAMPLE A.7. Let F be a finite algebraic number field. For any integral ideal $\mathfrak{n}$ of F, we define the norm $N_F(\mathfrak{n})$ by $N_F(\mathfrak{n}) = \#(\mathfrak{r}_F/\mathfrak{n})$. The norm N_F is naturally extended to a homomorphism from the group of fractional ideals to $\mathbb{Q}^{\times}$, which is called a **norm character**. The norm character N_F is an algebraic Hecke character of weight $\sum_{\sigma \in J_F} \sigma$, level 1.

In order to discuss the existence of algebraic Hecke characters other than the norm character, we recall the following fundamental result.

THEOREM A.8 (Theorem of Weil [**120**]). (1) *Let F be a totally real finite algebraic number field. Then any algebraic Hecke character on F is expressed as the product of an integral power N_F^r of the norm character N_F and an algebraic Hecke character of finite order.*

(2) *Let F be a finite algebraic number field and F^+ the largest totally real subfield of F. If there exists a subfield CM-field of F which contains F^+*[7]*, we denote it by F_0. Otherwise, we set $F_0 = F^+$. Then, any algebraic Hecke character η on F is expressed as $\eta = (\eta_0 \circ N_{F/F_0}) \cdot \eta_{\mathrm{fin}}$ where η_0 is an algebraic Hecke character on F_0 and η_{fin} is an algebraic Hecke character η_{fin} on F of finite order.*

Now, we want to discuss how much algebraic Hecke characters on F exist when F is a CM-field and we want to list them up. Below, we summarize some of known results according to [**21**, Remarque 8.2].

[6] Sometimes, it is also called a Grossencharacter of type A_0.

[7] We can easily check that such a CM-subfield is unique if it exists.

REMARK A.9. Let F be a CM-field which is of finite degree over $\mathbb{Q}$. We fix a CM-type Σ of F. Then the following statements hold (see [**102**, Chap. 0, §3] for example):

(1) Let η be an algebraic Hecke character on F of weight $\boldsymbol{n} = \sum_{\sigma \in J_F} n_\sigma \sigma$. Note that there exist unique elements $\boldsymbol{l} = \sum_{\sigma \in \Sigma} l_\sigma \sigma$ and $\boldsymbol{m} = \sum_{\overline{\sigma} \in \overline{\Sigma}} m_{\overline{\sigma}} \overline{\sigma}$ such that $\boldsymbol{n} = \boldsymbol{l} + \boldsymbol{m}$. Then we say that η has type $(\boldsymbol{l}, \boldsymbol{m})$.
(2) Let η be an algebraic Hecke character on F of type $(\boldsymbol{l}, \boldsymbol{m})$. Then the sum $l_\sigma + m_{\overline{\sigma}}$ is independent of $\sigma \in \Sigma$. We denote this value by $\sigma \in \Sigma$ and call it the absolute weight of η.
(3) Let us consider a pair $(\boldsymbol{l}, \boldsymbol{m})$ of elements $\boldsymbol{l} = \sum_{\sigma \in \Sigma} l_\sigma \sigma$ and $\boldsymbol{m} = \sum_{\overline{\sigma} \in \overline{\Sigma}} m_{\overline{\sigma}} \overline{\sigma}$ such that the sum $l_\sigma + m_{\overline{\sigma}}$ is independent of $\sigma \in \Sigma$. Then if we take a sufficiently small integral ideal $\mathfrak{f}$ of F, there exists an algebraic Hecke character on F of type $(\boldsymbol{l}, \boldsymbol{m})$, level $\mathfrak{f}$.

PROPOSITION A.10. *Let F be a finite algebraic number field, η an algebraic Hecke character on F, $\mathcal{K}$ a finite extension of $\mathbb{Q}_p$ which contains F^{gal} the Galois closure of F over $\mathbb{Q}$ and the values of η at every ideal of F. Then there exists a unique continuous Galois character $\eta^{\mathrm{gal}} : G_F \longrightarrow \mathcal{K}^\times$ such that we have $\eta^{\mathrm{gal}}(\mathrm{Frob}_{\mathfrak{l}}) = \eta(\mathfrak{l})$ for every prime ideal $\mathfrak{l}$ of F which does not divide the level of η nor (p).*

We give a rough sketch of the proof of the proposition.

PROOF. Let E be a finite algebraic number field which is obtained by adjoining the values of η at every ideal of F to F. Without losing the generality, we may assume that $\mathcal{K}$ is the completion of the image of E into $\overline{\mathbb{Q}}_p$ via the embedding ι'_p fixed in (3.76). We decompose the idèle group $\mathbb{A}_F^\times$ as

$$\mathbb{A}_F^\times = (F \otimes_{\mathbb{Q}} \mathbb{R})^\times \times (F \otimes_{\mathbb{Q}} \mathbb{Q}_p)^\times \times (\mathbb{A}_F^{(\infty p)})^\times,$$

where $\mathbb{A}_F^{(\infty p)}$ is the subgroup of $\mathbb{A}_F^\times$ obtained by removing factors over p, ∞. Further, we can decompose the group $(F \otimes_{\mathbb{Q}} \mathbb{R})^\times$ as

$$(F \otimes_{\mathbb{Q}} \mathbb{R})^\times = \{\pm 1\}^{r_1(F)} \times ((F \otimes_{\mathbb{Q}} \mathbb{R})^\times)_0$$

where $((F \otimes_{\mathbb{Q}} \mathbb{R})^\times)_0$ is the connected component of unity in $(F \otimes_{\mathbb{Q}} \mathbb{R})^\times$ and $r_1(F)$ is the number of real places of F. By extending trivially to finite places, we can regard $((F \otimes_{\mathbb{Q}} \mathbb{R})^\times)_0$ as a subgroup of $\mathbb{A}_F^\times$. Then, by definition, we have

$$\mathbb{A}_F^\times / ((F \otimes_{\mathbb{Q}} \mathbb{R})^\times)_0 \cong \{\pm 1\}^{r_1(F)} \times (F \otimes_{\mathbb{Q}} \mathbb{Q}_p)^\times \times (\mathbb{A}_F^{(\infty p)})^\times.$$

We define $\eta_\infty : \{\pm 1\}^{r_1(F)} \longrightarrow \{\pm 1\} \subset \mathcal{K}^\times$ by assigning -1 or $+1$ depending on whether η is ramified or unramified at each real infinite place. Recall that the group of fractional ideals of F is a quotient of $(F \otimes_{\mathbb{Q}} \mathbb{Q}_p)^\times \times (\mathbb{A}_F^{(\infty p)})^\times = (\mathbb{A}_F^{(\infty)})^\times$. Next, we denote by $\eta^{(\infty p)} : (\mathbb{A}_F^{(\infty)})^\times \longrightarrow E^\times$ the composition of a natural injection $(\mathbb{A}_F^{(\infty p)})^\times \hookrightarrow (\mathbb{A}_F^{(\infty)})^\times$, this quotient map and the character η. Suppose that the type of η is $(\boldsymbol{l}, \boldsymbol{m})$. Finally, we denote by $\eta_p : (F \otimes_{\mathbb{Q}} \mathbb{Q}_p)^\times \longrightarrow \mathcal{K}^\times$ the map induced by $F^\times \longrightarrow (F^{\mathrm{gal}})^\times$, $x \mapsto \prod_{\sigma \in \Sigma} (x^\sigma)^{-l_\sigma} (x^{\overline{\sigma}})^{-m_{\overline{\sigma}}}$. By using these maps η_∞, η_p and $\eta^{(\infty p)}$, we define the homomorphism

$$\hat{\eta} : \mathbb{A}_F^\times / ((F \otimes_{\mathbb{Q}} \mathbb{R})^\times)_0 \cong \{\pm 1\}^{r_1(F)} \times (F \otimes_{\mathbb{Q}} \mathbb{Q}_p)^\times \times (\mathbb{A}_F^{(\infty p)})^\times \longrightarrow \mathcal{K}^\times$$

by $(a, b, c) \mapsto \eta_\infty(a) \cdot \eta_p(b) \cdot \eta^{(\infty p)}(c)$. By construction, the map $\widehat{\eta}$ is trivial on the subgroup $F^\times \subset \mathbb{A}_F^\times$. Since the largest abelian quotient $(G_F)^{\mathrm{ab}}$ of the absolute Galois group G_F is isomorphic to $\mathbb{A}_F^\times/((F \otimes_{\mathbb{Q}} \mathbb{R})^\times)_0 F^\times$ by global class field theory, the map $\widehat{\eta}$ induces a Galois character $\eta^{\mathrm{gal}} : G_F \longrightarrow \mathcal{K}^\times$. Now, we have $\eta^{\mathrm{gal}}(\mathrm{Frob}_{\mathfrak{l}}) = \eta(\mathfrak{l})$ for every prime ideal $\mathfrak{l}$ of F which does not divide the level of η nor (p) by the definition of η^{gal} and by the compatibility of the global class field theory and the local class field theory. This completes the proof. □

We remark that the Galois character $(N_F)^{\mathrm{gal}}$ associated to the norm character N_F given in Example A.7 is equal to $\chi_{\mathrm{cyc}}^{-1}|_{G_F}$.

Finally we recall the definition of the L-function associated to an algebraic Hecke character.

Definition A.11. Let F be a finite algebraic number field, η an algebraic Hecke character on F. We associate the Dirichlet series $\sum_{\mathfrak{n} \subset \mathfrak{r}_F} \frac{\eta(\mathfrak{n})}{N(\mathfrak{n})^s}$ to η. This series is absolutely convergent for $\mathrm{Re}(s) \gg 0$ and is meromorphically continued to the whole $\mathbb{C}$-plane. It is also known that the series is holomorphic on the whole $\mathbb{C}$-plane if η is not a integral power of the norm character. We call this function the L-function of η and denote it by $L(\eta, s)$.

References

[1] Miho Aoki, *The Iwasawa main conjecture and Gauss sums*, J. Number Theory **89** (2001), no. 1, 151–164, DOI 10.1006/jnth.2001.2637. MR1838709

[2] M. F. Atiyah and I. G. Macdonald, *Introduction to commutative algebra*, Student economy edition, for the 1969 original see [MR0242802], Addison-Wesley Series in Mathematics, Westview Press, Boulder, CO, 2016. MR3525784

[3] Daniel Barsky, *Fonctions zeta p-adiques d'une classe de rayon des corps de nombres totalement réels* (French), Groupe d'Étude d'Analyse Ultramétrique (5e année: 1977/78), Secrétariat Math., Paris, 1978, pp. Exp. No. 16, 23. MR525346

[4] A. Besser, P. Buckingham, R. de Jeu, and X.-F. Roblot, *On the p-adic Beilinson conjecture for number fields*, Pure Appl. Math. Q. **5** (2009), no. 1, 375–434, DOI 10.4310/PAMQ.2009.v5.n1.a12. MR2520465

[5] Spencer Bloch and Kazuya Kato, *L-functions and Tamagawa numbers of motives*, The Grothendieck Festschrift, Vol. I, Progr. Math., vol. 86, Birkhäuser Boston, Boston, MA, 1990, pp. 333–400. MR1086888

[6] Armand Borel, *Cohomologie de* SL_n *et valeurs de fonctions zeta aux points entiers* (French), Ann. Scuola Norm. Sup. Pisa Cl. Sci. (4) **4** (1977), no. 4, 613–636. MR506168

[7] Nicolas Bourbaki, *Commutative algebra. Chapters 1–7*, Elements of Mathematics (Berlin), translated from the French; reprint of the 1989 English translation, Springer-Verlag, Berlin, 1998. MR1727221

[8] Armand Brumer, *On the units of algebraic number fields*, Mathematika **14** (1967), 121–124, DOI 10.1112/S0025579300003703. MR220694

[9] Winfried Bruns and Jürgen Herzog, *Cohen-Macaulay rings*, Cambridge Studies in Advanced Mathematics, vol. 39, Cambridge University Press, Cambridge, 1993. MR1251956

[10] Henri Carayol, *Sur les représentations l-adiques associées aux formes modulaires de Hilbert* (French), Ann. Sci. École Norm. Sup. (4) **19** (1986), no. 3, 409–468. MR870690

[11] Pierrette Cassou-Noguès, *Valeurs aux entiers négatifs des fonctions zêta et fonctions zêta p-adiques* (French), Invent. Math. **51** (1979), no. 1, 29–59, DOI 10.1007/BF01389911. MR524276

[12] John Coates, Takako Fukaya, Kazuya Kato, Ramdorai Sujatha, and Otmar Venjakob, *The* GL_2 *main conjecture for elliptic curves without complex multiplication*, Publ. Math. Inst. Hautes Études Sci. **101** (2005), 163–208, DOI 10.1007/s10240-004-0029-3. MR2217048

[13] J. Coates and R. Sujatha, *Cyclotomic fields and zeta values*, Springer Monographs in Mathematics, Springer-Verlag, Berlin, 2006. MR2256969

[14] J. Coates and A. Wiles, *On the conjecture of Birch and Swinnerton-Dyer*, Invent. Math. **39** (1977), no. 3, 223–251, DOI 10.1007/BF01402975. MR463176

[15] J. Coates and A. Wiles, *On p-adic L functions and elliptic units*, J. Austral. Math. Soc. Ser. A **26** (1978), no. 1, 1–25. MR510581

[16] Robert F. Coleman, *Division values in local fields*, Invent. Math. **53** (1979), no. 2, 91–116, DOI 10.1007/BF01390028. MR560409

[17] Robert F. Coleman, *Dilogarithms, regulators and p-adic L-functions*, Invent. Math. **69** (1982), no. 2, 171–208, DOI 10.1007/BF01399500. MR674400

[18] Pierre Colmez, *Résidu en* $s = 1$ *des fonctions zêta p-adiques* (French), Invent. Math. **91** (1988), no. 2, 371–389, DOI 10.1007/BF01389373. MR922806

[19] Pierre Colmez, *Fonctions L p-adiques* (French, with French summary), Séminaire Bourbaki, Vol. 1998/99, Astérisque **266** (2000), Exp. No. 851, 3, 21–58. MR1772669

[20] Pierre Deligne, *Formes modulaires et représentations l-adiques* (French), Séminaire Bourbaki. Vol. 1968/69: Exposés 347–363, Lecture Notes in Math., vol. 175, Springer, Berlin, 1971, pp. Exp. No. 355, 139–172. MR3077124

[21] P. Deligne, *Valeurs de fonctions L et périodes d'intégrales* (French), Automorphic forms, representations and L-functions (Proc. Sympos. Pure Math., Oregon State Univ., Corvallis, Ore., 1977), with an appendix by N. Koblitz and A. Ogus, Proc. Sympos. Pure Math., XXXIII, Amer. Math. Soc., Providence, R.I., 1979, pp. 313–346. MR546622

[22] P. Deligne and M. Rapoport, *Les schémas de modules de courbes elliptiques* (French), Modular functions of one variable, II (Proc. Internat. Summer School, Univ. Antwerp, Antwerp, 1972), Lecture Notes in Math., Vol. 349, Springer, Berlin, 1973, pp. 143–316. MR0337993

[23] Pierre Deligne and Kenneth A. Ribet, *Values of abelian L-functions at negative integers over totally real fields*, Invent. Math. **59** (1980), no. 3, 227–286, DOI 10.1007/BF01453237. MR579702

[24] Pierre Deligne and Jean-Pierre Serre, *Formes modulaires de poids* 1 (French), Ann. Sci. École Norm. Sup. (4) **7** (1974), 507–530 (1975). MR379379

[25] Ehud de Shalit, *Iwasawa theory of elliptic curves with complex multiplication: p-adic L functions*, Perspectives in Mathematics, vol. 3, Academic Press, Inc., Boston, MA, 1987. MR917944

[26] David Eisenbud, *Commutative algebra: With a view toward algebraic geometry*, Graduate Texts in Mathematics, vol. 150, Springer-Verlag, New York, 1995, DOI 10.1007/978-1-4612-5350-1. MR1322960

[27] L. J. Federer, *Regulator, Iwasawa modules, and the Main conjecture for* $p = 2$, Number theory related to Fermat's Last Theorem, Birkhäuser Verlag, pp. 289–296, 1982.

[28] Bruce Ferrero and Lawrence C. Washington, *The Iwasawa invariant* μ_p *vanishes for abelian number fields*, Ann. of Math. (2) **109** (1979), no. 2, 377–395, DOI 10.2307/1971116. MR528968

[29] Jean-Marc Fontaine and Jean-Pierre Wintenberger, *Le "corps des normes" de certaines extensions algébriques de corps locaux* (French, with English summary), C. R. Acad. Sci. Paris Sér. A-B **288** (1979), no. 6, A367–A370. MR526137

[30] K. Fujiwara, *Deformation rings and Hecke algebras in the totally real case*, arXiv:math/0602606 [math.NT]

[31] Roland Gillard, *Fonctions L p-adiques des corps quadratiques imaginaires et de leurs extensions abéliennes* (French), J. Reine Angew. Math. **358** (1985), 76–91, DOI 10.1515/crll.1985.358.76. MR797675

[32] Ralph Greenberg, *On a certain l-adic representation*, Invent. Math. **21** (1973), 117–124, DOI 10.1007/BF01389691. MR335468

[33] Ralph Greenberg, *On the Iwasawa invariants of totally real number fields*, Amer. J. Math. **98** (1976), no. 1, 263–284, DOI 10.2307/2373625. MR401702

[34] Cornelius Greither, *Class groups of abelian fields, and the main conjecture* (English, with English and French summaries), Ann. Inst. Fourier (Grenoble) **42** (1992), no. 3, 449–499. MR1182638

[35] Michel Gros, *Régulateurs syntomiques et valeurs de fonctions L p-adiques. I* (French), with an appendix by Masato Kurihara, Invent. Math. **99** (1990), no. 2, 293–320, DOI 10.1007/BF01234421. MR1031903

[36] Michel Gros, *Régulateurs syntomiques et valeurs de fonctions L p-adiques. II* (French), Invent. Math. **115** (1994), no. 1, 61–79, DOI 10.1007/BF01231754. MR1248079

[37] Benedict H. Gross, *A tameness criterion for Galois representations associated to modular forms (mod p)*, Duke Math. J. **61** (1990), no. 2, 445–517, DOI 10.1215/S0012-7094-90-06119-8. MR1074305

[38] Haruzo Hida, *Hecke algebras for* GL_1 *and* GL_2, Séminaire de théorie des nombres, Paris 1984–85, Progr. Math., vol. 63, Birkhäuser Boston, Boston, MA, 1986, pp. 131–163. MR897346

[39] Haruzo Hida, *Elementary theory of L-functions and Eisenstein series*, London Mathematical Society Student Texts, vol. 26, Cambridge University Press, Cambridge, 1993, DOI 10.1017/CBO9780511623691. MR1216135

[40] Haruzo Hida, *Anticyclotomic main conjectures*, Doc. Math. **Extra Vol.** (2006), 465–532. MR2290595

[41] Haruzo Hida, *Hilbert modular forms and Iwasawa theory*, Oxford Mathematical Monographs, The Clarendon Press, Oxford University Press, Oxford, 2006, DOI 10.1093/acprof:oso/9780198571025.001.0001. MR2243770

[42] Haruzo Hida, *Non-vanishing modulo p of Hecke L-values and application*, L-functions and Galois representations, London Math. Soc. Lecture Note Ser., vol. 320, Cambridge Univ. Press, Cambridge, 2007, pp. 207–269, DOI 10.1017/CBO9780511721267.007. MR2392356

[43] H. Hida and J. Tilouine, *Anti-cyclotomic Katz p-adic L-functions and congruence modules*, Ann. Sci. École Norm. Sup. (4) **26** (1993), no. 2, 189–259. MR1209708

[44] H. Hida and J. Tilouine, *On the anticyclotomic main conjecture for CM fields*, Invent. Math. **117** (1994), no. 1, 89–147, DOI 10.1007/BF01232236. MR1269427

[45] Ming-Lun Hsieh, *Eisenstein congruence on unitary groups and Iwasawa main conjectures for CM fields*, J. Amer. Math. Soc. **27** (2014), no. 3, 753–862, DOI 10.1090/S0894-0347-2014-00786-4. MR3194494

[46] Kenkichi Iwasawa, *A note on Kummer extensions*, J. Math. Soc. Japan **5** (1953), 253–262, DOI 10.2969/jmsj/00530253. MR62168

[47] Kenkichi Iwasawa, *On Γ-extensions of algebraic number fields*, Bull. Amer. Math. Soc. **65** (1959), 183–226, DOI 10.1090/S0002-9904-1959-10317-7. MR124316

[48] Kenkichi Iwasawa, *On a certain analogy between algebraic number fields and function fields* (Japanese), Sūgaku **15** (1963), 65–67. MR161850

[49] Kenkichi Iwasawa, *On some modules in the theory of cyclotomic fields*, J. Math. Soc. Japan **16** (1964), 42–82, DOI 10.2969/jmsj/01610042. MR215811

[50] Kenkichi Iwasawa, *On p-adic L-functions*, Ann. of Math. (2) **89** (1969), 198–205, DOI 10.2307/1970817. MR269627

[51] Kenkichi Iwasawa, *Lectures on p-adic L-functions*, Annals of Mathematics Studies, No. 74, Princeton University Press, Princeton, N.J.; University of Tokyo Press, Tokyo, 1972. MR0360526

[52] Kenkichi Iwasawa, *On the μ-invariants of Z_ℓ-extensions*, Number theory, algebraic geometry and commutative algebra, in honor of Yasuo Akizuki, Kinokuniya, Tokyo, 1973, pp. 1–11. MR0357371

[53] Kenkichi Iwasawa, *On $\mathbf{Z}_l$-extensions of algebraic number fields*, Ann. of Math. (2) **98** (1973), 246–326, DOI 10.2307/1970784. MR349627

[54] Kenkichi Iwasawa, *Riemann-Hurwitz formula and p-adic Galois representations for number fields*, Tohoku Math. J. (2) **33** (1981), no. 2, 263–288, DOI 10.2748/tmj/1178229453. MR624610

[55] Mahesh Kakde, *The main conjecture of Iwasawa theory for totally real fields*, Invent. Math. **193** (2013), no. 3, 539–626, DOI 10.1007/s00222-012-0436-x. MR3091976

[56] Kazuya Kato, *Euler systems, Iwasawa theory, and Selmer groups*, Kodai Math. J. **22** (1999), no. 3, 313–372, DOI 10.2996/kmj/1138044090. MR1727298

[57] Kazuya Kato, Nobushige Kurokawa, and Takeshi Saito, *Number theory. 1: Fermat's dream*, translated from the 1996 Japanese original by Masato Kuwata; Iwanami Series in Modern Mathematics, Translations of Mathematical Monographs, vol. 186, American Mathematical Society, Providence, RI, 2000, DOI 10.1090/mmono/186. MR1728620

[58] Kazuya Kato, Nobushige Kurokawa, and Takeshi Saito, *Number theory. 2: Introduction to class field theory*, translated from the 1998 Japanese original by Masato Kuwata and Katsumi Nomizu; Iwanami Series in Modern Mathematics, Translations of Mathematical Monographs, vol. 240, American Mathematical Society, Providence, RI, 2011, DOI 10.1090/mmono/240. MR2817199

[59] Nicholas M. Katz, *p-adic L-functions for CM fields*, Invent. Math. **49** (1978), no. 3, 199–297, DOI 10.1007/BF01390187. MR513095

[60] Nicholas M. Katz, *Another look at p-adic L-functions for totally real fields*, Math. Ann. **255** (1981), no. 1, 33–43, DOI 10.1007/BF01450554. MR611271

[61] Nicholas M. Katz and Barry Mazur, *Arithmetic moduli of elliptic curves*, Annals of Mathematics Studies, vol. 108, Princeton University Press, Princeton, NJ, 1985, DOI 10.1515/9781400881710. MR772569

[62] Nicholas M. Katz and William Messing, *Some consequences of the Riemann hypothesis for varieties over finite fields*, Invent. Math. **23** (1974), 73–77, DOI 10.1007/BF01405203. MR332791

[63] Yûji Kida, *l-extensions of CM-fields and cyclotomic invariants*, J. Number Theory **12** (1980), no. 4, 519–528, DOI 10.1016/0022-314X(80)90042-6. MR599821

[64] V. A. Kolyvagin, *Euler systems*, The Grothendieck Festschrift, Vol. II, Progr. Math., vol. 87, Birkhäuser Boston, Boston, MA, 1990, pp. 435–483. MR1106906

[65] Tomio Kubota and Heinrich-Wolfgang Leopoldt, *Eine p-adische Theorie der Zetawerte. I. Einführung der p-adischen Dirichletschen L-Funktionen* (German), J. Reine Angew. Math. **214(215)** (1964), 328–339. MR163900

[66] Masato Kurihara, *Some remarks on conjectures about cyclotomic fields and K-groups of* **Z**, Compositio Math. **81** (1992), no. 2, 223–236. MR1145807

[67] Masato Kurihara, *On the ideal class groups of the maximal real subfields of number fields with all roots of unity*, J. Eur. Math. Soc. (JEMS) **1** (1999), no. 1, 35–49, DOI 10.1007/PL00011159. MR1677689

[68] L. V. Kuz′min, *Some duality theorems for cyclotomic Γ-extensions over algebraic number fields of CM-type* (Russian), Izv. Akad. Nauk SSSR Ser. Mat. **43** (1979), no. 3, 483–546, 733–734. MR541659

[69] Serge Lang, *Cyclotomic fields I and II*, 2nd ed., with an appendix by Karl Rubin, Graduate Texts in Mathematics, vol. 121, Springer-Verlag, New York, 1990, DOI 10.1007/978-1-4612-0987-4. MR1029028

[70] Heinrich Wolfgang Leopoldt, *Eine p-adische Theorie der Zetawerte. II Die p-adische Γ-Transformation* (German), J. Reine Angew. Math. **274(275)** (1975), 224–239, DOI 10.1515/crll.1975.274-275.224. MR379446

[71] Hideyuki Matsumura, *Commutative ring theory*, 2nd ed., translated from the Japanese by M. Reid, Cambridge Studies in Advanced Mathematics, vol. 8, Cambridge University Press, Cambridge, 1989. MR1011461

[72] B. Mazur, *Deforming Galois representations*, Galois groups over **Q** (Berkeley, CA, 1987), Math. Sci. Res. Inst. Publ., vol. 16, Springer, New York, 1989, pp. 385–437, DOI 10.1007/978-1-4613-9649-9_7. MR1012172

[73] Barry Mazur and Karl Rubin, *Kolyvagin systems*, Mem. Amer. Math. Soc. **168** (2004), no. 799, viii+96, DOI 10.1090/memo/0799. MR2031496

[74] B. Mazur and J. Tilouine, *Représentations galoisiennes, différentielles de Kähler et "conjectures principales"* (French), Inst. Hautes Études Sci. Publ. Math. **71** (1990), 65–103. MR1079644

[75] B. Mazur and A. Wiles, *Class fields of abelian extensions of* **Q**, Invent. Math. **76** (1984), no. 2, 179–330, DOI 10.1007/BF01388599. MR742853

[76] James S. Milne, *Étale cohomology*, Princeton Mathematical Series, No. 33, Princeton University Press, Princeton, N.J., 1980. MR559531

[77] Toshitsune Miyake, *Modular forms*, translated from the Japanese by Yoshitaka Maeda, Springer-Verlag, Berlin, 1989, DOI 10.1007/3-540-29593-3. MR1021004

[78] David Mumford, *Abelian varieties*, Tata Institute of Fundamental Research Studies in Mathematics, vol. 5, Published for the Tata Institute of Fundamental Research, Bombay by Oxford University Press, London, 1970. MR0282985

[79] Jan Nekovář, *Selmer complexes* (English, with English and French summaries), Astérisque **310** (2006), viii+559. MR2333680

[80] Jürgen Neukirch, *Algebraic number theory*, Grundlehren der mathematischen Wissenschaften [Fundamental Principles of Mathematical Sciences], vol. 322, translated from the 1992 German original and with a note by Norbert Schappacher; with a foreword by G. Harder, Springer-Verlag, Berlin, 1999, DOI 10.1007/978-3-662-03983-0. MR1697859

[81] Jürgen Neukirch, Alexander Schmidt, and Kay Wingberg, *Cohomology of number fields*, Grundlehren der mathematischen Wissenschaften [Fundamental Principles of Mathematical Sciences], vol. 323, Springer-Verlag, Berlin, 2000. MR1737196

[82] Tadashi Ochiai, *A generalization of the Coleman map for Hida deformations*, Amer. J. Math. **125** (2003), no. 4, 849–892. MR1993743

[83] Tadashi Ochiai, *Euler system for Galois deformations* (English, with English and French summaries), Ann. Inst. Fourier (Grenoble) **55** (2005), no. 1, 113–146. MR2141691

[84] Tadashi Ochiai and Kazuma Shimomoto, *Bertini theorem for normality on local rings in mixed characteristic (applications to characteristic ideals)*, Nagoya Math. J. **218** (2015), 125–173, DOI 10.1215/00277630-2891620. MR3345626

[85] Masami Ohta, *Companion forms and the structure of p-adic Hecke algebras*, J. Reine Angew. Math. **585** (2005), 141–172, DOI 10.1515/crll.2005.2005.585.141. MR2164625

[86] Masami Ohta, *Companion forms and the structure of p-adic Hecke algebras. II*, J. Math. Soc. Japan **59** (2007), no. 4, 913–951, DOI 10.2969/jmsj/05940913. MR2369998

[87] Bernadette Perrin-Riou, *Théorie d'Iwasawa p-adique locale et globale* (French), Invent. Math. **99** (1990), no. 2, 247–292, DOI 10.1007/BF01234420. MR1031902

[88] Bernadette Perrin-Riou, *Représentations p-adiques ordinaires* (French), with an appendix by Luc Illusie; Périodes p-adiques (Bures-sur-Yvette, 1988), Astérisque **223** (1994), 185–220. MR1293973

[89] Bernadette Perrin-Riou, *La fonction L p-adique de Kubota-Leopoldt* (French, with French summary), Arithmetic geometry (Tempe, AZ, 1993), Contemp. Math., vol. 174, Amer. Math. Soc., Providence, RI, 1994, pp. 65–93, DOI 10.1090/conm/174/01852. MR1299735

[90] Bernadette Perrin-Riou, *Fonctions L p-adiques* (French), Proceedings of the International Congress of Mathematicians, Vol. 1, 2 (Zürich, 1994), Birkhäuser, Basel, 1995, pp. 400–410. MR1403940

[91] Bernadette Perrin-Riou, *Systèmes d'Euler p-adiques et théorie d'Iwasawa* (French, with English and French summaries), Ann. Inst. Fourier (Grenoble) **48** (1998), no. 5, 1231–1307. MR1662231

[92] Mic Raynaud, *Schémas en groupes de type* $(p,\dots,p)$ (French), Bull. Soc. Math. France **102** (1974), 241–280. MR419467

[93] Kenneth A. Ribet, *A modular construction of unramified p-extensions of* $Q(\mu_p)$, Invent. Math. **34** (1976), no. 3, 151–162, DOI 10.1007/BF01403065. MR419403

[94] Kenneth A. Ribet, *Galois representations attached to eigenforms with Nebentypus*, Modular functions of one variable, V (Proc. Second Internat. Conf., Univ. Bonn, Bonn, 1976), Lecture Notes in Math., Vol. 601, Springer, Berlin, 1977, pp. 17–51. MR0453647

[95] Kenneth A. Ribet, *Report on p-adic L-functions over totally real fields*, Journées Arithmétiques de Luminy (Colloq. Internat. CNRS, Centre Univ. Luminy, Luminy, 1978), Astérisque, vol. 61, Soc. Math. France, Paris, 1979, pp. 177–192. MR556672

[96] K. A. Ribet, *On modular representations of* $\mathrm{Gal}(\overline{\mathbf{Q}}/\mathbf{Q})$ *arising from modular forms*, Invent. Math. **100** (1990), no. 2, 431–476, DOI 10.1007/BF01231195. MR1047143

[97] Jürgen Ritter and Alfred Weiss, *Toward equivariant Iwasawa theory*, Manuscripta Math. **109** (2002), no. 2, 131–146, DOI 10.1007/s00229-002-0306-8. MR1935024

[98] Jürgen Ritter and Alfred Weiss, *On the "main conjecture" of equivariant Iwasawa theory*, J. Amer. Math. Soc. **24** (2011), no. 4, 1015–1050, DOI 10.1090/S0894-0347-2011-00704-2. MR2813337

[99] Serge Lang, *Cyclotomic fields I and II*, 2nd ed., with an appendix by Karl Rubin, Graduate Texts in Mathematics, vol. 121, Springer-Verlag, New York, 1990, DOI 10.1007/978-1-4612-0987-4. MR1029028

[100] Karl Rubin, *The "main conjectures" of Iwasawa theory for imaginary quadratic fields*, Invent. Math. **103** (1991), no. 1, 25–68, DOI 10.1007/BF01239508. MR1079839

[101] Karl Rubin, *Euler systems*, Hermann Weyl Lectures, The Institute for Advanced Study, Annals of Mathematics Studies, vol. 147, Princeton University Press, Princeton, NJ, 2000, DOI 10.1515/9781400865208. MR1749177

[102] Norbert Schappacher, *Periods of Hecke characters*, Lecture Notes in Mathematics, vol. 1301, Springer-Verlag, Berlin, 1988, DOI 10.1007/BFb0082094. MR935127

[103] Leila Schneps, *On the* μ*-invariant of p-adic L-functions attached to elliptic curves with complex multiplication*, J. Number Theory **25** (1987), no. 1, 20–33, DOI 10.1016/0022-314X(87)90013-8. MR871166

[104] A. J. Scholl, *Motives for modular forms*, Invent. Math. **100** (1990), no. 2, 419–430, DOI 10.1007/BF01231194. MR1047142

[105] René Schoof, *Catalan's conjecture*, Universitext, Springer-Verlag London, Ltd., London, 2008. MR2459823

[106] J. P. Serre, *Classes des corps cyclotomiques*, Seminaire Bourbaki, exposé 174, (1958-1959)

[107] Jean-Pierre Serre, *Abelian l-adic representations and elliptic curves*, 2nd ed., with the collaboration of Willem Kuyk and John Labute, Advanced Book Classics, Addison-Wesley Publishing Company, Advanced Book Program, Redwood City, CA, 1989. MR1043865

[108] Jean-Pierre Serre, *Local fields*, translated from the French by Marvin Jay Greenberg, Graduate Texts in Mathematics, vol. 67, Springer-Verlag, New York-Berlin, 1979. MR554237

[109] Jean-Pierre Serre, *Cohomologie galoisienne* (French), 5th ed., Lecture Notes in Mathematics, vol. 5, Springer-Verlag, Berlin, 1994, DOI 10.1007/BFb0108758. MR1324577

[110] Romyar Sharifi, *A reciprocity map and the two-variable p-adic L-function*, Ann. of Math. (2) **173** (2011), no. 1, 251–300, DOI 10.4007/annals.2011.173.1.7. MR2753604

[111] G. Shimura, *An l-adic method in the theory of automorphic forms*, the text of a lecture at the conference Automorphic functions for arithmetically defined groups, Oberwolfach, Germany, 1968. (published in Collected papers. II. 1967–1977. Springer-Verlag, New York, 2002.)

[112] Goro Shimura, *Introduction to the arithmetic theory of automorphic functions*, Kanô Memorial Lectures, No. 1, Publications of the Mathematical Society of Japan, No. 11, Iwanami Shoten Publishers, Tokyo; Princeton University Press, Princeton, N.J., 1971. MR0314766

[113] W. Sinnott, *On the μ-invariant of the Γ-transform of a rational function*, Invent. Math. **75** (1984), no. 2, 273–282, DOI 10.1007/BF01388565. MR732547

[114] C. Soulé, *Perfect forms and the Vandiver conjecture*, J. Reine Angew. Math. **517** (1999), 209–221, DOI 10.1515/crll.1999.095. MR1728540

[115] J. T. Tate, *p-divisible groups*, Proc. Conf. Local Fields (Driebergen, 1966), Springer, Berlin, 1967, pp. 158–183. MR0231827

[116] Richard Taylor and Andrew Wiles, *Ring-theoretic properties of certain Hecke algebras*, Ann. of Math. (2) **141** (1995), no. 3, 553–572, DOI 10.2307/2118560. MR1333036

[117] Lawrence C. Washington, *Introduction to cyclotomic fields*, 2nd ed., Graduate Texts in Mathematics, vol. 83, Springer-Verlag, New York, 1997, DOI 10.1007/978-1-4612-1934-7. MR1421575

[118] Charles A. Weibel, *An introduction to homological algebra*, Cambridge Studies in Advanced Mathematics, vol. 38, Cambridge University Press, Cambridge, 1994, DOI 10.1017/CBO9781139644136. MR1269324

[119] Charles Weibel, *Algebraic K-theory of rings of integers in local and global fields*, Handbook of K-theory. Vol. 1, 2, Springer, Berlin, 2005, pp. 139–190, DOI 10.1007/3-540-27855-9_5. MR2181823

[120] André Weil, *On a certain type of characters of the idèle-class group of an algebraic number-field*, Proceedings of the international symposium on algebraic number theory, Tokyo & Nikko, 1955, Science Council of Japan, Tokyo, 1956, pp. 1–7. MR0083523

[121] André Weil, *Basic number theory*, 3rd ed., Die Grundlehren der mathematischen Wissenschaften, Band 144, Springer-Verlag, New York-Berlin, 1974. MR0427267

[122] A. Wiles, *On p-adic representations for totally real fields*, Ann. of Math. (2) **123** (1986), no. 3, 407–456, DOI 10.2307/1971332. MR840720

[123] A. Wiles, *On ordinary λ-adic representations associated to modular forms*, Invent. Math. **94** (1988), no. 3, 529–573, DOI 10.1007/BF01394275. MR969243

[124] A. Wiles, *The Iwasawa conjecture for totally real fields*, Ann. of Math. (2) **131** (1990), no. 3, 493–540, DOI 10.2307/1971468. MR1053488

[125] Andrew Wiles, *Modular elliptic curves and Fermat's last theorem*, Ann. of Math. (2) **141** (1995), no. 3, 443–551, DOI 10.2307/2118559. MR1333035

[126] Jean-Pierre Wintenberger, *Le corps des normes de certaines extensions infinies de corps locaux; applications* (French), Ann. Sci. École Norm. Sup. (4) **16** (1983), no. 1, 59–89. MR719763

[127] Malte Witte, *On a noncommutative Iwasawa main conjecture for varieties over finite fields*, J. Eur. Math. Soc. (JEMS) **16** (2014), no. 2, 289–325, DOI 10.4171/JEMS/434. MR3161285

[128] Rodney I. Yager, *On two variable p-adic L-functions*, Ann. of Math. (2) **115** (1982), no. 2, 411–449, DOI 10.2307/1971398. MR647813

[129] Rodney I. Yager, *p-adic measures on Galois groups*, Invent. Math. **76** (1984), no. 2, 331–343, DOI 10.1007/BF01388600. MR742854

[130] *Two hours with Kenkichi Iwasawa* (Japanese), Sūgaku **45** (1993), no. 4, 366–372. MR1256476

[131] Gary Cornell and Joseph H. Silverman (eds.), *Arithmetic geometry*, papers from the conference held at the University of Connecticut, Storrs, Connecticut, July 30–August 10, 1984, Springer-Verlag, New York, 1986, DOI 10.1007/978-1-4613-8655-1. MR861969

[132] M. Rapoport, N. Schappacher, and P. Schneider (eds.), *Beilinson's conjectures on special values of L-functions*, Perspectives in Mathematics, vol. 4, Academic Press, Inc., Boston, MA, 1988. MR944987

[133] Gary Cornell, Joseph H. Silverman, and Glenn Stevens (eds.), *Modular forms and Fermat's last theorem*, papers from the Instructional Conference on Number Theory and Arithmetic Geometry held at Boston University, Boston, MA, August 9–18, 1995, Springer-Verlag, New York, 1997, DOI 10.1007/978-1-4612-1974-3. MR1638473

[134] Kenkichi Iwasawa, *Collected papers. Vol. I, II*, edited and with a preface by Ichiro Satake, Genjiro Fujisaki, Kazuya Kato, Masato Kurihara and Shoichi Nakajima; with an appreciation of Iwasawa's work in algebraic number theory by John Coates, Springer-Verlag, Tokyo, 2001. MR1851503

[135] J. Coates and R. Sujatha, *The main conjecture*, Guwahati Workshop on Iwasawa Theory of Totally Real Fields, Ramanujan Math. Soc. Lect. Notes Ser., vol. 12, Ramanujan Math. Soc., Mysore, 2010, pp. 113–140. MR2759412

Index

Selected Published Titles in This Series

272 **Tadashi Ochiai,** Iwasawa Theory and Its Perspective, Volume 1, 2023
271 **Hideto Asashiba,** Categories and Representation Theory, 2022
270 **Leslie Hogben, Jephian C.-H. Lin, and Bryan L. Shader,** Inverse Problems and Zero Forcing for Graphs, 2022
269 **Ralph S. Freese, Ralph N. McKenzie, George F. McNulty, and Walter F. Taylor,** Algebras, Lattices, Varieties, 2022
268 **Ralph S. Freese, Ralph N. McKenzie, George F. McNulty, and Walter F. Taylor,** Algebras, Lattices, Varieties, 2022
267 **Dragana S. Cvetković Ilić,** Completion Problems on Operator Matrices, 2022
266 **Kate Juschenko,** Amenability of Discrete Groups by Examples, 2022
265 **Nabil H. Mustafa,** Sampling in Combinatorial and Geometric Set Systems, 2022
264 **J. Scott Carter and Seiichi Kamada,** Diagrammatic Algebra, 2021
263 **Vugar E. Ismailov,** Ridge Functions and Applications in Neural Networks, 2021
262 **Ragnar-Olaf Buchweitz,** Maximal Cohen–Macaulay Modules and Tate Cohomology, 2021
261 **Shiri Artstein-Avidan, Apostolos Giannopoulos, and Vitali D. Milman,** Asymptotic Geometric Analysis, Part II, 2021
260 **Lindsay N. Childs, Cornelius Greither, Kevin P. Keating, Alan Koch, Timothy Kohl, Paul J. Truman, and Robert G. Underwood,** Hopf Algebras and Galois Module Theory, 2021
259 **William Heinzer, Christel Rotthaus, and Sylvia Wiegand,** Integral Domains Inside Noetherian Power Series Rings, 2021
258 **Pramod N. Achar,** Perverse Sheaves and Applications to Representation Theory, 2021
257 **Juha Kinnunen, Juha Lehrbäck, and Antti Vähäkangas,** Maximal Function Methods for Sobolev Spaces, 2021
256 **Michio Jimbo, Tetsuji Miwa, and Fedor Smirnov,** Local Operators in Integrable Models I, 2021
255 **Alexandre Boritchev and Sergei Kuksin,** One-Dimensional Turbulence and the Stochastic Burgers Equation, 2021
254 **Karim Belabas and Henri Cohen,** Numerical Algorithms for Number Theory, 2021
253 **Robert R. Bruner and John Rognes,** The Adams Spectral Sequence for Topological Modular Forms, 2021
252 **Julie Déserti,** The Cremona Group and Its Subgroups, 2021
251 **David Hoff,** Linear and Quasilinear Parabolic Systems, 2020
250 **Bachir Bekka and Pierre de la Harpe,** Unitary Representations of Groups, Duals, and Characters, 2020
249 **Nikolai M. Adrianov, Fedor Pakovich, and Alexander K. Zvonkin,** Davenport–Zannier Polynomials and Dessins d'Enfants, 2020
248 **Paul B. Larson and Jindrich Zapletal,** Geometric Set Theory, 2020
247 **István Heckenberger and Hans-Jürgen Schneider,** Hopf Algebras and Root Systems, 2020
246 **Matheus C. Bortolan, Alexandre N. Carvalho, and José A. Langa,** Attractors Under Autonomous and Non-autonomous Perturbations, 2020
245 **Aiping Wang and Anton Zettl,** Ordinary Differential Operators, 2019
244 **Nabile Boussaïd and Andrew Comech,** Nonlinear Dirac Equation, 2019
243 **José M. Isidro,** Jordan Triple Systems in Complex and Functional Analysis, 2019

For a complete list of titles in this series, visit the
AMS Bookstore at **www.ams.org/bookstore/survseries/**.